Genetics, Embryology, and Development of Auditory and Vestibular Systems

Editor-in-Chief for Audiology
Brad A. Stach, PhD

Genetics, Embryology, and Development of Auditory and Vestibular Systems

Sherri M. Jones
Timothy A. Jones

SAN DIEGO
OXFORD
BRISBANE

5521 Ruffin Road
San Diego, CA 92123

e-mail: info@pluralpublishing.com
Web site: http://www.pluralpublishing.com

49 Bath Street
Abingdon, Oxfordshire OX14 1EA
United Kingdom

Copyright © by Plural Publishing, Inc. 2011

Typeset in 10½/13 Palatino by Flanagan's Publishing Services, Inc.
Printed in Malaysia

All rights, including that of translation, reserved. No part of this publication may be reproduced, stored in a retrieval system, or transmitted in any form or by any means, electronic, mechanical, recording, or otherwise, including photocopying, recording, taping, Web distribution, or information storage and retrieval systems without the prior written consent of the publisher.

For permission to use material from this text, contact us by
Telephone: (866) 758-7251
Fax: (888) 758-7255
e-mail: permissions@pluralpublishing.com

Every attempt has been made to contact the copyright holders for material originally printed in another source. If any have been inadvertently overlooked, the publishers will gladly make the necessary arrangements at the first opportunity.

Library of Congress Cataloging-in-Publication Data:
Jones, Sherri M.
  Genetics, embryology, and development of auditory and vestibular systems / Sherri M. Jones and Timothy A. Jones.
     p. ; cm.
  Includes bibliographical references and index.
  ISBN-13: 978-1-59756-201-0 (alk. paper)
  ISBN-10: 1-59756-201-7 (alk. paper)
 1. Ear—Growth. 2. Auditory pathways—Growth. 3. Developmental genetics. I. Jones, Timothy A. (Timothy Arthur), 1948- II. Title.
  [DNLM: 1. Ear—growth & development. 2. Auditory Pathways—growth & development. 3. Ear—embryology. 4. Genetic Processes. 5. Vestibule, Labyrinth—growth & development. WV 201]
  QP461.J66 2010
  612.8'5—dc22

2010054527

Contents

Foreword by Robert F. Burkard, PhD — *vii*
Preface — *ix*

1. Introduction — 1
2. Nature of Cells and Tissues — 27
3. Basic Concepts in Genetics — 53
4. Basic Concepts in Embryology — 89
5. Embryogenesis of the Outer and Middle Ear — 131
6. Embryogenesis of the Inner Ear — 153
7. Emergence of Inner Ear Function — 207

Index — 262

Foreword

I love getting a new book. It is almost like Christmas morning when I was a child, looking at the tree and wondering what was in those packages left for me. This particular book was sent to me piecemeal as separate chapters in prepublication form, but I have looked forward to seeing what was inside a book that covered materials not often included in an audiology (or hearing science) curriculum. I know a fair amount about the anatomy and physiology of hearing, and have at least some working knowledge of vestibular function. I also have known Sherri and Tim Jones for many years. I can think of few others who could write a book on development of the peripheral auditory and vestibular system from a molecular and genetic framework *and* make it understandable to an audiology/hearing scientist audience. Their work has included the anatomy and physiology of the auditory and vestibular systems, often focusing on development. Their more recent research has focused on the genetics of the auditory and vestibular systems. It doesn't hurt that Sherri Jones is a clinical audiologist as well as a hearing/vestibular scientist.

The book has seven chapters. I know Tim and Sherri took the lead on different chapters, and yet, in terms of writing style, the book is written in a single voice. The consistent writing style is of huge benefit for the reader, and the authors are to be commended. It makes this book easier to read. This writing style is concise and matter of fact. Although the book is not particularly long, it is data rich. The authors do not compromise accuracy for simplicity. In my quick read of each chapter, there is no way I can possibly remember all of the facts that were new to me (these days, I am happy at the end of the day to remember where I parked my truck that morning). This book will have to be read several times to adequately digest the information, and will need to sit on the shelf as a ready reference for those who purchase it. Some terms will be new for the readers. I had to look up at least one term (I Googled the term "anlagen"), and so access to a medical dictionary (or the electronic equivalent) will make progress through the book easier. The references refer to literature that extends one full century (from 1910 through 2010). Some classical work in the area of auditory/vestibular development was performed early in the 20th century. The cellular/molecular biology and molecular genetics work is much more recent, as these tools have only been available the last few decades, and are in continuous development. As acknowledged by the authors, the molecular chapters will need to be updated in future editions of the book, as the book is no doubt already out of date in the molecular aspects.

Some features of the book are helpful to the reader. Starting each chapter with "Key Points" will help the reader know what they should learn in each chapter. Ending each chapter with suggestions of other books to review (under "General

Resources"), for the reader who craves more information on the chapter's topics, is also a great feature. The authors put a lot of thought and effort into the figures and tables. Many of the figures were created by the authors themselves (some of which are shown on the book cover) and are a real strength of the book. They graphically depict in a fairly straightforward fashion what are (in many cases) very complex biological processes. The tables also are very useful. Many of the tables summarize information that will no doubt be out of date in the near future. For example, listings of genes known to cause hearing loss or vestibular dysfunction (such as Table 6–7) will likely be changing almost continuously. Future editions of this book can add to the tables, and hence be used to update our current knowledge of such factors.

This book attempts to make the reductionistic world of molecular/cellular biology and molecular genetics available to audiologists and hearing scientists. It succeeds in this attempt, but students will have to work hard to reap the benefits of this effort. For the cellular/molecular biologist or molecular geneticist, this book is a good summary of the current state of knowledge (or ignorance) of development of the peripheral auditory and vestibular systems. The book makes it clear when knowledge of human development is inferred from animal data (often mouse or chick data). The authors, in an often understated way, point to areas where additional study is required. Many of these areas would make excellent dissertation topics for the basic science students reading this book. This book does not provide any experimental details of the studies reviewed, and if the reader is looking for a primer on experimental approaches in cellular/molecular biology and molecular genetics they will have to look elsewhere. Similarly, the reader will have to look elsewhere for summaries of the development of the central auditory and vestibular systems, as this book focuses on the peripheral systems, up to and including the auditory and vestibular nerves.

This book opens to the reader a brave new molecular world. When I was a student, development was mostly discussed in terms of anatomy and electrophysiology. Developmental neurobiology was in its infancy. Our recent investment of billions of dollars to map the human genome has not yet, as promised, led to any cures of human diseases or genetic disorders, but has given us a glimpse into the complexity of human development, and provided us with an incomplete understanding as to how these sensory systems develop. Coming back to the "Christmas morning" analogy I started this Foreword with, reading this book reminds me that there is still magic in our world, and that the purpose of science is to try and demystify at least portions of this world. In the case of this book, our understanding of auditory/vestibular development is still in its infancy, and there is still much for us to learn. I hope the next edition can review some initial progress into treatments for the genetic and other developmental disorders identified in this book.

Robert F. Burkard, PhD
Professor and Chair
Department of Rehabilitation Sciences
University at Buffalo,
   State University of New York

Preface

This text was written for first-year students of audiology who may or may not have a significant undergraduate training in the basic sciences. The authors' goal was to provide an introductory presentation regarding basic science concepts in genetics and embryology and then apply those concepts to the development of the auditory and vestibular systems during the embryonic and fetal periods of development. We attempted to cover many of the genes and processes that are important for development of the ear, but recognize that there is much yet to be learned. Some of the information may well be out of date or revised by the time this book is in print.

The general concepts covered in each chapter are discussed more extensively in the resources listed at the end of each chapter. We hope that this text will prepare each student to begin his or her own journey into the scientific literature addressing normal development of the ear as well as genetic hearing loss and vestibular impairment. The authors thank Dr. Brad Stach for the invitation to write such a text, and special thanks to Plural Publishing, Dr. Sadanand and Angie Singh, Casey Stach, Sandy Doyle and Stephanie Meissner for patience and encouragement throughout this long endeavor.

1

Introduction

> **KEY POINTS**
>
> 1. Ontogeny is the series of steps that leads to a mature state.
> 2. The human being is precocious with regard to hearing. Although humans may be considered precocious with respect to vestibulo-ocular reflexes and other vestibular reflexes, we certainly are altricial with respect to upright posture and balance.
> 3. Several animal models have been used to understand molecular, structural, and functional development of auditory and vestibular systems including Drosophilia, zebrafish, mouse, and humans. Even yeast genome has revealed information about conserved biological pathways.
> 4. The mature inner ear is a complex system of chambers consisting of one auditory and five vestibular end organs, each specialized to encode sound or head motion, respectively.
> 5. Sensory neuroepithelia consist of specialized non-neural mechanoreceptors innervated by afferent and efferent neurons.
> 6. Supporting cells and ancillary structures are critical for inner ear homeostasis.

ONTOGENY

To understand the meaning of ontogeny, we need ask only what it means to be mature. If we accept what the mature state is, then we can define ontogeny as the series of steps that ultimately leads to the mature state. This seems rather unremarkable until one realizes that, in all cases, the series of steps begins with one cell, the fertilized ovum, and ends, in most cases, when this one cell becomes billions of cells organized into a fully functional, complex organism. To further underscore how remarkable this is, consider that the series of steps can lead to hundreds of thousands of different organisms each with a different ontogenetic plan. Fortunately, given the diverse number of different organisms, the ontogenetic plans across the organisms have many steps in common or that occur in analogous ways. We can recognize these steps as milestones of ontogeny across classes of animals and generalize how each organism has successfully built its own plan based on a modification of a preceding plan. Ontogeny is the process of developing from a single cell to a mature organism. Understanding how ontogeny works is the focus of developmental sciences. In this text, we will study the ontogeny of inner ear sensors. For obvious reasons, we will not attempt to consider all possible organisms. Instead, we focus on a select few organisms for which we have the most information about the ontogeny of hearing and vestibular senses.

PRECOCIAL VERSUS ALTRICIAL

The human neonate is able to hear normal ambient sounds at birth. This is not the case for many mammals. The neonatal kitten is essentially deaf and blind at birth and does not begin to hear appreciably until weeks later. The words precocial and altricial traditionally were used to indicate the maturity of birds at hatch where altricial birds are helpless, naked, and blind (e.g., pigeon), and precocial birds are down-covered, can see, and are mobile at birth (e.g., chicken). The neonatal chicken is precocial and can likely hear even before hatching. In contrast, the pigeon is altricial and has a delayed onset of sensitive hearing until well after hatch. These examples serve to illustrate that the status of the ear and hearing at birth varies considerably across animal classes as well as across species within classes of animals.

The words altricial and precocial (or altricious and precocious), although originally used for birds, are used to characterize neonatal mammals in today's hearing and vestibular literature. Hence, the human may be considered precocial with respect to hearing whereas the kitten and mouse are altricial (blind and deaf at birth). Although humans may be precocious with regard to the vestibulo-ocular and perhaps other vestibular reflexes, they certainly are altricial in terms of upright balance and postural behaviors.

ANIMAL MODELS FOR ONTOGENY

Much of our knowledge regarding normal molecular, structural, and physiological development of the ear as well as genetics of hearing and vestibular impairment has come from studies using animal models. Indeed, some of the genes now known to be critical for development of the ear were first discovered in fruit flies (Drosophila). Table 1–1 compares the genomes for several organisms. The choice of model organism for any given study is driven by the research question(s) under study. Indeed, some cellular and molecular processes can be elucidated by studying plants or single cell organisms such as yeast. Who would think that studies of yeast genes might actually apply to human beings? In fact, mechanisms for cell division (discussed in Chapter 3) are well understood thanks to studies of brewer's yeast. Indeed, basic cellular processes and the proteins needed for those processes are conserved in yeast, worms,

Table 1–1. Comparison of Genomes for Yeast, Worms, Flies, Zebrafish, Chicken, Mice, and Humans

Species	Number of Chromosomes (2n)	Genome Size (base pairs)	Number of Genes
Brewer's Yeast	32	12,162,996	6.532
Worm (C. Elegans)	12	103,021,882	20,158
Fruit Fly	8	168,736,537	14,141
Zebrafish	50	1,563,441,531	20,870
Chicken	78	1,050,947,331	15,926
Mouse	40	3,420,842,930	22,974
Human	46	3,272,480,989	22,258

Source: Information from Ensembl Genome Browser (http://www.ensembl.org).

flies, zebrafish, mice, and humans. Furthermore, research in yeast set the stage for scientists to be able to study how the expression of the entire genome (i.e., all of the organism's genes) changes in development, across the life span or in various experimental conditions (see Howard Hughes Medical Institute [HHMI], 2001).

Genomic, proteomic (expression of proteins), and transcriptomic (expression of transcripts) studies are now commonplace in a variety of organisms. Such studies have tremendous implications for human health and disease.

For inner ear function and the genetics associated with development and dysfunction, the mouse has become a widely used model; 97% of the human genome is similar to the mouse. The mouse and human have very similar middle and inner ear anatomy. A wealth of phenotypic and genotypic data are available for mouse strains that allow us to understand genetic aspects regarding development of the ear as well as disorders of the ear including otitis media and other middle ear disorders, inner ear dysfunction that accompanies many diseases, and syndromic and nonsyndromic hearing loss. The mouse genome can be readily manipulated to test hypotheses about gene function, to understand the role(s) for particular genes in development of the ear, and to develop potential therapies for genetic hearing loss and vestibular impairment. Mice breed readily and often produce large litters providing sufficient numbers of genetic controls and experimental models. Mice have a short gestation (18 to 21 days) and a relatively short life span (2 to 2.5 years); therefore, studies covering embryology to gerontology can be accomplished in about two to three years (Kispert & Gossler, 2004). Use of the mouse model does not negate other models (e.g., zebrafish and bird), which also have been instrumental in studies of ontogeny; and the reader will see references to various species throughout the text. Table 1–2 lists developmental age comparisons for chick, mouse, and human.

Table 1–2. General Developmental Age Comparisons for Chicken, Mouse, and Human

Chick	*Mouse*	*Human*
0–3 days	0–7.5 dpc	0–3 weeks
3–6 days	8–15 dpc	4–8 weeks
7–11 days	15–20 dpc	9–12 weeks
12–15 days	21 dpc (birth)–7dpn	13–20 weeks
16–18 days	8–9 dpn	21–28 weeks
19–20 days	10–11 dpn	30–35 weeks
20–21 days (birth)	11–14 dpn	36–40 weeks (birth)

Sources: Information from: Hamburger and Hamilton (1951); Mullen, Li, and Ryan (2004); and Carlson (2004).

Note: Age correlations across species for specific organ systems may vary considerably from ages listed here. For the chicken, days indicate number of days of incubation for mouse days postconception (dpc) or days postnatal (dpn), and for human weeks of gestation are indicated. Notice that humans and chickens are precocious relative to the mouse at birth. The mouse is born unable to see or hear. The chick generally is more mature at birth than the human.

PHILOSOPHY OF SENSORY PROCESSING

It is the "miracle" of the brain that transforms mere action potentials into the wealth of auditory, visual, somatosensory, or olfactory experiences we attribute to the senses. These wonderful and utterly distinct sensations or sensory qualities are spectacular inventions of the brain. This is a hard lesson to appreciate, but it is that very difficult concept that so impresses and amazes us. After all, the sensation of a sound is nothing like an action potential, or so it seems. Although we may cleverly reveal sensory illusions of all sorts by manipulating the nature of stimuli, there is no more profound illusion than that which defines elemental sense qualities and the awareness of them.

Hearing is defined by the *Random House Unabridged Dictionary* (Flexner & Hauck, 1993) as follows: "the faculty or sense by which sound is perceived"; "the act of perceiving sound" (p. 882). To hear: "to perceive by the ear"; "to be capable of perceiving sound by the ear; have the faculty of perceiving sound vibrations," or "the faculty or sense by which sound is perceived" (p. 882, Flexner & Hauck, 1993). We should add the definition of perceive and perceiving from the same source, that is, "to become aware of, know or identify by means of the senses" (p. 1437). Contrast these definitions with those for vision: "the act or power of sensing with the eyes; sight" (p. 2126, Flexner & Hauck, 1993) and from Soukhanov (1992), "the faculty of sight" (vision, p. 1997), "the ability to see," (sight, p. 1678) and "to perceive with the eyes" (see, p. 1632).

Perception is defined as "a single unified awareness derived from sensory processes while a stimulus is present" (p. 1437,

Flexner & Hauck, 1993). As defined, a perception is an interpretation. Why? Sensory processes involve neural codes, which are action potentials. In order to experience or "be aware" of a sensation, the brain must "interpret" action potentials. This is an interpretation because action potentials are not sounds (or other physical stimuli) nor does it seem likely that action potentials per se are "awareness." The action potentials are representations of the physical stimuli. As for "awareness," despite how familiar the idea is to us, it is not at all clear what awareness is. However, it is clear that "awareness" requires neural circuitry; at least we know this to be the case in the human. Hence, to enjoy a "single unified awareness" of a stimulus, many neurons (a circuit) and presumably many action potentials are required. Furthermore, "awareness" resides in only some circuits mediating action potentials, those of the neocortex. Thus, logically, given our limited understanding, awareness is an invention of neural activity in some special circuits in the brain. How it is that these "special" circuits give rise to human awareness remains to be shown. Ask yourself what it means to be aware of a sound such as a "click," a color such as red, or the presence of something on the skin, such as a touch?

THE MATURE EAR

The mature ear includes two major sensory systems each mediating the distinct sensory qualities of hearing and vestibular sense. These two modalities are mediated by distinct structures in the ear, the cochlea (hearing) and the labyrinth (vestibular sense). As an introduction to the mature ear, it is important to come to some kind of understanding about what these two sensory systems are. To this end, we offer the following definition of hearing and vestibular sensation.

Hearing and vestibular sensation are unique sensory modalities defined by their adequate stimuli (sound or head motion, respectively), by their respective sensors (the cochlea, maculae, and cristae), and distinct neural pathways in the brain as well as by unique sensory qualities (the sensation experienced when one hears a sound or moves the head). Both are associated with separate neuroanatomical systems termed the auditory and vestibular systems. We define hearing as the sense by which sound is identified, experienced, and perceived. Vestibular sensation is defined as the sense by which head motion and orientation are identified, experienced, and perceived. Both modalities involve physiological processes that provide the brain with a neural representation of the sound environment or head motion. Hearing provides the basis upon which our appraisal of the sound environment is made. Vestibular sensation provides the basis upon which we are aware of head position and motion. Both modalities include the detection and encoding of information in the peripheral sensors as well as the analysis and processing of that encoded information in the central nervous system.

Auditory and vestibular sensory processes involve all levels of the neuraxis from the periphery to the neocortex. Normal sensation depends on a peripheral sensory system that detects stimuli and encodes the spectral content and dynamics of the stimuli continuously. This begins in the sensory end organ, the ear. There are several excellent texts on auditory and vestibular anatomy (see General Resources). Anatomy and physiology are covered briefly here as a reference for the mature ear.

Outer Ear

The human outer ear is an acoustic chamber consisting of the pinna and external auditory meatus. It funnels sound to the tympanic membrane and functions as a passive amplifier for frequencies from about 1500 Hz to 5000 Hz. The pinna is comprised of epithelium tightly bound to cartilage, which is held to the temporal bone by three muscles: (1) the anterior auricularis; (2) superior auricularis; and (3) posterior auricularis. Several identifying folds and ridges are visible on the pinna (Figure 1–1), the most notable of which are the helix (superior and posterior perimeter of the pinna), concha (deep bowl-shaped depression surrounding the entrance to the ear canal), anithelix (ridge surrounding the concha), and crua of the anithelix (superior and inferior crua) that form the boundaries of the triangular fossa. Finally, the tragus and antitragus are two flap-like structures anterior and posterior to the external auditory meatus, respectively, while the earlobe forms the most inferior portion of the pinna. Sensory innervation to the pinna is provided by branches of the trigeminal, vagus, and a small portion of the facial nerves, as well as the cervical plexus. The glossopharyngeal nerve also may provide some sensory innervation. Motor innervation, which is minor in the human, is supplied by the facial nerve. Blood supply to the pinna is via branches of the external carotid.

The human external auditory meatus (EAM) is a tube approximately 2.5 to 2.7 cm long and approximately 1 cm in diameter that is open to air on one end and closed by the tympanic membrane at the other end (Figure 1–2). Epithelium lines the cartilaginous and bony portions as well as the lateral layer of the tympanic membrane. Sebaceous glands along the EAM secrete cerumen, which helps maintain skin moisture and protect the ear canal from foreign debris. The juncture between the cartilaginous and bony portions narrows in diameter and is called the isthmus.

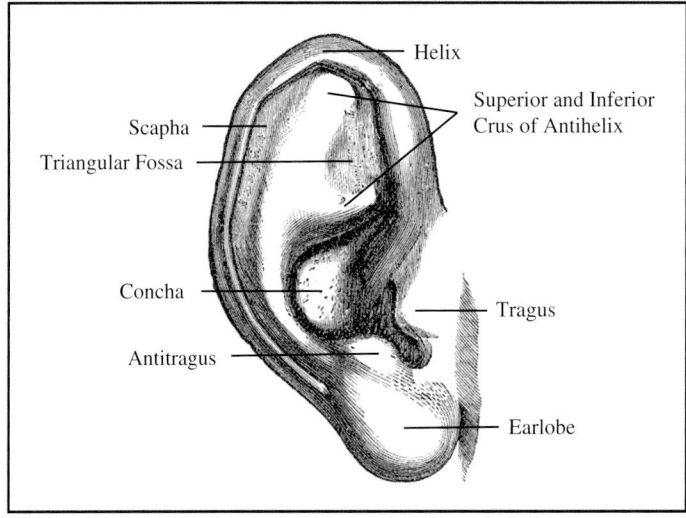

Figure 1–1. Illustration of the pinna showing major visible landmarks. Adapted from Gray (1918). Used with permission.

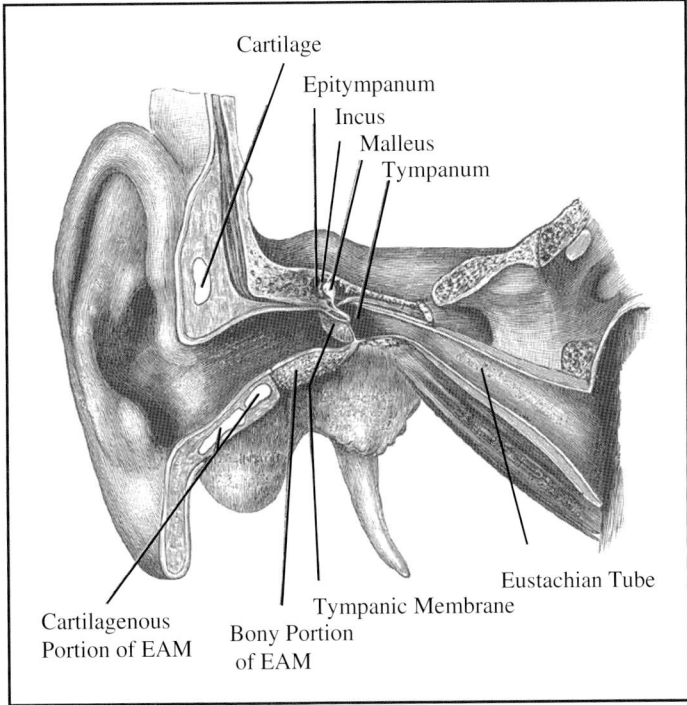

Figure 1–2. Coronal section through the external acoustic meatus (EAM) and middle ear highlighting the cartilaginous and bony portions of the EAM along with several middle ear structures. Adapted from Gray (1918). Used with permission.

Innervation is supplied by the trigeminal, facial, glossopharyngeal, and vagus nerves. Branches of the external carotid provide the vascular supply.

Middle Ear

The middle ear is an air-filled cavity, lined with mucous membrane that extends from the tympanic membrane laterally to the oval window medially and from the epitympanum (or attic) superiorly to the Eustachian tube antero-inferiorly (see Figures 1–2 and 1–3). The middle ear is critical for transmission of airborne sound to the fluid-filled cochlea and functions as an impedance matching transformer. Impedance matching ensures that sound energy reaching the entrance to the middle ear (the tympanic membrane) passes to the cochlea rather than being reflected back out of the ear.

The middle ear cavity is surrounded by the temporal bone; superior, posterior, and inferior by the mastoid portion, superior and medial by the petrous portion, and anterior by the tympanic portion. At the antero-inferior wall is the opening to the Eustachian tube, which travels to the nasopharynx and functions to maintain normal air pressure in the middle ear cavity.

Several structures within the middle ear are critical for sound transmission including the tympanic membrane, ossicles (malleus, incus, stapes), and middle

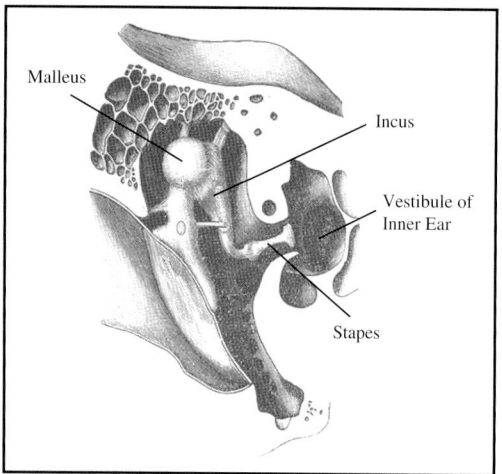

Figure 1–3. *Ossicles of the middle ear span the cavity from the tympanic membrane laterally to the oval window medially. The oval window opens into the vestibule of the inner ear. Ligaments suspend the ossicles within the cavity. Adapted from Gray (1918). Used with permission.*

ear muscles (stapedius, tensor tympani) (Figures 1–3 and 1–4). The tympanic membrane consists of three layers: (1) the outer layer is continuous with the epidermis of the ear canal; (2) the inner layer is continuous with the middle ear mucosa; and (3) a layer of radial and circular fibers is sandwiched between. The ossicles form a chain spanning the cavity from the tympanic membrane laterally to the oval window medially. The manubrium of the malleus is attached to the tympanic membrane, the stapes footplate sits within the oval window, and the incus is suspended between the two at joints encapsulated with connective tissue. Two muscles, the tensor tympani and stapedius, attach to the malleus and stapes, respectively, via tendons. They function in the middle ear acoustic reflex pathway, modifying the

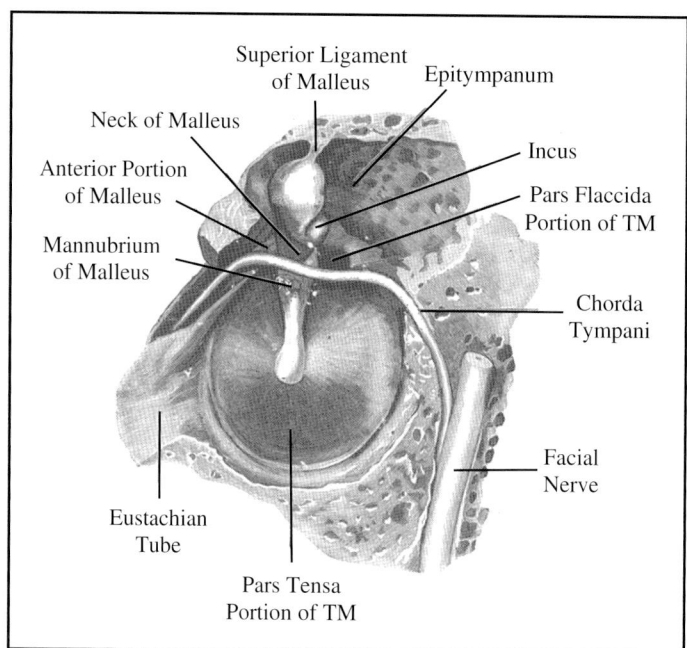

Figure 1–4. *View of the structures of the middle ear from the medial side (i.e., from the oval window looking laterally). A branch of the facial nerve (the chorda tympani) serving sensation to a portion of the tongue traverses the middle ear space. Adapted from Gray (1918). Used with permission.*

stiffness of the ossicular chain and thereby controlling the amplitude of sound transmitted to the cochlea. The trigeminal nerve innervates the tensor tympani whereas the facial nerve innervates the stapedius muscle. The trigeminal nerve also innervates the middle ear mucosa along with the glossopharyngeal and the vagus nerves. The vagus nerve innervates the lateral surface of the tympanic membrane along with small branches of the glossopharyngeal and facial nerves. The vascular supply to the middle ear is via the internal and external carotid arteries.

INNER EAR

The inner ear is contained within the temporal bone of the skull (the petrous portion). The bony labyrinth is a complex of channels in the temporal bone that form the osseous walls of the semicircular canals, a common connecting atrium referred to as the vestibule and the cochlea (Figure 1–5). The bony labyrinth is fluid-filled and lined with a periosteal (perilymphatic) membrane. Within the lumen of the channels formed by the inner surface of the periosteum lies a second inner compartment called the membranous labyrinth, which extends from the termination in the cochlear apex, via the vestibule, to the endolymphatic sac located on the cranial side of the vestibular aqueduct (Figure 1–6). The membranous labyrinth forms a complex of tubes and chambers that follows the general shape and outline of the bony labyrinth. The space between the periosteum of the bony labyrinth and the membranous labyrinth is filled with perilymph. The membranous labyrinth is filled with endolymph. The endolymphatic fluid space does not freely exchange fluid constituents with plasma, perilymph, or cerebrospinal fluid as there are no known

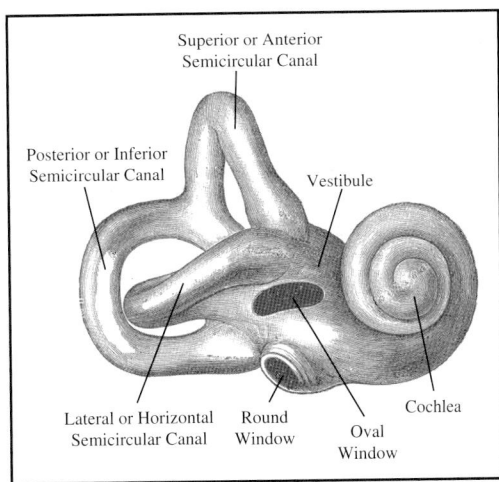

Figure 1–5. The bony labyrinth is a series of canals and chambers within the petrous portion of the temporal bone and is filled with perilymph. Adapted from Gray (1918). Used with permission.

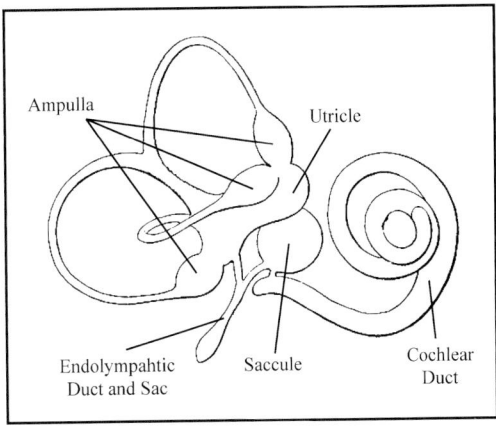

Figure 1–6. The membranous labyrinth follows the contours of the bony labyrinth, is filled with endolymph, and houses the sensory receptor organs. Adapted from Gray (1918). Used with permission.

channels, ducts, or vessels that would permit bulk flow between compartments. Furthermore, the endolymph extends to and bathes all sensory receptor surfaces.

There are six receptor organs in the inner ear. One receptor organ for hearing is in the cochlea (organ of Corti), three sensory organs for detecting head rotation are in the ampullae of the semicircular canals (cristae ampullaris), and two sensors for detecting linear motion (translation) of the head and head tilt in relation to the gravity vector of Earth are in the vestibule (otoconial or macular organs: the saccule and utricle).

All inner ear sense organs depend on hair cell receptors for the detection and encoding of sensory information. Hair cells are specialized non-neural mechanoreceptors (Figure 1–7). They are organized into sensory patches that form a single-layered receptor sheet overlying supporting cells and neural dendrites. At the apical (top) end of each hair cell, highly organized stereocilia emerge from the cuticular plate and form a rising staircase made of rows of successively taller stereocilia (see Figure 1–7B). Stereocilia are specialized microvilli. Together the stereocilia form a bundle called the stereociliary bundle. Immediately adjacent to the tallest row of stereocilia in the hair bundle is a single true cilium called the kinocilium. Mature cochlear hair cells lack a kinocilium but retain a remnant known as the basal body. The stereociliary bundle, with or without a prominent kinocilium, forms the sensory transduction apparatus. Transduction occurs when shearing forces bend the hair bundle toward the kinocilium or toward the tallest stereocilia (see Figure 1–7). This movement stretches stereociliary tip links spanning the distance between neighboring rows of stereocilia causing transduction channels to open, which in turn causes the hair cell to depolarize (Hudspeth, 1989a, 1989b; Hudspeth & Logothetis, 2000; Pickles & Corey, 1992). Bending the hair bundle away from the kinocilium (or tallest stereocilia in the cochlea) reduces tension in tip links, closes transduction channels, and hyperpolarizes the hair cell. This bidirectional action operates best along a "polarization" line or vector. Shearing forces in a direction perpendicular to the cells polarization vector have little or no effect on the hair cell. Therefore, hair cells are said to be functionally polarized and the preferred direction can be characterized as a morphological polarization vector (MPV, Figure 1–8).

Tight junctions between hair cell apical cuticular plates and supporting cells form a transport barrier between the endolymphatic and perilymphatic spaces thus preventing significant free exchange of constituents (Figure 1–9). Below the transport barrier at the apical surface of hair cells, perilymphatic fluids mix with extracellular interstitial fluids to bathe hair cells, supporting cells, and neural dendrites that make up the sensory neuroepithelium. The sensory neuroepithelium itself is firmly attached to the periosteum and wall of the bony labyrinth for the vestibular organs; the neuroepithelium is suspended across the cochlear partition by the basilar membrane in the cochlea. Neural dendrites innervating hair cells of the various sensory epithelia are the most distal extension of processes originating from primary afferent neurons in the vestibular (Scarpa's) and cochlear ganglia.

Together, each primary afferent and all hair cells that it innervates form a sensory unit (see Figure 1–8B) and this represents the fundamental unit of information input to the brain. Information about head motion or sound must reach the brain and to do that it ultimately must be encoded as neural discharge patterns in individual sensory units. This begins with the stimulus-induced shearing forces that act on the transduction channels to cause changes in the hair cell membrane potential.

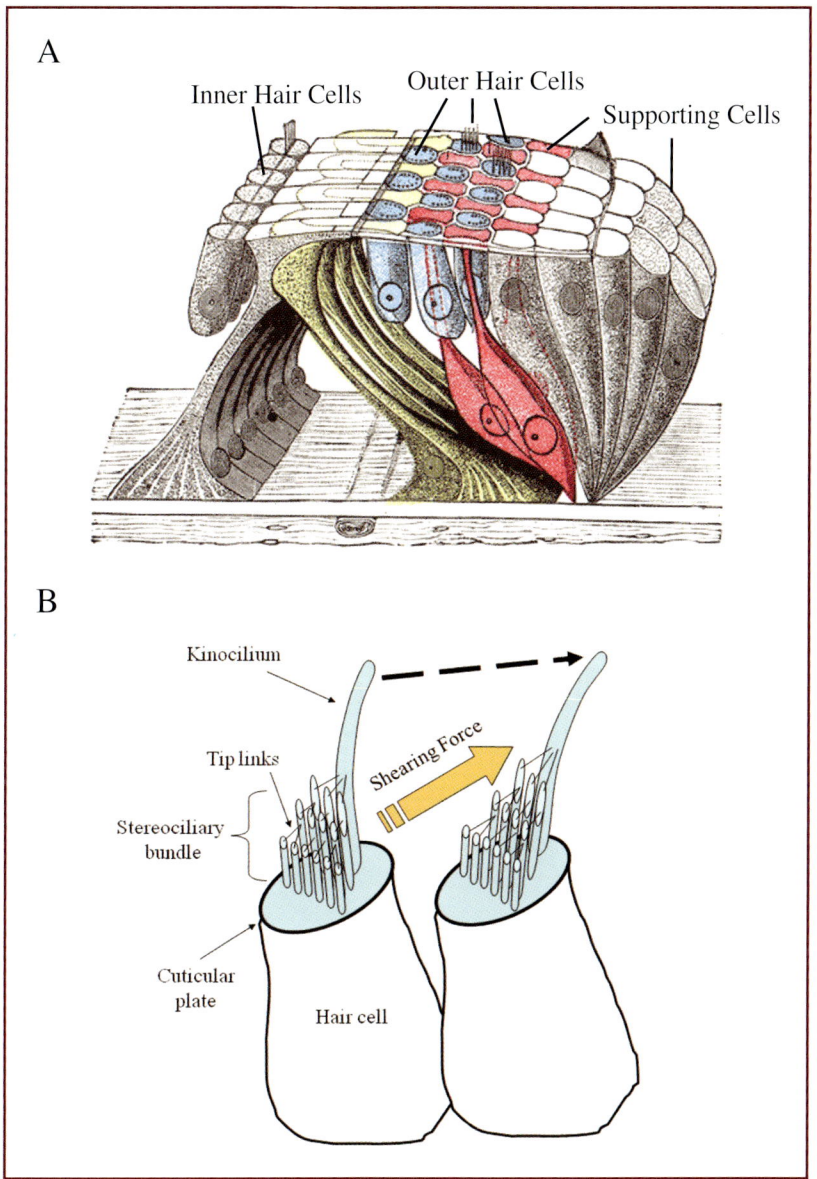

Figure 1–7. *The inner ear sensory receptors are hair cells that have stereocilia protruding from the top. Panel A shows the organization of the inner and outer hair cells in the organ of Corti where the stereocilia are shown explicitly on one IHC and two OHC, and appear as dots on the remaining hair cells. Panel B shows an example of vestibular hair cells, which have a kinocilium. The stereociliary bundles atop the hair cells consist of progressively taller rows as shown in panel B. Movement of the bundle toward the tallest row depolarizes the hair cell whereas movement in the opposite direction hyperpolarizes the cell. Panel A is adapted from Gray (1918). Used with permission. Panel B is from Jones and Jones (2007). Used with permission from Lippincott, Williams and Wilkins.*

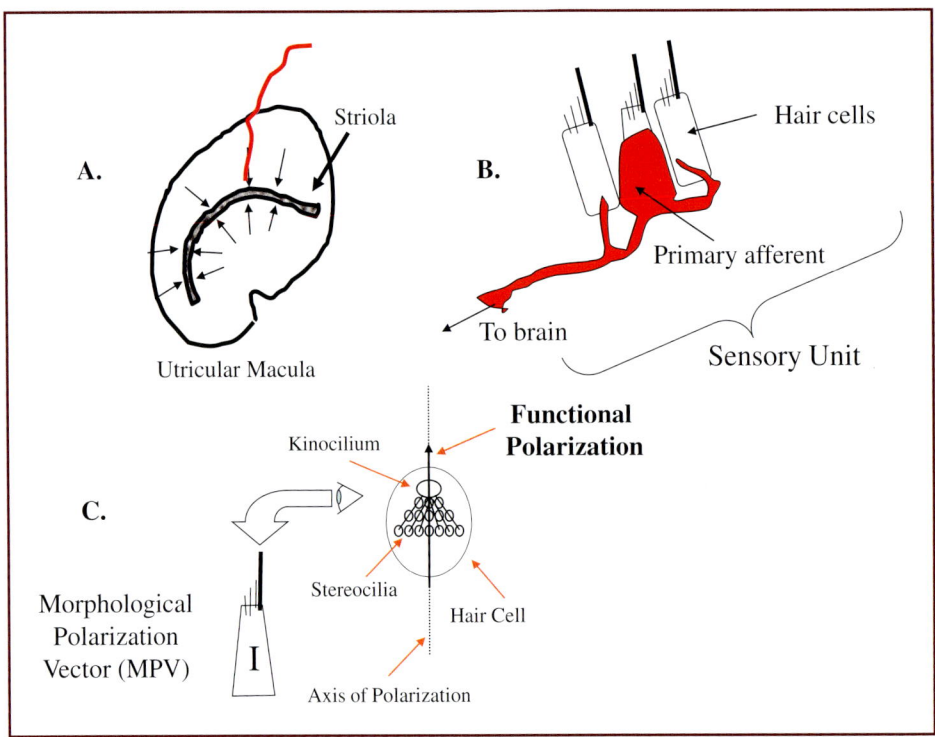

Figure 1–8. Using the utricle as an example, the orientation (or polarization) of stereociliary bundles is depicted across the epithelia surface (**A**), within a given hair cell (**B** and **C**). From Jones and Jones (2007). Used with permission from Lippincott, Williams and Wilkins.

Changes in hair cell membrane potential modulate the release of neurotransmitter from the base of the hair cell. The release of transmitter modulates the discharge rate of the postsynaptic primary afferent neuron and thus initiates neural activity corresponding to the head motion or sound stimulus. The primary afferent in turn transmits the encoded information to the brain.

The blood supply to the inner ear is mediated by the internal auditory artery, a branch of the anterior inferior cerebellar artery or the basilar artery. Postganglionic autonomic neurons arise from the superior cervical ganglion and innervate blood vessels within the inner ear (Shibamori, Tamamaki, Saito, & Nojyo, 1994; Spoendlin, 1981).

Potassium (K^+) is a key ion in the inner ear. It is found at unusually high concentrations in the extracellular fluid of the endolymph and contributes to the endolymphatic (endocochlear) potential that is about +80 mV in the scala media and +20 mV or so in the vestibular labyrinth (Salt, Melichar, & Thalmann, 1987; Wangeman, 2002). Potassium is the primary current carrier during transduction of sound and head motion. Potassium diffuses from the endolymph into hair cells through transduction channels in stereocilia (Figure 1–10). It then diffuses out of the basolateral surfaces of the hair cell, through the supporting cells, to connective tissue structures lining the endolymphatic fluid compartment (e.g., the spiral

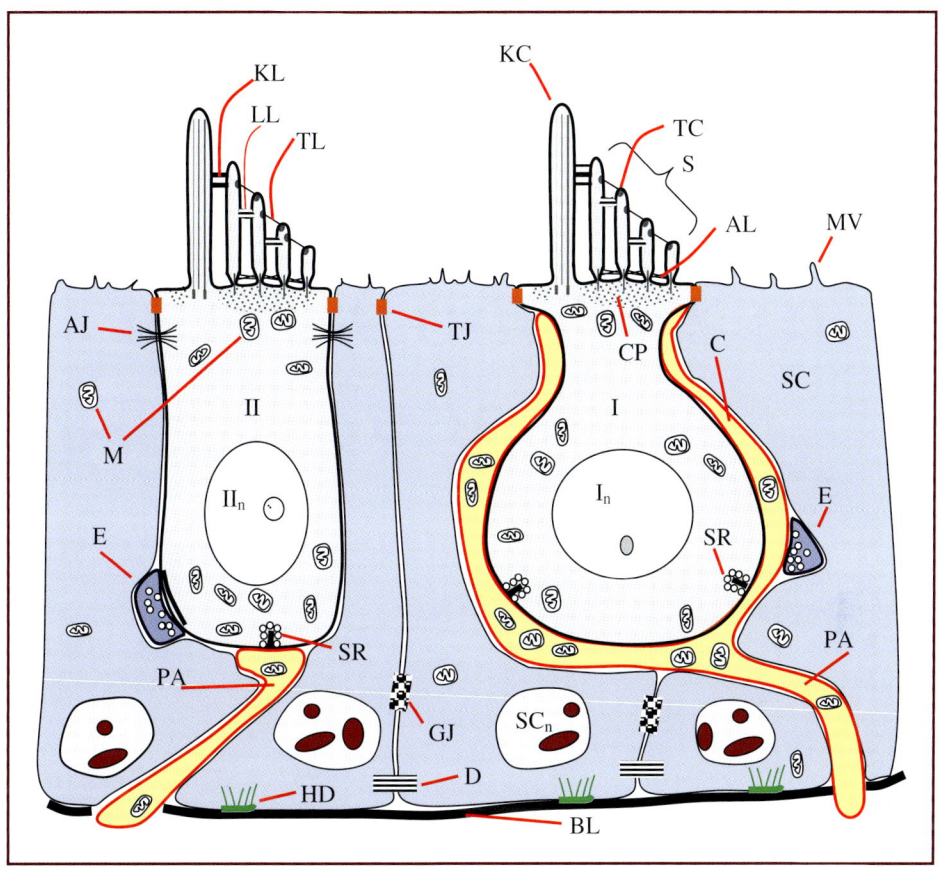

Figure 1–9. Summary schematic of major features of cellular elements of the sensory neuroepithelium. Type I and II vestibular hair cells are used in this example.

AJ, adherens junction
AL, ankle link
BL, basal lamina
C, afferent calyx
CP, cuticular plate
D, desmosome
E, efferent terminal
GJ, gap junction
HD, hemidesmosome
I, type I hair cell
I_n, nucleus, type I cell
II, type II hair cell
II_n, nucleus, type II cell
KC, kinocilium
KL, kinocilial link
LL, lateral links
M, mitochondria
MV, microvilli
PA, primary afferent
S, stereocilia
SC, supporting cell
SC_n, nucleus, supporting cell
SR, synaptic ribbon
TC, transduction channel
TL, tip link
TJ, tight junction

ligament of the cochlea). Potassium pumps in supporting cells and connective tissue of the spiral ligament move K$^+$ from the interstitial fluids into cells and help main- tain steep diffusion gradients along the intracellular path to the stria vascularis. Ultimately, K$^+$ diffuses to the dark cells in the vestibular labyrinth and to the cells

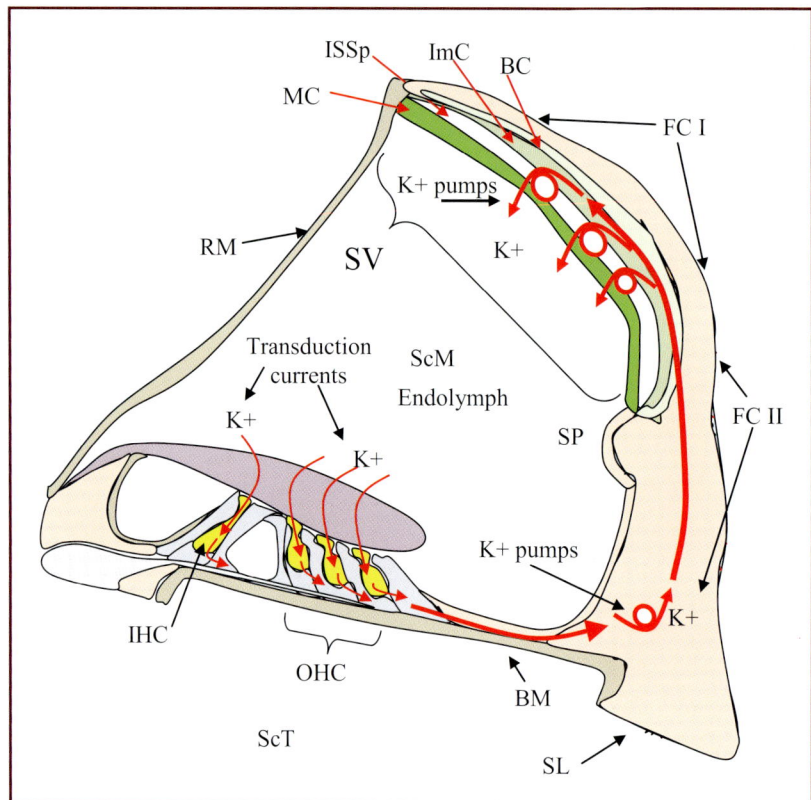

Figure 1–10. *Schematic cross section through the cochlea showing the organ of Corti with its inner (IHC) and outer (OHC) hair cells (yellow), the scala media (ScM) containing endolymph and the stria vascularis (SV) serving as the source of potassium flow into the ScM. The SV consists of basal cells (BC), intermediate cells (ImC), and marginal cells (MC). Potassium (K^+) flows from the endolymph into hair cells during transduction of sound. K^+ is recycled back to the marginal cells of the stria vascularis through gap junctions located in supporting cells of the organ of Corti (follow red arrows), fibrocytes (FCI & FCII) of the spiral ligament (SL), and BC, and ImC of the stria vascularis. K^+ pumps in marginal cells mediate the secretion of K^+ back into the endolymph. K^+ pumps in supporting cells, marginal cells and type II fibrocytes (FCII) maintain steep electrochemical diffusion gradients for K^+ cycling. Other structures highlighted include the basilar membrane (BM), intrastrial space (ISSp), scala tympani (ScT), and spiral prominence (SP).*

of the stria vascularis in the cochlea. These epithelial specializations function to maintain potassium concentrations in the endolymph (see Figure 1–10). The diffusion pathway through supporting cells and connective tissues utilizes gap junctions (Zhao, Kikuchi, Ngezahayo, & White, 2006), and once K^+ reaches the dark cell complex or stria vascularis, active K^+ transport pumps, K^+ cotransporters, and specialized K^+ channels (e.g., KCNE1, KCNQ1) function to return the K^+ back to the endolymphatic compartment (Lang, Vallon, Knipper, & Wangemann, 2007). The cycling of K⁺ is sometimes referred to as the K⁺ shuttle and it serves to conserve K⁺ and maintain high concentration of K⁺ in endolymphatic fluid. High concentrations of endolymphatic K⁺ and the endocochlear potential are critical for normal hair cell transduction.

Cochlea

Sound arrives in the cochlear fluids of the scala vestibuli via mechanical vibrations of the stapes footplate in the oval window. Footplate velocities produce pressure differences across the cochlear partition, which separates the scala vestibuli from the scala tympani. The pressure gradients induce vibrations in the cochlear partition and basilar membrane. The cochlear partition contains the sensory cells responsible for transducing vibrations of the basilar membrane into biological signals. Cochlear hair cells are separated into two types, inner (3000 in number) and outer hair cells (12,000 to 15,000 in number). Inner hair cells are heavily innervated by type I primary afferent neurons whereas outer hair cells are innervated by relatively few primary afferent neurons (mainly type II) but receive a dense innervation by efferent neurons (Figure 1–11; Spoendlin, 1966). The functional role of each hair cell type also appears to be distinct in that inner hair cells mediate the primary sensory input to the central nervous system regarding ambient sounds, whereas outer hair cells are involved in shaping the tuning characteristics of the basilar membrane and cochlear partition itself. Thus, inner hair cells provide the signals that are reported to the brain about vibrations on the

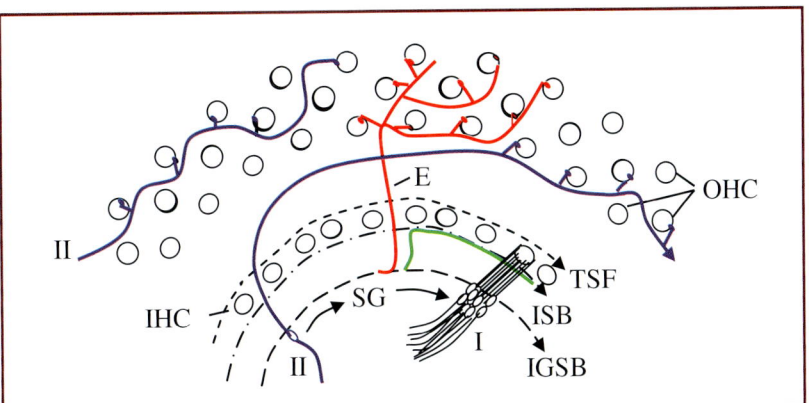

Figure 1–11. Schematic drawing of the neural connections to cochlear inner (IHC) and outer (OHC) hair cells. Afferents: Type I radial afferents to inner hair cells (I). Type II afferents (blue) innervate outer hair cells (II). Efferents: Efferent fibers to outer hair cells (E) are part of the medial olivocochlear projections (red). Efferent terminals on primary afferent terminals (green) are part of the lateral olivocochlear projections and pass through the intraganglionic spiral bundle (IGSB) to the inner spiral bundle (ISB) to reach IHC region. Other structures identified include spiral ganglion (SG) and tunnel spiral fibers (TSF). Modified from Spoendlin (1985), Anatomy of Cochlear Innervation, American Journal of Otolaryngology, 6, pages 453–467 with permission from Elsevier, Ltd.

cochlear partition and outer hair cells work in the cochlear partition to sharpen tuning and enhance frequency analysis of these vibrations.

Signals reported to the brain are mediated by thousands of neurons called primary afferents or cochlear ganglion cells (approximately 30,000 afferents in the human cochlear nerve). Each of these neurons terminates forming a synapse on only one inner hair cell. Each inner hair cell is innervated by 10 to 30 of these primary afferent neurons. Thus, each inner hair cell is associated with a cohort of neurons, where each neuron encodes signals as action potentials. The pattern of action potentials reflects the activity of the hair cell. This is remarkable as inner hair cells lie adjacent to one another and collectively form a single line extending along the cochlear partition from base to apex. Thus, each inner hair cell occupies a unique position along the length of the cochlear partition and encodes information about vibrations at its particular location. This systematic spatial arrangement of hair cells forms the basis for the cochleotopic map and for the cochleotopic projections of primary afferent neurons to the central nervous system.

There are two sources of efferent axons innervating the adult mammalian cochlea (Figures 1–12 and 1–13; Guinan, 2006; Iurato, 1974): the lateral olivocochlear bundle (LOCB) and the medial olivocochlear bundle (MOCB). The LOCB originates in the lateral portion of the superior olivary nuclei (LSO) and projects to the cochlea onto the dendritic terminals of afferents contacting IHCs. The MOCB originates near the medial superior olivary nuclei, projects to, and synapses directly on OHCs. The fiber projections of both OCBs join in the roots of the vestibular component of the VIIIth nerve on their path to the cochlea. The OCB is segregated out and leaves the inferior vestibular nerve as a discrete efferent bundle called the vestibulocochlear anastomosis (or anastomosis of Oort, see Figure 1–13; Arnesen & Oson, 1984; Tian, Xu, Huang, Liao, & Huang, 2008). This fiber bundle enters the modiolus basally and then turns radially to reach the lateral edge of the spiral ganglion where it turns again to run spirally to the base and apex along the edge of the spiral ganglion. This spiraling efferent bundle is called the intraganglionic spiral bundle (IGSB). Along the length of the cochlear spiral, radial efferent fibers arise from the IGSB and project across the osseous spiral lamina into the organ of Corti (see Figure 1–11). The major neurotransmitters thought to mediate efferent actions are acetylcholine and gamma aminobutyric acid (GABA) for the LOCB and acetylcholine for the MOCB.

The cochlear partition is mechanically tuned to a range of vibration frequencies. The best frequency for a given cochlear position gradually changes along the length of the cochlea such that the highest frequency vibrations displace the base and lowest frequencies preferentially displace the apex. The systematic spatial distribution of best frequencies forms the basis for tonotopy in the cochlea. Cochlear tonotopy is represented by the inner hair cells by virtue of their respective positions along the length of the cochlea. Tonotopy is preserved by the discrete groups of primary afferent neurons innervating hair cells and this neurotopic organization is maintained in neural relays and projections of the auditory system all the way to the cortex. The cochleotopic organization of neural projections forms the cochlear "labeled lines" that reach the cerebral auditory cortex, any of which when activated will result in the sensation or perception of a sound. In the mature animal,

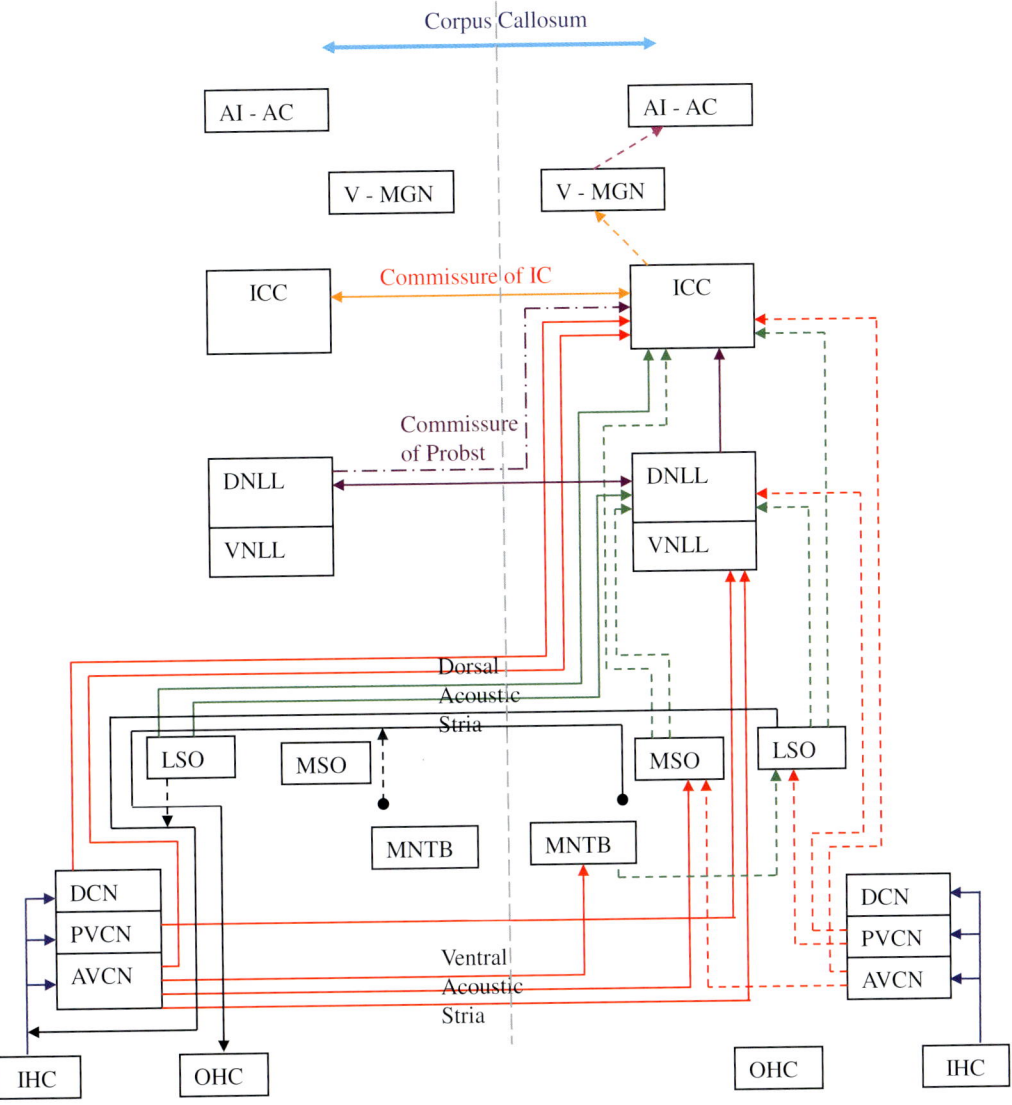

Figure 1–12. Afferent auditory pathways course through several relay nuclei in the brainstem before converging at the thalamus after which the neurons project to the primary auditory cortex where perception occurs. Efferent cell bodies (large black dots) located near the LSO and MSO project axons back to the cochlea via the lateral and medial olivocochlear bundles. IHC, inner hair cell; OHC, outer hair cell; AVCN, anteroventral cochlear nucleus; PVCN, posteroventral cochlear nucleus; DCN, dorsal cochlear nucleus; MNTB, medial nucleus of the trapezoid body; MSO, medial superior olive; LSO, lateral superior olive; VNLL, ventral nucleus of the lateral lemniscus; DNLL, dorsal nucleus of the lateral lemniscus; ICC, inferior colliculus complex; V-MGN, ventral medial geniculate nucleus; AC, auditory cortex

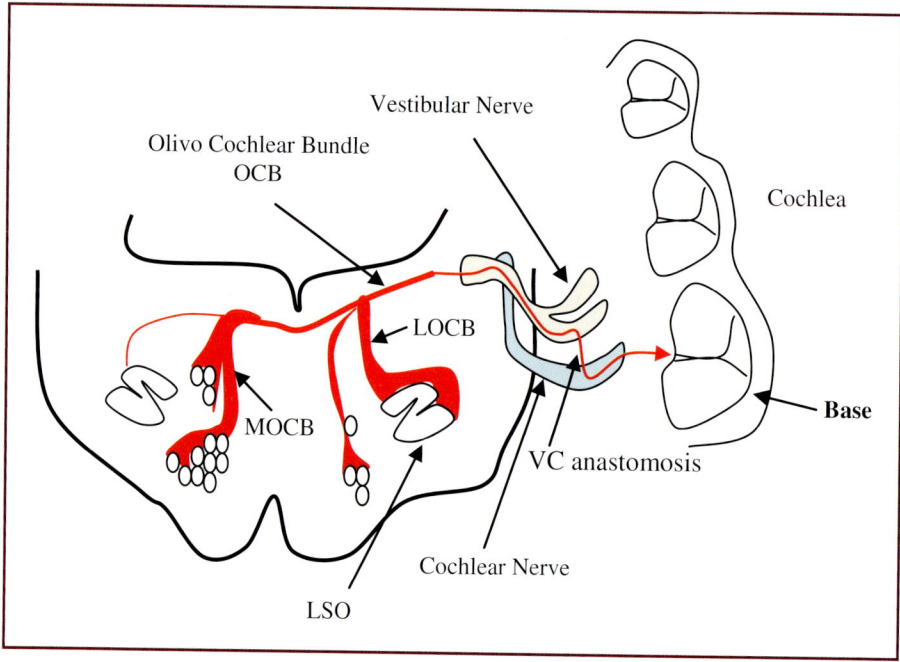

Figure 1–13. Projections of the olivocochlear efferent system. Lateral olivocochlear bundle (LOCB). Medial olivocochlear bundle (MOCB). Vestibulocochlear anastomosis (VC). Lateral superior olivary complex (LSO). Efferent projections of these cells pass from the brainstem nuclei via the olivocochlear bundle (OCB) to the cochlea as illustrated in red.

afferent and efferent systems operate together to ensure a precisely adjusted stream of sensory information to the brain.

Semicircular Canals

The three semicircular canals each lie predominantly in their own distinct plane and at an angle of approximately 90 degrees relative to the planes of the other two canals (see Figures 1–5 and 1–6). Each is named according to its anatomical position: the posterior canal, the anterior or superior canal, and the lateral or horizontal canal. The fact that the canals are not exactly at 90-degree angles virtually guarantees that any angular rotation will stimulate all three to some extent.

One end of each canal dilates to form a swelling called the ampulla. Near the central region of the ampulla, the lumen of the membranous duct is occluded by a structure called the cupula, which encounters the free flow of endolymph in either direction through the lumen of the duct (Figure 1–14). The cupula overlies the sensory epithelium (hair cells and supporting cells) of the crista ampullaris. The mechanical coupling of the cupula to the crista provides the critical link between head rotation, flow of endolymphatic fluid within the canal, and mechanical shearing action on hair cell stereociliary bundles. The stereociliary bundle and kinocilium extend into the cupula and movements of the cupula shear the stereociliary bundle to initiate transduction. The sensory

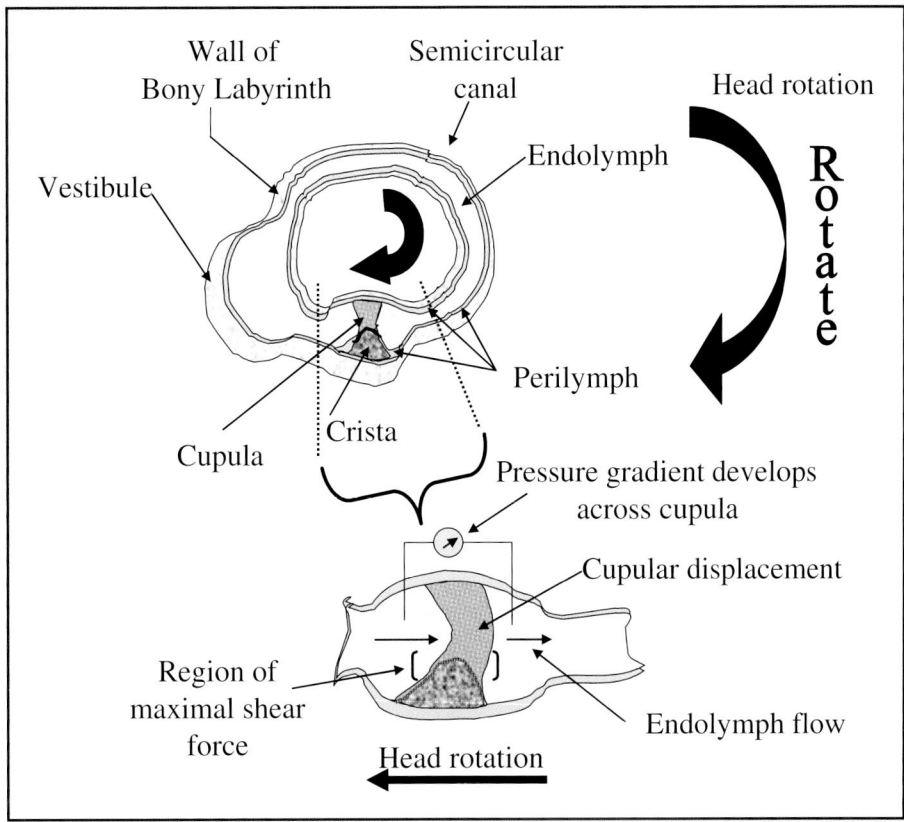

Figure 1–14. *Schematic showing the effect of head rotation on the movement of endolymph within the semicircular canal and eventual displacement of the cupula. Cupular displacement shears the stereociliary bundle and will depolarize or hyperpolarize the hair cells depending on the direction of shearing. From Jones and Jones (2007). Used with permission of Lippincott, Williams and Wilkins.*

receptors (about 4000 for each crista based on a surface area of 0.4 mm² and hair cell densities from Merchant, Velazquez-Villasenor, Tsuji, Glynn, Wall, et al., [2000]) consist of type I and type II hair cells (see Figure 1–9) with type I cells predominant at the crest of each crista. With respect to any given cross section of a canal, endolymph can undergo bulk flow in only two directions along the duct. The polarization vectors of individual hair cells in a given cristae are uniformly oriented in the same direction such that there is a collective increase or decrease in afferent activity with endolymphatic flow in one direction or the other.

The adequate stimulus for canal sensors is rotational movement of the head. When the head is turned or rotated, the temporal bone and bony labyrinth undergo an identical angular acceleration. This results in a relative movement (displacement) between the bony labyrinth and the endolymphatic fluid. With this displacement, pressures are developed across the cupula, which in turn cause the cupula to be deformed (see Figure 1–14). The cupular displacements introduce shearing motion in hair cell stereociliary bundles that ultimately lead to the modulation of activity in primary afferent neurons.

The canal receptors are preferentially activated by angular acceleration in planes that approximate the anatomical plane of the canal (horizontal, anterior or posterior). Each canal gives a unique response to a given rotation and the three canals together provide a three-dimensional vector representation of the motion (Rabbitt, 1999). Each of the three components is represented to the brain by the primary afferent discharge patterns of the three respective ampullary nerves. This vectored information is used by the brain to: (1) perceive head motion and position at any given point in time; (2) control the position of the eyes and head as well as tone in postural muscles; and (3) anticipate and implement necessary changes in postural motor outflow to support ongoing volitional motor programs.

Otoconial Organs

The two otoconial organs, the saccule and utricle, are present in the vestibule of the labyrinth. These sensors serve to transduce and encode linear acceleration of the head during translational head movements as well as detect gravity and encode head tilt. Together, the two kinds of linear acceleration stimuli (gravity and linear translation) often are referred to as gravitoinertial stimuli. The otoconial organs often are simply referred to as gravity receptors.

The neuroepithelium of the saccule and utricle is attached firmly to the membranous labyrinth, the periosteum and bony wall of the vestibule. Each macula contains around 3000 to 3500 type I and type II hair cells (see Figure 1–9, estimates based on a surface area of 0.5 mm^2 and densities from Merchant, et al., 2000). The sensory sheet (macula) of the utricle lies in a plane that approaches the horizontal plane, whereas the saccule is nearly perpendicular to the utricle and therefore is aligned closely with the vertical plane (Figure 1–15A). Like the cochlea and cristae, the gravity receptors have a membrane lying atop the sensory receptors into which the stereociliary bundle projects. The gravity receptors have an otoconial membrane (Figure 1–15B), a gelatinous matrix into which is incorporated high-density calcium carbonate crystals called otoconia. The otoconia produce an overall membrane density that is considerably greater than the tissues and fluids surrounding the membrane. The presence of otoconia ensures that the membrane will undergo movement relative to the surrounding tissue when stimulated by linear acceleration during head motion or head tilt.

The polarization vectors for individual macular hair cells are systematically placed such that the vectors are organized across the epithelium to produce particular surface patterns (see Figure 1–8). Both maculae have a centrally located striola, which is characterized in part by a line of polarization vector reversal. In the utricle, for example, the line of polarity reversal separates a polarization field pointing largely medial from a field pointing largely lateral (see Figure 1–15). Along the striola, adjacent hair cells on either side of the reversal line have stereociliary bundles oriented in opposite directions. The polarization vectors for the maculae and cristae are highly significant because they provide the fundamental basis for encoding stimulus direction in the vestibular system. Thus, hair cells not only detect and encode the amplitude and dynamics of a

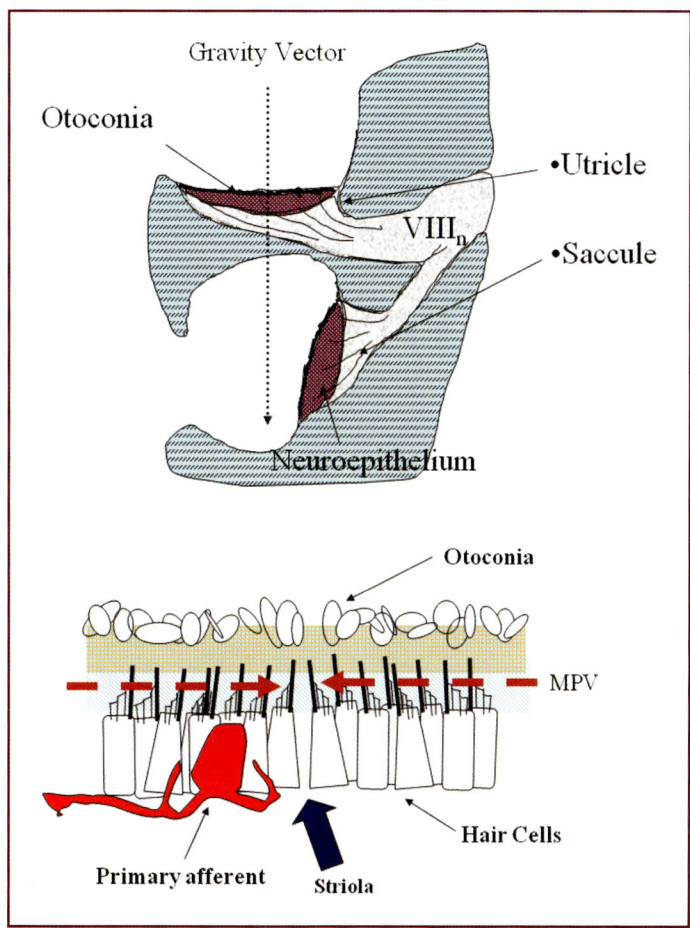

Figure 1–15. **A.** *The utricle and saccule are oriented orthogonal to one another with the utricle oriented horizontally and the saccule predominately vertically.* **B.** *Schematic showing a cross section through the macula highlighting organization of polarization vectors across the line of polarity reversal (striola). Movement of the otoconial matrix will stimulate (i.e., depolarize) a subset of hair cells while simultaneously hyperpolarizing another subset due to the arrangement of morphological polarization vectors across the epithelium. Both panels are from Jones and Jones (2007). Used with permission of Lippincott, Williams and Wilkins.*

stimulus but collectively encode the direction of a stimulus.

Two divisions, superior and inferior, make up Scarpa's ganglion (the vestibular branch of the VIIIth nerve). At birth in the human, about 13,000 cells comprise the superior division and 8500 for the inferior division (Velazquez-Villasenor, Merchant, Tsuji, Glynn, Wall, et al., 2000). The peripheral dendrites form three kinds of afferent terminals (see Figures 1–8 and 1–9). Calyx afferents innervate type I hair cells only.

Bouton afferents innervate type II hair cells only. Dimorphic afferents consist of both calyx and bouton synapses and innervate both types of hair cells. Afferents project to the vestibular nuclei in the brainstem and to the cerebellum. The vestibular nuclei in turn send controlling signals to the eyes, spinal cord, and brainstem motor nuclei. They also send signals to the cerebellum, which in turn relays back information to refine vestibular adjustments for smooth and accurate action. Neural projections from the vestibular nuclei project through the thalamus to the cortex for perception of head position and motion (Figure 1–16).

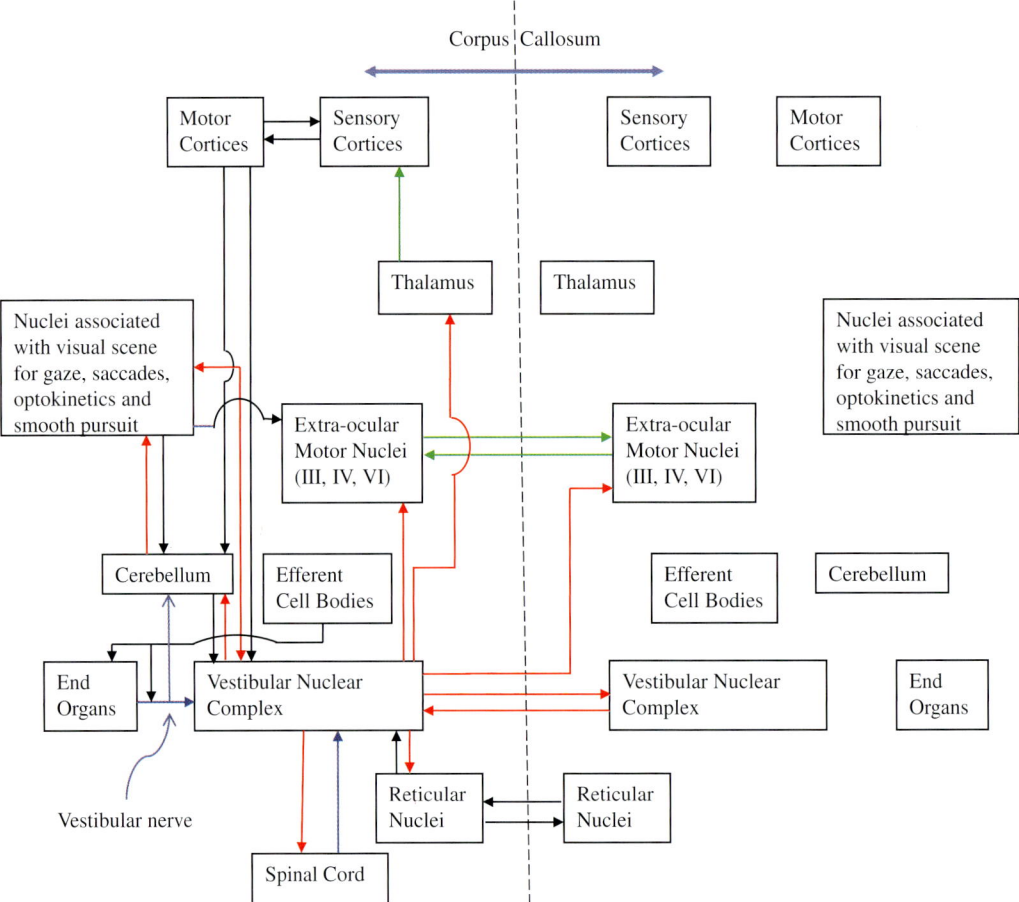

Figure 1–16. *Afferent vestibular pathways serve several functions. The primary sensory input is via the vestibular nerve to the vestibular nuclear complex, then through the thalamus to the sensory cortex where perception occurs. Reflex output motor pathways exist from the vestibular nuclear complex to the cranial nerve nuclei controlling the eye muscles and to motor neuron pools in the spinal cord controlling the lower limbs and neck muscles. Cerebellar control as well as input from the visual system modulates output of the vestibular nuclear complex. Efferent cell bodies located near the vestibular nuclear complex project to the labyrinth to regulate hair cell and primary afferent activity. From Jones, Jones, Mills, and Gaines (2009). Used with permission of Thieme Medical Publishers, Inc.*

Vestibular efferent fibers arise from neuron cell bodies located bilaterally in the dorsal brainstem near the motor nuclei of the abducens (VIth) cranial nerve (Goldberg, Brichta, & Wackym, 2000; Simmons, 2002). Vestibular efferent cells project to the vestibule via the inferior vestibular nerve and distribute their terminals within the neuroepithelia of all vestibular sensory organs. As with the cochlea, the vestibular afferent and efferent systems operate together to ensure a precisely adjusted stream of sensory information to the brain.

CONCLUSIONS

This brief description of the mature ear underscores a number of key steps or requirements for normal hearing and vestibular sensation to occur:

1. Vibrations (sound or head motion) must reach the inner ear sensory organs.
2. Vibrations must be transduced into biological signals.
3. Attributes of vibrations (e.g., amplitude, frequency, etc.) must be encoded.
4. The encoded information must be reliably reported to central relays.
5. Central relays must reliably transfer and analyze information from both ears in order to create a three-dimensional map or "picture" of the environment.

These key elements are not present at the beginning of the life of an animal. Indeed, starting as a single cell clearly requires that all of these key elements be formed and implemented before hearing and vestibular sensation can take place. This process of formation represents the ontogeny of hearing and vestibular sense. We next examine how these key elements for hearing and vestibular function come into being during ontogeny.

GENERAL RESOURCES

Geisler, D. (1998). *From sound to synapse.* New York, NY: Oxford University Press.

Gulick, W. L., Gescheider, G. A., & Frisina, R. D. (1989). *Hearing: Physiological acoustics, neural coding, and psychoacoustics.* New York, NY: Oxford University Press.

Highstein, S. M., Fay, R. R., & Popper, A. N. (2004). *The vestibular system.* New York, NY: Springer-Verlag.

Martin, F. (1981). *Medical audiology: Disorders of hearing.* Englewood Cliffs, NJ: Prentice-Hall.

Pickles, J. O. (2009). *Introduction to auditory physiology* (3rd ed.). New York, NY: Academic Press.

Webster, D. B., Popper, A. N., & Fay, R. R. (1992). *The mammalian auditory pathway: Neuroanatomy.* Berlin, Germany: Springer-Verlag.

LITERATURE CITED

Arneson, A. R., & Oson, K. K. (1984). Fibre population of the vestiubulocochlear anastomosis in the cat. *Acta Otolaryngologica, 98,* 255–269.

Carlson, B. M. (2004). *Human embryology and developmental biology* (pp. xiii–xv). Philadelphia, PA: Mosby.

Flexer, S. B., & Hauck, L. C. (1993). *Random House unabridged dictionary.* New York, NY: Random House.

Goldberg, J. M., Brichta, A. M., & Wackym, P. A. (2000). Efferent vestibular system: Anatomy, physiology, and neurochemistry. In A. J. Beitz & J. H. Anderson (Eds.), *Neurochemistry of the vestibular system* (pp. 61–94). Boca Raton, FL: CRC Press.

Gray, H. (1918). *Anatomy of the human body.* Philadelphia, PA: Lea & Febiger. Retrieved July 28, 2009 from Bartleby.com, 2000. http://www.bartleby.com/107/

Guinan, J. J. (2006). Olivocochlear efferents: Anatomy, physiology, function, and the measurement of efferent effects in humans. *Ear and Hearing, 27,* 589–607.

Hamburger, V., & Hamilton, H. L. (1951). A series of normal stages in the development of the chick embryo. *Journal of Morphology, 88,* 44–92.

Howard Hughes Medical Institute. (2001). *The genes we share with yeast, worms, flies and mice: New clues to human health and disease,* Chevy Chase, MD: Howard Hughes Medical Institute.

Hudspeth, A J. (1989a). How the ear's works work. *Nature, 341,* 397–404.

Hudspeth, A. J. (1989b). Mechanoelectrical transduction by hair cells of the bullfrog's sacculus. In J. H. J. Allum & M. Hulliger (Eds.), *Afferent control of posture and locomotion* (pp. 129–135). Amsterdam, The Netherlands: Elsevier.

Hudspeth, A. J., & Logothetis, N. K. (2000). Sensory systems. *Current Opinion in Neurobiology, 10,* 631–641.

Iurato, S. (1974). Efferent innervation of the cochlea. In W. D. Keidel & W. D. Neff (Eds.), *Auditory system: Anatomy and physiology. Handbook of sensory physiology* (Vol. V1, pp. 261–282). New York, NY: Springer-Verlag.

Jones, S. M., Jones, T. A., Mills, K. N., & Gaines, C. G. (2009). Anatomical and physiological considerations in vestibular dysfunction and compensation. *Seminars in Hearing, 30,* 231–241.

Jones, T. A., & Jones, S. M. (2007). Vestibular evoked potentials. In R. F. Burkard, M. Don, & J. J. Eggermont (Eds.), *Auditory evoked potentials: Basic principles and clinical applications* (pp. 622–650). Baltimore, MD: Lippincott Williams & Wilkins.

Kispert, A., & Gossler, A. (2004). Introduction to early mouse development. In H. Hedrich (Ed.), *The laboratory mouse* (pp. 175–191). Amsterdam, The Netherlands: Elsevier.

Lang, F., Vallon, V., Knipper, M., & Wangemann, P. (2007). Functional significance of channels and transporters expressed in the inner ear and kidney. *American Journal of Physiology Cell Physiology, 293,* C1187–C1208.

Merchant, S. N., Velazquez-Villasenor, L., Tsuji, K., Glynn, R. J., Wall, C., & Rauch, S. D. (2000). Temporal bone studies of the human peripheral vestibular system. Normative vestibular hair cell data. *Annals of Otology, Rhinology, and Laryngology Supplement, 181,* 3–13.

Mullen, L. M., Li, Y., & Ryan, A. F. (2004). Normal development of the ear in the human and mouse. In P. J. Willems (Ed.), *Genetic hearing loss* (pp. 1–131), New York, NY: Marcel Dekker.

Pickles, J. O., & Corey, D. P. (1992). Mechano-electrical transduction by hair cells. *Trends in Neuroscience, 15,* 254–259.

Rabbitt, R. D. (1999). Directional coding of three-dimensional movements by the vestibular semicircular canals. *Biological Cybernetics, 80,* 417–431.

Salt, A. N., Melichar, I., & Thalmann, R. (1987). Mechanisms of endocochlear potential generation by stria vascularis. *Laryngoscope, 97,* 984–991.

Shibamori, Y., Tamamaki, N., Saito, H., & Nojyo, Y. (1994). The trajectory of the sympathetic nerve fibers to the rat cochlea as revealed by anterograde and retrograde WGA-HRP tracing. *Brain Research, 646,* 223–229.

Simmons, D. D. (2002). Development of the inner ear efferent system across vertebrate species. *Journal of Neurobiology, 53,* 228–250.

Soukhanov, A. H. (1992). *The American Heritage dictionary of the English language.* Boston, MA: Houghton-Mifflin.

Spoendlin, H. (1966). *The organization of the cochlear receptor.* Basel, Switzerland: Karger.

Spoendlin, H. (1985). Anatomy of cochlear innervation. *American Journal of Otolaryngology, 6,* 453–467.

Spoendlin, H. (1981). Autonomic innervations of the inner ear. *Advances in Oto-Rhino-Laryngology, 27,* 1–13

Tian, G., Xu, D., Huang, D., Liao, H., & Huang, M. (2008). The topographical relationships and anastomosis of the nerves in the human internal auditory canal. *Surgical Radiology and Anatomy, 30,* 243–247.

Velazquez-Villasenor, L., Merchant, S. N., Tsuji, K., Glynn, R. J., Wall, C., & Rauch, S. D. (2000). Temporal bone studies of the human peripheral vestibular system 2. Normative Scarpa's ganglion cell data. *Annals of Otology, Rhinology, and Laryngology, 109,* 14–19.

Wangemann, P. (2002). K+ cycling and the endocochlear potential. *Hearing Research, 165,* 1–9.

Zhao, H. B., Kikuchi, T., Ngezahayo, A., & White, T. W. (2006). Gap junctions and cochlear homeostasis. *Journal of Membrane Biology, 209,* 177–186.

2

Nature of Cells and Tissues

> **KEY POINTS**
>
> 1. Within each cell is a complex environment containing many organelles to carry out the cell's functions.
> 2. The human body contains 80 to 100 trillion cells, most of which are immersed in a fluid environment that is strictly regulated to maintain homeostasis.
> 3. Cells combine to form connective, epithelial, neural, and muscle tissues. Tissues combine to form the organ systems in the human body.
> 4. Cell signaling pathways control the development of tissues and organs. Such pathways include hedgehog, notch, and Wnt signaling.
> 5. Polarity of individual cells is controlled by molecular signals intrinsic to the cells.
> 6. Planar cell polarity (PCP, specifically Wnt-PCP pathway) defines the axes for a tissue sheet.
> 7. Convergent extension relies on PCP to properly elongate a structure along one dimension while shortening it in another dimension.

THE CELL

The human body is comprised of some 80 to 100 trillion individual cells. Most cells must be immersed in a special aqueous environment in order to survive. Indeed, our bodies are 60% fluid. Extracellular fluid (20%) includes interstitial fluid that surrounds and bathes cells as well as plasma that surrounds blood cells. Extracellular fluids contain the necessary electrolytes, biochemical fuels, and building blocks (carbohydrates, amino acids, lipids, etc.), as well as oxygen necessary for cell survival. These fluids also provide a reservoir for the removal of waste produced during the course of cellular metabolism. Intracellular fluid (40%), called cytosol, is contained inside the cell and surrounds the cell's organelles. The cell membrane is a phospholipid bilayer that forms a selective barrier between the extra- and intracellular fluid spaces. Pores, channels, and other specialized transport mechanisms exist at the cell membrane to control movement of substances into and out of the cell. Therefore, substances are not uniformly distributed across cell membranes. The aqueous environment must be strictly regulated, its constituents controlled and kept within certain limits. The maintenance of the cellular environment is called homeostasis.

Inside the cell are many organelles and associated key molecular components that perform necessary cell functions (Figure 2–1). Three organelles, the nucleus, the endoplasmic reticulum, and the mitochondria are particularly important in the context of genetics and protein synthesis (Chapter 3). Two key macromolecules associated with genetics and protein synthesis are the chromosome and the ribosome, respectively. The cytoskeleton is a fifth organelle that contributes to the cell's strength, shape, mobility, and polarity.

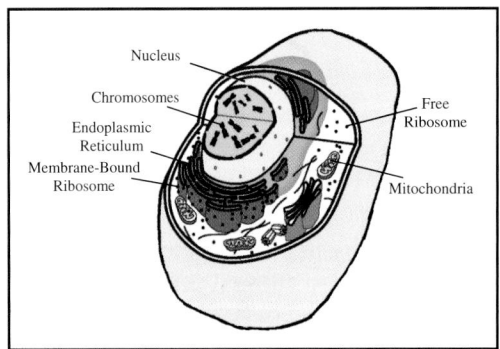

Figure 2–1. *The cell and its constituents. Modified from a figure courtesy of the National Human Genome Research Institute.*

Every cell in the human body has at least one nucleus at some stage in its formation. For example, red blood cells have a nucleus early in their development, but have none at maturity. Skeletal muscle cells have many nuclei and these cells are called multinucleated. Within each nucleus one finds the chromosomes, which contain the genetic material (i.e., deoxyribonucleic acid or DNA) that makes us who we are.

Mitochondria are organelles that synthesize most of a cell's adenosine triphosphate (ATP), the molecule that is used to supply the chemical energy needed for cellular metabolism, biochemical synthesis, and chemical/mechanical work. In response to cellular stress, mitochondria also may generate reactive oxygen species (i.e., potentially damaging hydrogen peroxide and the superoxide free radical) and can initiate apoptosis (cell death). Cells need energy to perform a multitude of functions and the mitochondria are well equipped to convert the molecules derived from the food we eat into the necessary energy molecules. The number of mitochondria in a given cell depends on the cell's energy requirements. Generally, the fraction of cell volume given to mitochondria is considerable. The size of mitochondria varies between 0.5 to 1 micron in diameter. Some cells are sufficiently large and create such a high energy demand that extremely high numbers of mitochondria are present (e.g., liver cells contain 1000 or more mitochondria). Cells in the ear utilize the chemical energy from ATP for a number of functions such as maintaining potassium levels in endolymphatic fluids.

Ribosomes are responsible for protein synthesis. Two populations of ribosomes exist in the cell cytosol. Some ribosomes are attached or bound to the membrane of endoplasmic reticulum and are called membrane-bound ribosomes. Free ribosomes float freely in the intracellular fluid. Whether a ribosome is membrane-bound or free at any given moment in time is driven by the particular protein that the ribosome is synthesizing.

The endoplasmic reticulum (ER) plays important roles in production, packaging and transport of proteins and lipids and contributes to calcium homeostasis within the cell. The ER works together with several other organelles such as the Golgi complex to maintain exocytotic (secretory) and endocytotic pathways that move constituents out of and into the cell, respectively. The ER has a rough or smooth appearance, which is due to ribosomes associated with the ER. Rough ER has ribosomes bound to the ER membrane, which are major sites for protein synthesis. Smooth ER has no ribosomes and plays a role in lipid metabolism.

The cytoskeleton coordinates cell shape, organizes and establishes mechanical properties across the cell, mediates linkage between cells, and is important for cell division. The cytoskeleton plays a major role in mechanisms generating cell polarity and is under the control of molecular signals such as the Wnt-PCP pathways described later in the chapter. Cells are often oriented such that some consistent landmark or function occupies a particular location within the cell. To orient a cell, there must be at least one internal or external distinguishing feature that determines how the cell is "pointed." Such features serve to define the "cell polarity" (Figure 2–2). Individually, epithelial cells including hair cells of the inner ear generally have intracellular features that reflect cellular polarization. These include specific locations for particular membrane transport systems, the location of cell-to-cell junctions, the membrane position of protein signaling complexes, the surface containing the mechanoreceptor apparatus (stereocilia and kinocilium), and so on. All of the internal asymmetric organization is supported and maintained by the cytoskeleton.

The cytoskeleton itself is composed of actin filaments, microtubules, and intermediate filaments and is highly complex and dynamic (Figure 2–3). Actin and tubulin filaments are formed by assembling together protein subunits. Actin filaments form arrays or networks on the inner surface of the cell membrane adding strength and shape to the surface membrane. They provide a support framework for the cell against an underlying surface and form the infrastructure of lamellipodia and filopodia in migrating cells. They are the primary structural element of stereocilia in inner ear hair cells. Actin filaments anchor various surface structures to the cell interior including adherens junctions between epithelial cells and junctions with the extracellular matrix. Actin filaments also form part of the contractile apparatus in muscle cells.

Generally, microtubules radiate out from a single structural element known as the centrosome. Microtubule radiations essentially extend to all regions of the cytosol from this central point and they serve to determine the location of various organelles within the cell and direct intracellular communications and vesicle transport. Notably, microtubules form the core structure of the hair cell kinocilium as well as true cilia and flagella. A single tubulin filament is relatively unstable when free in the cytosol, but forms stable structures as microtubules. A microtubule is a thin rigid tube composed of tubulin filaments. Tubulin subunits are added to the growing end of the microtubule during polymerization, whereas tubulin subunits are removed from the end of the microtubule when it is disassembled. The filaments within the microtubule are highly dynamic and there is a constant turnover of subunits. In the normal cell, the rates of

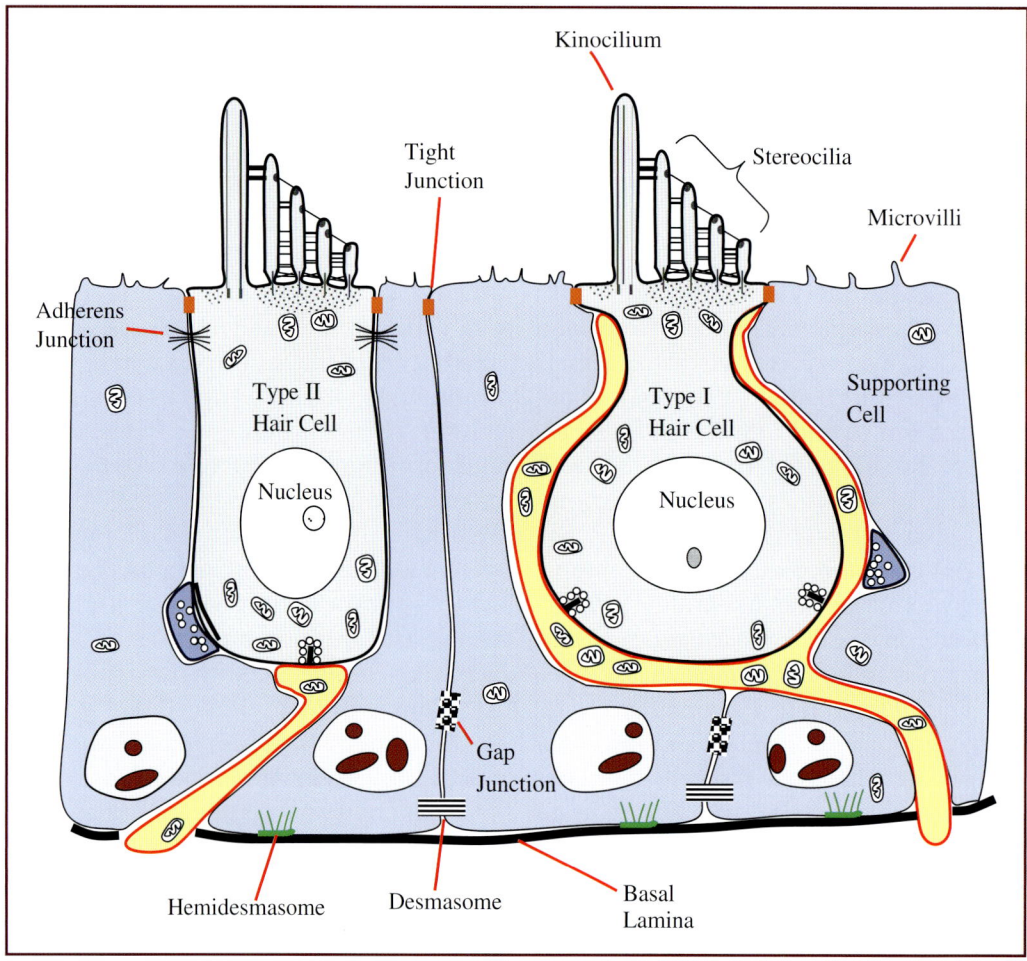

Figure 2–2. Schematic of inner ear hair cells highlighting the various junctions between cells and cellular organelles that define apical (e.g., stereocilia, microvilli, tight junctions) versus basal (e.g., basal lamina, synaptic contacts, desmasomes) polarity of a given cell.

growth and disassembly of microtubules are controlled by signaling pathways. Stable polymerization of tubulin filaments also requires a stable site to grow. There are two major sites that favor stable assembly of tubulin subunits. One has already been mentioned, the tip of a microtubule. A second site is known as the microtubule organizing center (MTOC). The centrosome is the MTOC in animal cells. The MTOC has initiation sites or "seeds" for polymerization of tubulin subunits, thus greatly increasing the rate of growth of stable microtubules. This process of stable polymerization is called nucleation and the centrosome is referred to as a nucleation site. A similar process exists for actin polymerization. Nucleation sites regulate where and how many microtubules and filaments can be formed in the cytosol.

There are a number of different classes of intermediate filaments. Those found in inner ear tissue include keritins, vimentin, and neurofilaments (Deschesne,

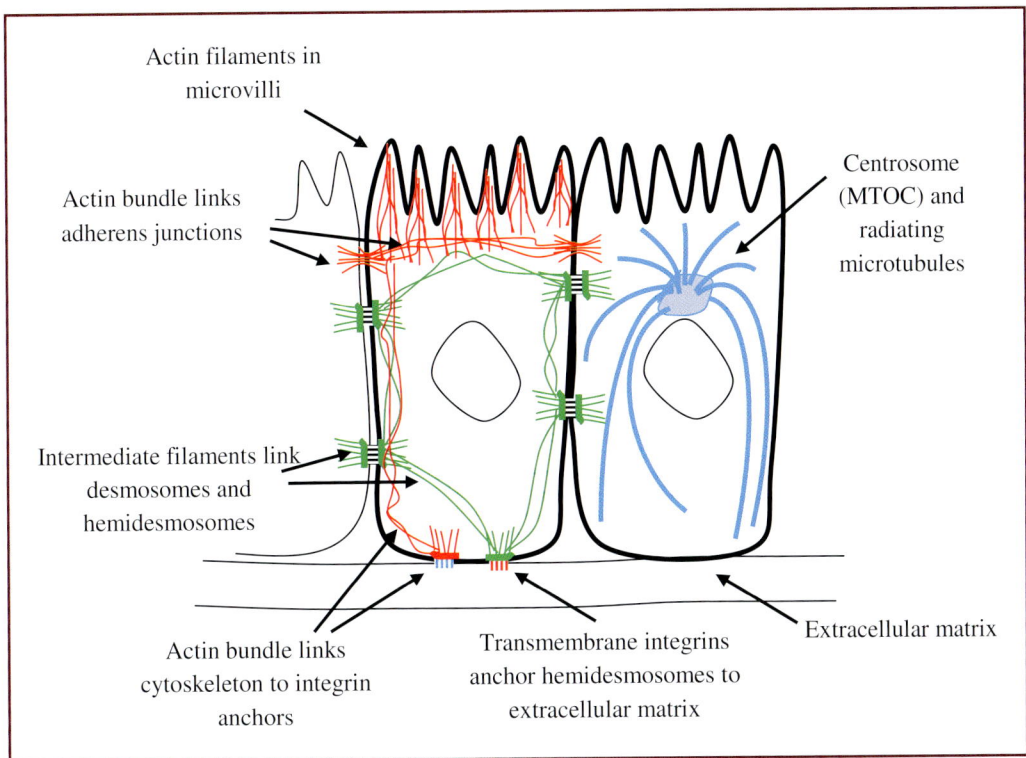

Figure 2–3. The cytoskeleton of a cell.

1992). These filaments form part of the structural framework of cells and many serve to anchor the intercellular adhesions such as desmosomes and hemidesmosomes to the cell interior.

The above discussion touched on only a few of the key organelles and structures within the cell. Normal cell function requires many organelles working together. Indeed, all of a cell's organelles create a dynamic environment in order to maintain homeostasis of the cell and ultimately homeostasis of the tissues and organism. Abnormal development or genetic mutations that affect the organelles can cause acquired or inherited hearing loss and/or vestibular deficits. We will learn more about chromosomal abnormalities and genetic mutations in Chapter 3.

TISSUES

Although many types of cells exist in the body (e.g., blood cells, sensory cells, germ cells, fibroblasts, macrophages, etc.), they combine in various ways to form tissues. Four types of tissues are found in vertebrates: (1) connective tissue, (2) epithelium, (3) nerves, and (4) muscle. Each is briefly described below and highlighted by examples in the auditory system.

Connective Tissue

Connective tissue is composed largely of an extracellular matrix that provides support, mechanical strength, and structure to an organism. Some examples include cartilage, bone, and subdermis under the skin. It also provides support and nourishment to epithelial tissues. The cellular component of connective tissue includes fibroblasts, osteoblasts (bone cells), chondrocytes (cartilage cells), adipocytes (fat cells), and smooth muscle cells. Connective tissue forms from the embryonic mesoderm. Blood vessels also form from mesoderm and are considered a component of connective tissue. One major constituent of this extracellular matrix is collagen, a protein that is secreted by connective tissue cells and distributed throughout the extracellular matrix (Figure 2–4). There are many types of collagen and cells secrete the most appropriate form depending on the amount of rigid support that may be required. For example, the bones of the skeleton are quite rigid whereas the connective tissue associated with the skin is more flexible. Other protein components of connective tissue include fibronectin, elastin, and laminin. When connective tissue is developed in close association with epithelia, the epithelia are distinguished from connective elements by referring to them as parenchyma (poured in beside), whereas the connective tissue is termed stroma (something to sit or lay on).

In the inner ear, examples of connective tissue include the cartilage of the pinna and ear canal, the basilar membrane, extracellular matrices of the maculae and cristae, spiral ligament, spiral limbus, and spiral lamina. Collagen types I, II, III, IV, V, and IX (among others) are present in the connective tissues of the ear, some demonstrating tissue specific localization (Cosgrove, Samuelson, & Pinnt, 1996; Khetarpal, Robertson, Yoo, & Morton, 1994; Nehrlich, 1995). Laminin and fibronectin are also found in the extracellular matrix. Cochlin, the most

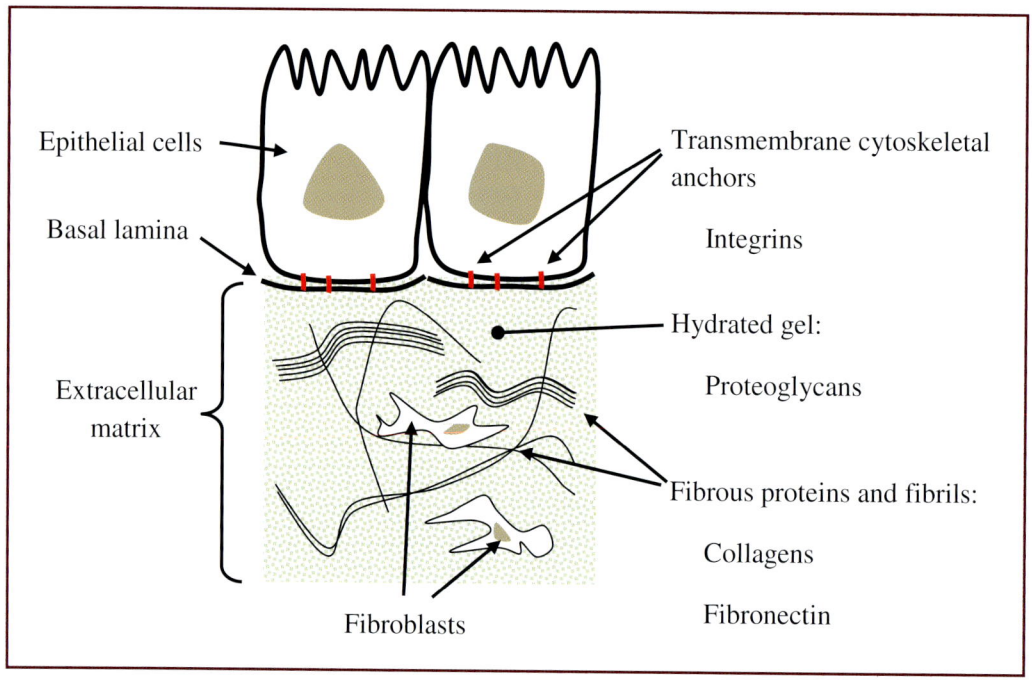

Figure 2–4. Extracellular matrix.

prevalent protein in the inner ear, is highly expressed in the connective tissues (Robertson et al., 2001). The basilar membrane is connective tissue that suspends the organ of Corti within the cochlea and this matrix is specialized in the cochlea. The thickness and width of the basilar membrane vary systematically from the base to apex of the cochlea contributing to the mechanical tuning of the cochlear partition. Vibrations reaching the cochlea travel along the basilar membrane resulting in stimulation of the hair cells. Hence, this specialized extracellular matrix plays a critical role in hearing.

Epithelium

Epithelia are coverings or linings that are composed of cells joined together by cell to cell junctions in a single layer (simple epithelium) or multiple layers (stratified epithelium). The shapes of the cells that form the epithelium provide additional descriptors such as columnar, cuboidal, or squamous (Figure 2–5). Epithelia form barriers that selectively control what may pass through. Epithelial membranes are supported by connective tissue as epithelia are avascular and must receive oxygen and nutrients from and give up carbon dioxide and other wastes to nearby capillaries in the connective tissue. General examples of epithelium include the skin and mucosal lining of the gut. Examples in the ear include the mucosal lining of the middle ear, the sensory epithelium of the inner ear (organ of Corti), and the stria vascularis.

All epithelial tissues, whether they line the gut or the organ of Corti, have an asymmetry that marks and distinguishes the apical and basal surfaces (see Figure 2–2). The apical surface forms microvilli and is a free surface that typically faces an air or fluid environment. Proteins near the apical end form tight junctions between cells to create a restrictive barrier. The basal surface is distinguished by the basal lamina (basement membrane), which is a thin (40 to 120 nanometers thick), tough, flexible sheet of extracellular matrix composed of noncellular protein secretions including nidogen, perlecan, laminin, and type IV collagen. The basal lamina is

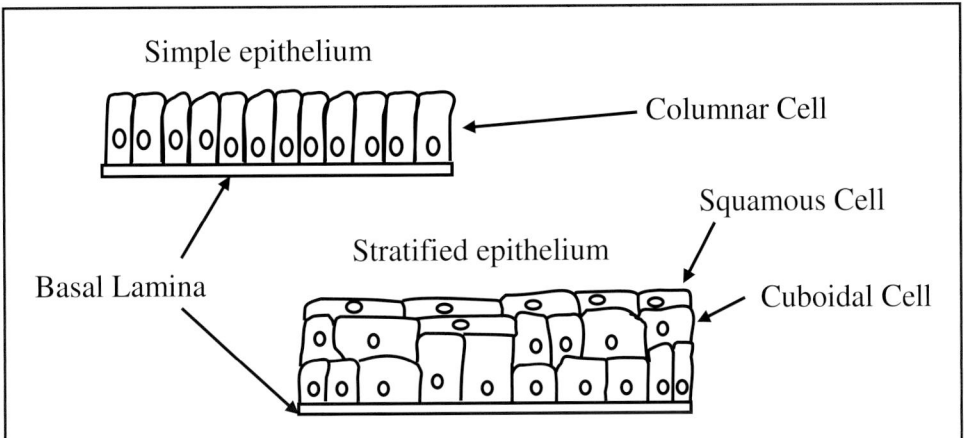

Figure 2–5. Schematic showing simple and stratified epithelium.

a small component of the extracellular matrix associated with the epithelial sheet. The basal lamina is formed from secretions of the epithelium and other cells lying within the supporting connective tissue or "stroma." Epithelial cells form attachments to the basal lamina via integrin receptor proteins. Integrin receptors are associated with cell-to-matrix adhesions or with hemidesmasomes, which are cell structures at the basal end of epithelial cells (see Figure 2–2). The basal lamina often forms the principal mechanical coupling between epithelial cells. In the case of inner ear sensory epithelia, epithelial supporting cells of the organ of Corti form attachments to the basilar lamina which is embedded in the extracellular matrix of the basilar membrane.

As previously stated, epithelial sheets serve as barriers controlling the transport and exchange of substances between compartments (by virtue of the tight junctions near the apical surface). The compositions of the two compartments on either side of the epithelial sheet generally are very different and the epithelium itself often is responsible for maintaining such differences. Epithelia create special environments to serve particular functions by partitioning structures and lining structural surfaces to enclose internal regions. This is well illustrated by the sensory epithelia of the inner ear. The apical side is exposed to the endolymph whereas the basal side is anchored to the extracellular matrix (i.e., basilar membrane) that is bathed in perilymph.

Neural Tissue

Individual nerve cells are called neurons (Figure 2–6). We are most familiar with the collective action of neurons; that is,

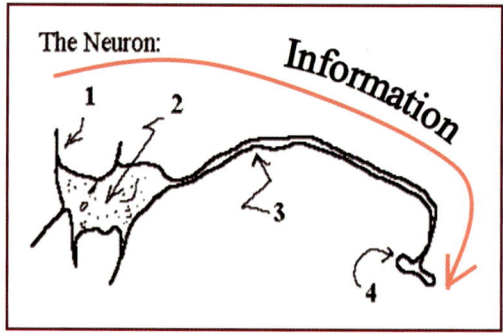

Figure 2–6. Schematic showing the major portions of a neuron: (1) dendrites, (2) soma or cell body, (3) axon, (4) axon terminal.

most people understand that the brain is responsible for our physical and mental activities (i.e., our behaviors). Neurons act to measure the environment and pass on the information to the central nervous system. The brain then interprets the information and responds appropriately. The single neuron then must be able to receive information from and relay information to other neurons or nonneural cells. Communication between nerve cells occurs at the synapse, which is literally a very narrow "space" (approximately 20 nm) that separates cells where they make contact. The apposed membranes and intracellular spaces immediately adjacent to the synapse are specialized for the transfer of chemical signals (chemical synaptic transmission).

Most sensory neurons have cell bodies in peripheral ganglia and display a polarization such that information (i.e., neural impulses) will "flow" in a certain direction. Anatomically the neuron receives signals at synapses located on dendrites (see Figure 2–6). The information is then transmitted from dendrites to the axon and axon terminal. The axon terminal forms a synaptic contact on another cell (postsynaptic cell), where the information can be passed on via synaptic transmission. Normal direction of information flow is typically: (1) dendrite, (2) cell body, (3) axon, and (4) axon terminal.

Neurons are said to be excitable and collectively they represent one of the two excitable tissues (i.e., nerve and muscle tissues) in the body. The word excitable refers to the fact that nerve cells (neurons) and muscle cells can initiate action potentials (Figure 2–7). Action potentials often

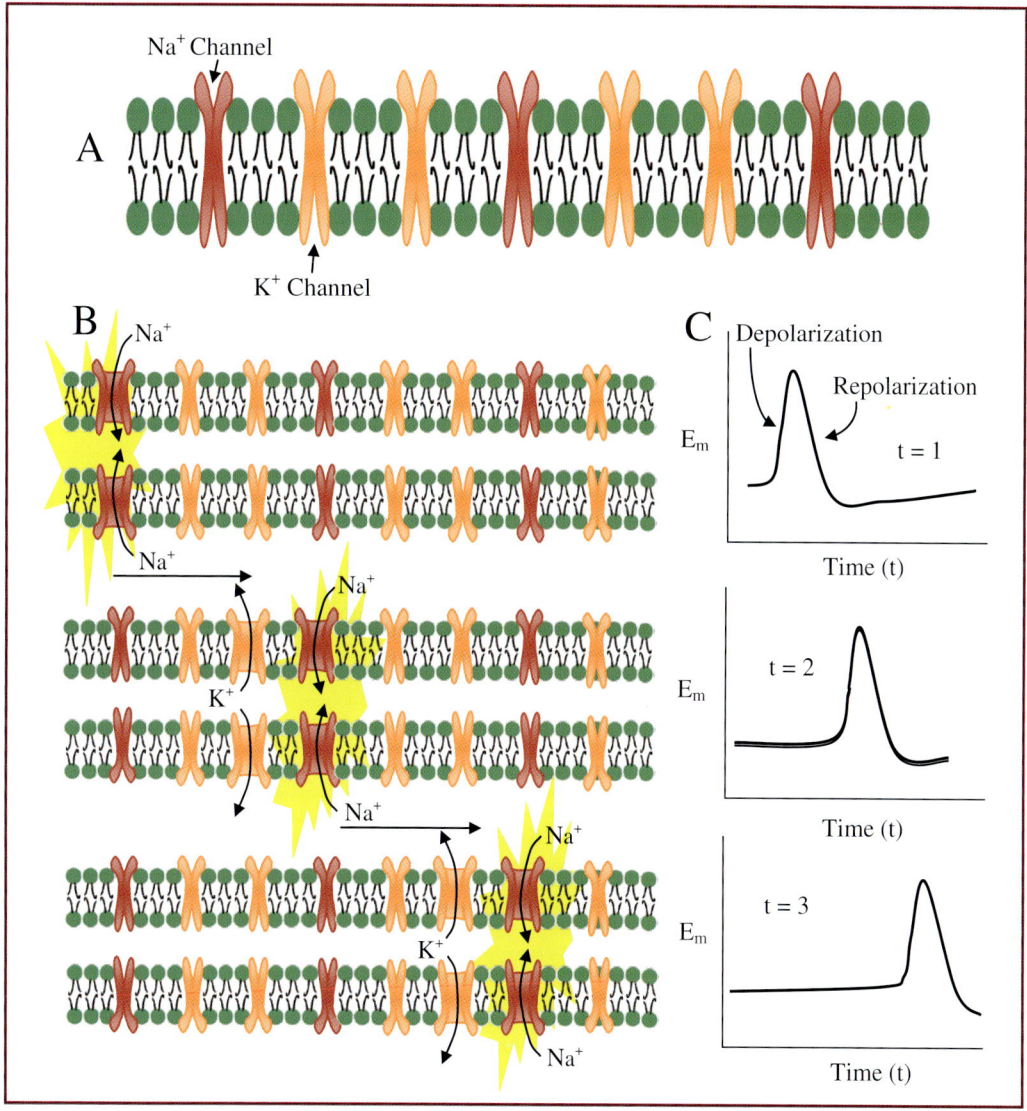

Figure 2–7. Neural tissue is excitable. The phospholipid bilayer of a neural cell membrane (**A**) contains ion channels that when activated allow Na^+ and K^+ ions to diffuse into or out of the cell, respectively (**B**). The influx of Na^+ produces the depolarization phase of an action potential while the efflux of K^+ repolarizes the membrane (E_m) to resting voltage levels (**C**). Depolarization at one location on the neuron activates neighboring regions of cell membrane, thus propagating the action potential along the entire length of the neuron.

are referred to as neural impulses or spikes. Once an action potential is initiated or triggered, it travels or propagates by conduction along the cell membrane to all parts of the cell. In a neuron, the action potential provides a signal that can travel without degradation for long distances over the entire cell (some neurons are several meters long). Action potential signals in neurons therefore can be used to transmit information reliably. The information can be represented by a code of action potential impulses. The pattern or code then is passed between cells giving each cell in turn the opportunity to examine, modify, or simply pass on the impulse code.

The excitability of a neuron, its resting state, and the action potential it produces depend on electrochemical gradients that are established across the cell membrane. Proteins bound by the cell membrane serve as channels for ion diffusion or as transporters for ions and nutrients. The numbers and types of channels present in the membrane determine the behavior of the cell. Active transporters (e.g., Na^+/K^+ ATPase) are responsible for establishing and maintaining ionic gradients (electrochemical gradients) that exist across the cell membrane. Voltage-sensitive channels are critical for generation and propagation of action potentials. In neurons, the principal channels for the production of action potentials are Na^+ and K^+, where Na^+ is the dominant depolarizing ion and voltage sensitive Na^+ channels control action potential initiation. The makeup of the population of membrane channels may change through development and these changes can transform the function of the cell. One example is the inner hair cell, and although the hair cell is not a neuron, it behaves like one during development. Before hearing begins in the neonate, the inner hair cell generates Ca^{++} dependent action potentials. This capability disappears as hearing begins and the disappearance is associated with the appearance of a very fast K^+ conductance that prevents the Ca^{++} regenerative depolarization from reaching threshold (Kros, Ruppersberg, & Rusch, 1998).

The nervous system is made up of billions of individual neurons and has three functional components: (1) the input, (2) the processor, and (3) the output (Figure 2–8). In biological terms, the input systems represent afferent or sensory systems. The output systems act on the world and are called efferent or motor systems. The processor evaluates information provided to it by the sensory input systems

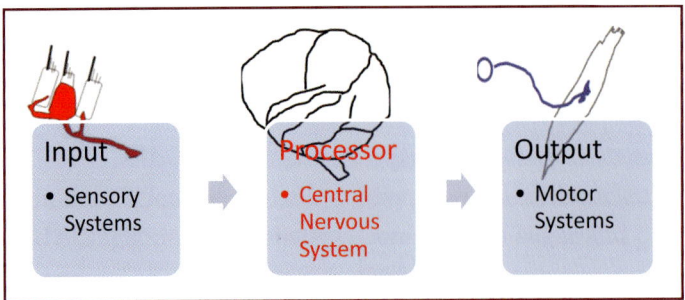

Figure 2–8. General organization of the nervous system. Peripheral sensory input is processed and integrated in the central nervous system to produce final output to muscles and glands.

and determines an output plan, which it then implements through the motor systems. The overall organization of the nervous system is such that the outside world and many internal variables are monitored by peripheral sensors, which translate events of these worlds into afferent or sensory signals that travel in the form of action potentials to the brain and spinal cord. Sensory information passes from the periphery via primary afferent neurons into the various laminae of the dorsal horn of the spinal cord or directly into sensory relay nuclei in the brainstem. The primary afferent neurons synapse with second order neurons, which in turn pass information more rostrally to higher nuclear assemblies. In most sensory modalities, second (or higher) order neurons pass with or without intervening synapses to the contralateral side and ultimately to the thalamus. Thalamic relays in turn pass information to the primary receiving cortices on the side opposite the original stimulus. This is the simplified projection scheme that summarizes the sensory modalities of hearing, vision, vestibular sense, touch, and taste. The one exception to this scheme is found in the olfactory system, which bypasses the thalamus and reaches the olfactory cortices in pathways coursing from the olfactory mucosa through the olfactory bulb.

We will be concentrating on sensory systems of the inner ear, that is, the vestibular and auditory systems. The sensory pathways of each follow the general scheme outlined above for the nervous system. Both have quite complex processing pathways within the brain and have efferent (output) pathways that regulate the sensory end organs (reviewed in Chapter 1).

Muscle

Forty to fifty percent of the body mass is muscle. Muscles are classified as either striated or smooth (Figure 2–9) and are composed of many cells that collectively

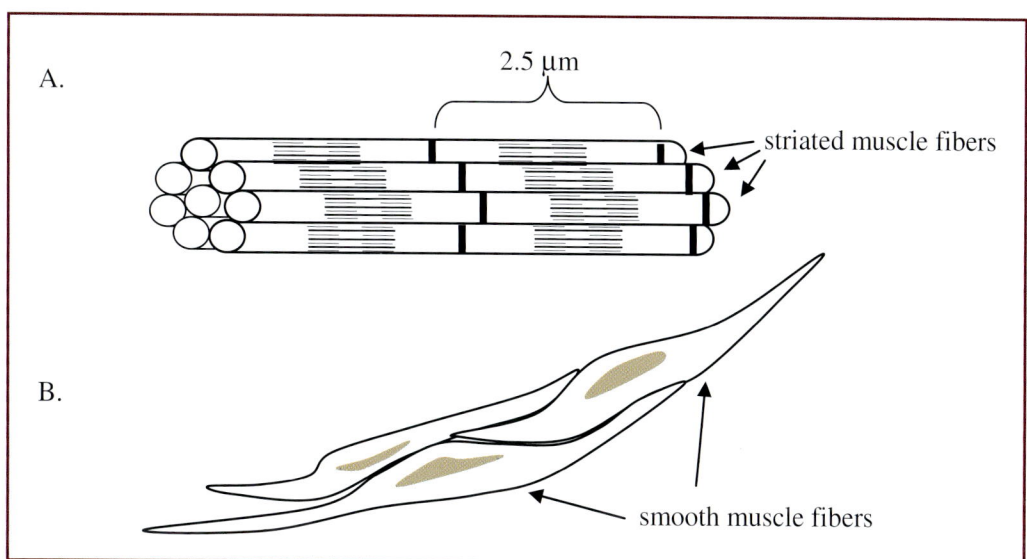

Figure 2–9. Schematic contrasting striated and smooth muscle tissues.

contract to produce movement or develop tension. Muscles sometimes are called effectors because they implement the instructions signaled by neurons serving to control them. Skeletal and cardiac muscle cells are striated whereas smooth muscle has no striations.

Skeletal muscle cells usually are very large cells that are multinucleated and have a striking banding pattern (striations) due to highly organized subcellular structures. Each muscle fiber is composed of hundreds to thousands of myofibrils that are arranged in parallel and run the length of the cell. Myofibrils are composed of myofilaments, which are segmented into discrete sarcomeres that contain contractile proteins actin and myosin. In muscle, the passing of an action potential causes the muscle cells to contract, thus providing a means to move and develop force. Skeletal muscle is innervated by alpha motor neurons, the cell body of which is located in the central nervous system (ventral spinal cord and brainstem). Skeletal muscles are under conscious voluntary control from neurons in the cerebral motor cortex, are responsible for the movements of our joints and limbs, and form the basis of our somatic musculature. Skeletal muscles act on the environment by developing forces and creating motion thus providing the work necessary to implement our behaviors. In contrast, cardiac and smooth muscle are innervated by postganglionic neurons of the autonomic motor system and generally are not under conscious voluntary control. Cardiac muscle is, of course, the muscle of the heart and is responsible for circulating blood through the vascular system. Smooth muscle is responsible for peristalsis (rhythmic contractions) of hollow viscera including the intestines, the contraction of blood vessels, and dilation of the pupils among many other things.

The middle ear muscles are skeletal muscle. The vasculature of the ear contains smooth muscle.

CELL SIGNALING

The development of tissues and organs in multicellular organisms is controlled by several signaling pathways that work together. A few examples of these pathways include hedgehog signaling, Notch signaling, and Wnt signaling. There are many others, but these have been well studied in development of the ear and are reviewed briefly here.

Signaling molecules are secreted from cells and act on neighboring cells. Signaling molecules are ligands that generally must bind to membrane receptors, which then initiate a cascade of actions within the receiving cell. The receptors for secreted ligands may be contained on or within cells in the region (paracrine signaling) or on the secreting cell itself (autocrine signaling). The cascade is referred to as a signal transduction pathway, where transduction is the process of activating the pathway. Some ligands can also diffuse within a cell to find receptor sites (binding cites or domains on large molecules) within different cellular compartments. In other cases, the ligand may be a molecule that is expressed by a cell and then translocated to the surface of the cell membrane where it remains firmly attached to the cell yet it is exposed to the extracellular space. In this case, the action of the ligand can occur only when the cell comes into contact with a cognate receptor (high affinity binding site) on the surface of a neighboring cell. Thus, cell-to-cell contact causes binding to occur and initiates signal transduction leading to a response in the adjacent neighboring cell.

Hedgehog Signaling Pathway

The hedgehog (Hh) proteins are secreted signaling molecules originally discovered in *Drosophila* (a fly) where there is only one form. In vertebrates, there are three forms of hedgehog proteins: (1) Sonic (Shh), (2) Desert (Dhh) and (3) Indian (Ihh). When Hh binds with receptors (named patched and smoothened) the corresponding Hh pathway is activated (Hooper & Scott, 2005; Figure 2–10). This in turn leads to the activation of Hh target genes, which encode gene regulatory proteins (Gli1, Gli2, Gli3). Gli1 and Gli2 are thought to be transcription activators (i.e., activate transcription of genes) whereas Gli3 may act as either a repressor (suppress transcription) or activator. Transcription and transcriptional regulation are discussed in Chapter 3. The hedgehog signaling pathway is thought to play a role in defining the prosensory domain in the otocyst

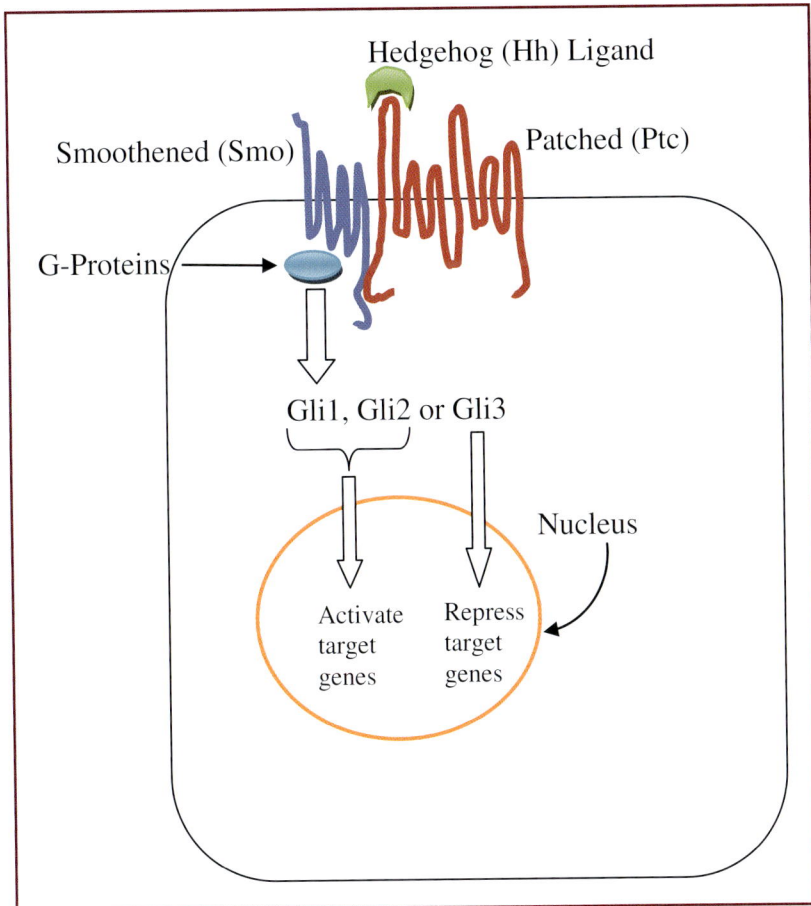

Figure 2–10. Sonic hedgehog signaling pathway.

from which the sensory epithelium will develop (Driver et al., 2008). Mutations in *Gli3*, a downstream target of Shh signaling cause Pallister Hall syndrome, which has a predominately low-frequency sensorineural hearing loss as part of the phenotype (Kang, Graham, Olney, & Biesecker, 1997).

Notch Signaling Pathway

Notch is an example of a surface receptor that interacts with surface ligands such as Jagged and Delta. Mammals have four different Notch receptors, two Jagged (Jag) ligands, and three Delta-like (Dll) ligands (Radtke, Schweisguth, & Pear, 2005). The Notch receptor has two portions: (1) an extracellular portion (or domain) that binds with the ligand; and (2) an intracellular domain (Notch Intracellular Domain, abbreviated NICD) that will move into the nucleus when the receptor and the Notch pathway are activated (Figure 2–11). When NICD moves into the nucleus it binds with a transcription factor known as CSL (C-Promotor Binding Factor 1, CBF1, in humans) increasing the transcription of Notch target genes such as the "hairy and enhancer of split" (HES) family of genes.

The Notch pathway mediates a process of lateral inhibition. Lateral inhibition involves the expression of a ligand in one cell that activates receptors on adjacent cells. Activation of the receptors on adjacent cells in turn prevents (or inhibits) the adjacent cells from adopting a particular developmental fate. More succinctly, ligands expressed on one cell bind with and activate receptors on surrounding cells. Activation of the receptors on surrounding cells inhibits their differentiation. One example of this occurs with the activation of the Notch1 receptors on

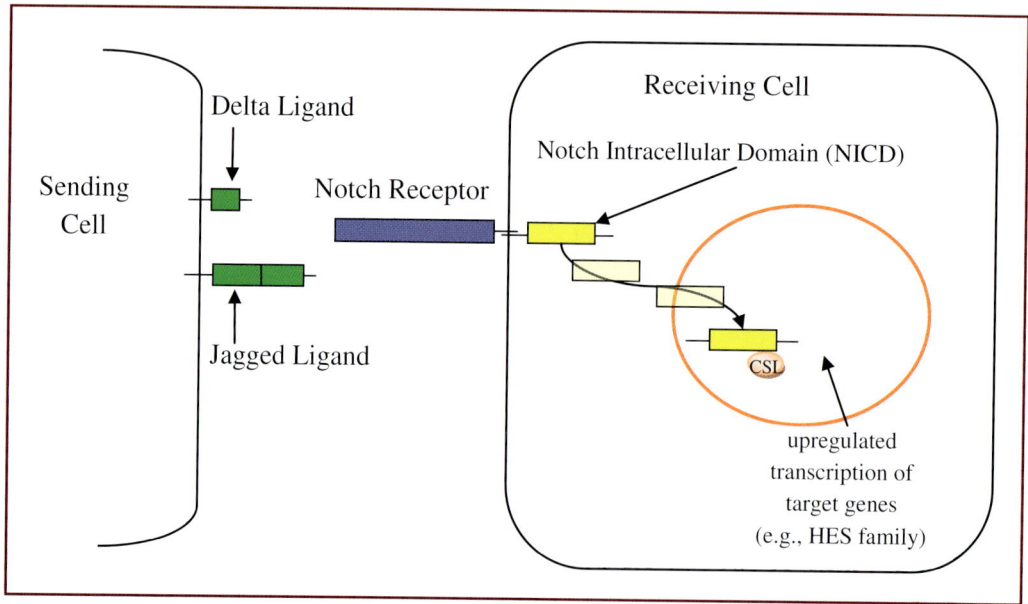

Figure 2–11. Notch signaling pathway.

prosensory cells (precursors to the sensory epithelium) of the cochlear duct (see Figure 6–6, Kelley, 2007; Lanford et al., 1999). The prosensory cells are destined to become either hair cells or supporting cells. Surface ligands (Jag2 or Dll1) appear on the surface of one cell. When the ligand binds with the Notch1 surface receptor located on an adjacent cell, the receptor is activated thus initiating a series of events within the adjacent cell. These events lead to the upregulation of Notch target genes (e.g., *Hes1* or *Hes5*), which prevents the cell from becoming a hair cell and the cell adopts a supporting cell fate instead. Therefore, this lateral inhibition may contribute to the production of a mosaic of one hair cell surrounded by supporting cells, a pattern that is repeated throughout the epithelium (reviewed by Bryant, Goodyear, & Richardson, 2002; Kelly & Chen, 2009).

Wnt Signaling Pathway

In *Drosophila* the Wingless (*Wg*) gene encodes Wnt proteins. Wnt is the nomenclature for Wingless-type MMTV Integration Site Family. Wingless was first described in 1973. In 1982, the gene *Int1* (integration 1) was described for the mouse in relation to mouse mammary tumor virus (hence, MMTV in the full name) and later found to be identical to Wingless (Klaus & Birchmeier, 2008). The name Wnt was derived from both Wingless (Wg) and Int1 to become Wnt. The name Wnt refers to either the gene or the protein. There are 19 human WNT genes and corresponding proteins. They serve different but commonly overlapping functions (Logan & Nusse, 2004; Montcouquiol, Crenshaw, & Kelley, 2006; Nusse, 2005). Wnt proteins activate primarily three signaling pathways: (1) the canonical Wnt pathway (Wnt/β-catenin pathway); (2) the planar cell polarity pathway (Wnt-PCP pathway); and (3) the Wnt/Ca^{2+} pathway. All three pathways are mediated by the Frizzled (Fzd) family of surface receptors (seven members in humans). In the canonical signal pathway, when Wnt binds to Fzd receptors, the protein Disheveled (Dvl) is activated and coupled to a complex of proteins, which acts to cause an accumulation of β-catenin in the cytosol. This ultimately upregulates transcription of *Fzd* target genes and stimulates cell growth and proliferation (Figure 2–12).

The Wnt-PCP and Wnt-Ca^{2+} pathways are called noncanonical Wnt signaling. Figure 2–13A illustrates the Wnt-Ca^{2+} pathway. Ca^{2+} is a ubiquitous intracellular signaling molecule and its concentrations normally are maintained at very low levels in the cytosol. The Wnt-Ca^{2+} pathway provides a means for Wnt signaling to influence Ca^{2+} levels in the cell. Wnt binds to Fzd receptors and activates Dvl, which leads to increased Ca^{2+} levels. Cytosol concentrations of Ca^{2+} regulate numerous cellular processes including exocytosis, gene expression, and protein phosphorylation. Figure 2–13A illustrates two downstream target proteins Ca^{2+}/calmodulin-dependent kinase II (CaMKII) and protein kinase C (PKC), which are activated by increased Ca^{2+} levels.

The key molecular players mediating the Wnt-PCP pathway are schematically illustrated in Figure 2–13B. Wnt signals act through Fzd receptors and the protein Dvl to regulate the formation and maintenance of cytoskeletal actin filaments and microtubules (e.g., via Rho/Rac GTPases). Interference with Wnt-PCP signaling alters PCP mechanisms (discussed below) causing serious abnormalities in development.

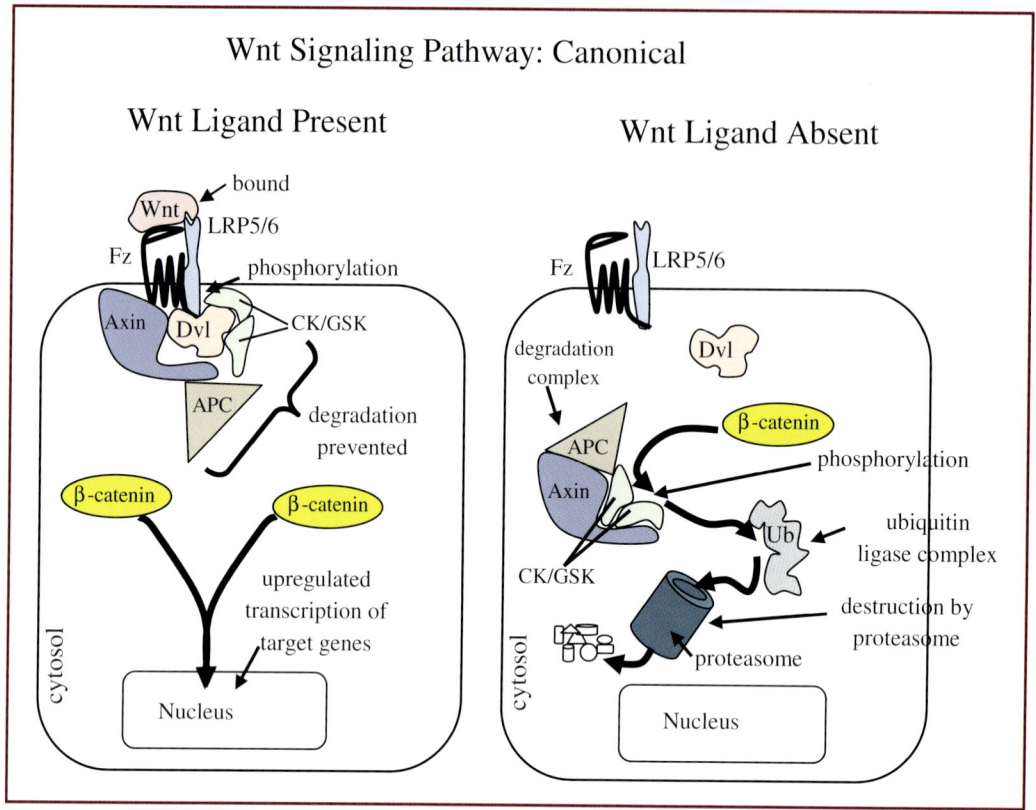

Figure 2–12. Canonical Wnt signaling pathway.

PLANAR CELL POLARITY

Planar cell polarity (PCP) was originally identified simply as a tissue "polarity" (Gubb & Garcia-Bellido, 1982) and it refers to the uniform orientation of individual cells within a tissue sheet. Thus, PCP refers to how cells orient in relation to other cells. First, we must explore how it is that a single cell has an innate (autonomous or intrinsic) internal spatial organization that distinguishes its top, bottom, and sides.

Polarization of Individual Cells

Many cells can develop and maintain polarization in complete isolation from other cells (intrinsic or cell autonomous polarization). This indicates that all necessary molecular components are available in these cells and they can execute the necessary polarization programs in isolation. Some cells contain all the molecular machinery necessary to polarize but may not do so unless coaxed into polarizing by the presence of other cells. These cells require cooperative polarization to form planar cell polarity. Epithelial cells can act in this way such that in isolation an individual epithelial cell may show no evidence of polarization. However, after dividing and producing numerous additional cells, the cells may spontaneously assemble to form an epithelial cluster or colony, complete with an interior central fluid-filled extracellular compartment (Figure 2–14). Under close examination,

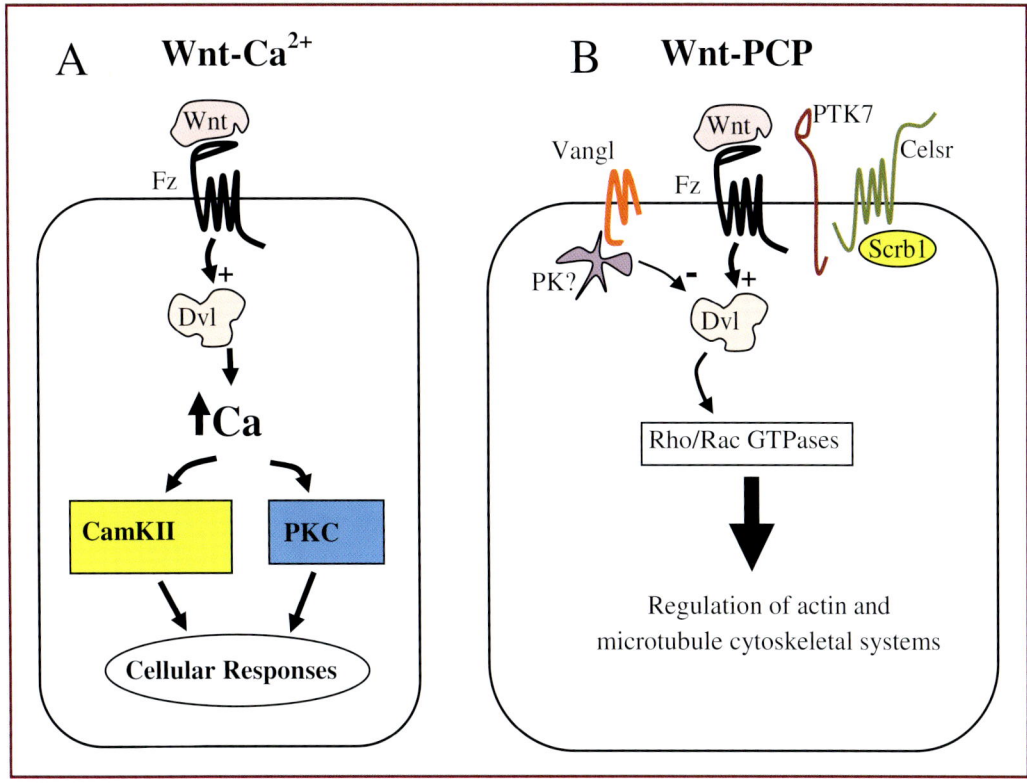

Figure 2–13. *Noncanonical Wnt signaling pathways include Wnt-Ca^{2+} (**A**) and Wnt-PCP (**B**).*

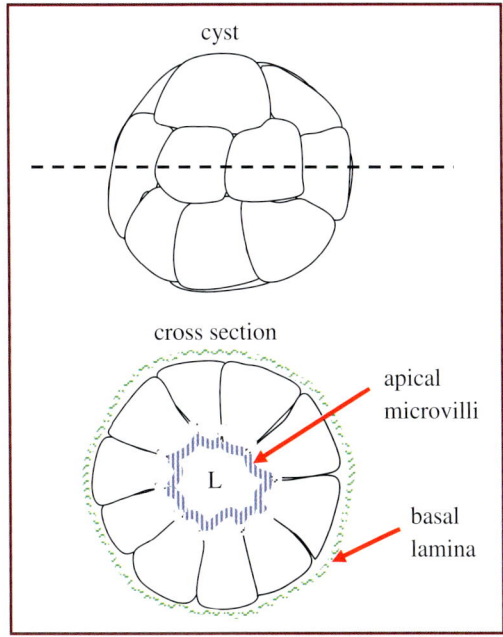

Figure 2–14. *Self-assembling polarization.*

you would likely see that the central extracellular compartment is lined exclusively by apical microvilli. Moreover, the cells would be joined together by apico-lateral tight junctions, and the exterior surface of the cluster would be lined with basal cell surfaces attached to a basal lamina. Thus, during the spontaneous assembly, the cells underwent normal cooperative polarization, alignment, and attachment. (O'Brien et al., 2001).

The key genes and their associated proteins required for polarization of individual epithelial cells (apex versus base) include the "partitioning defective" (*Par*) family of genes. There are at least six members of this family. The proteins Par3 and Par6 along with "atypical protein kinase C" (aPKC) form a complex with tight junctions located in the lateral wall near the apical border of each cell (Figure 2–15). These scaffold proteins have binding sites for two GTPases, Rac and Cdc42 (regulators of actin assembly). The GTPases play an important role in establishing where the Par3-Par6-aPKC complex is located

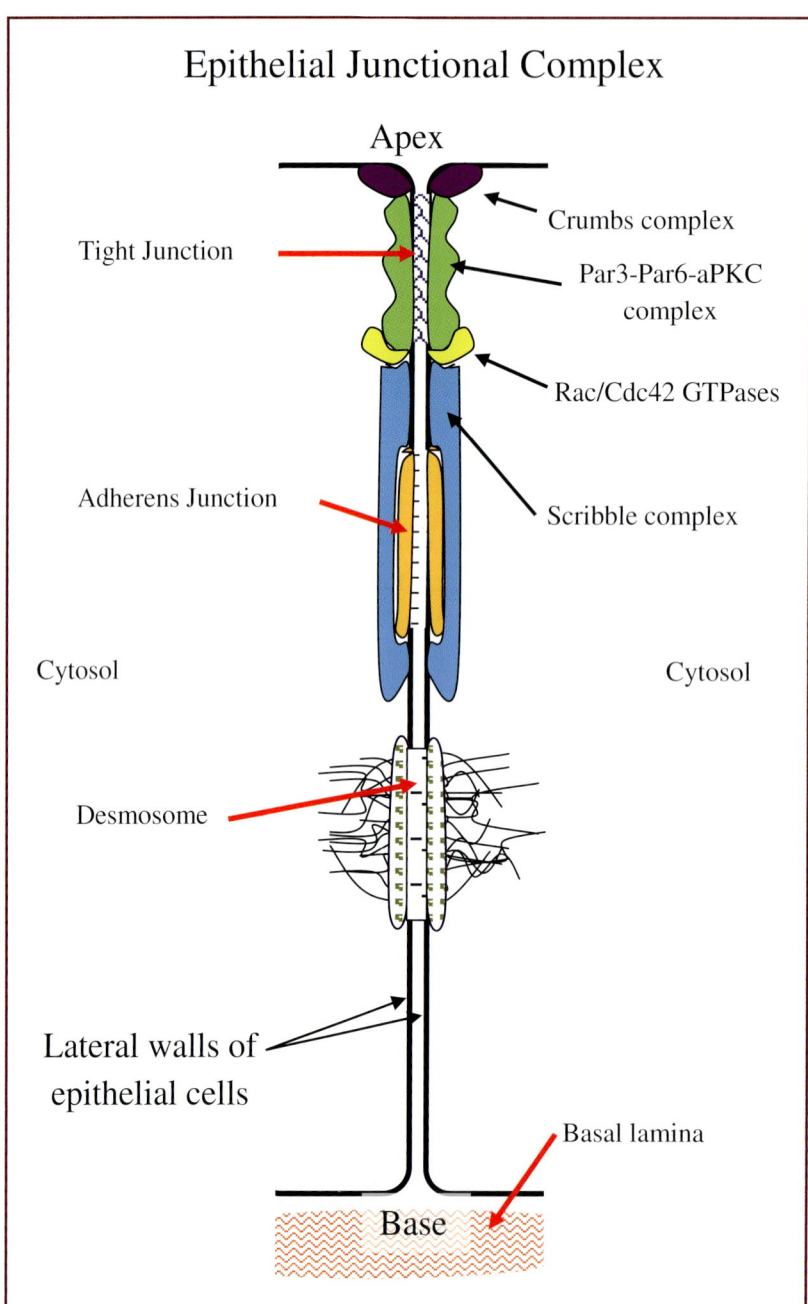

Figure 2–15. Epithelial junctional complex along the apical to basal length of the cell.

and hence where the apical pole of the cell will emerge. Placement of this complex affects the placement of two other key components: (1) the "Crumbs complex" at the apex; and (2) the "Scribble complex" toward the base of cells. Under the influence of the apically located Rac protein, laminin and other components of the basal lamina are secreted and deposited on the exterior of the cell base. Constituents of the deposited basal lamina, including laminin, in turn act on basal regions of the source cell as well as neighboring cells via integrin receptors (transmembrane protein receptors) to facilitate the formation of basal characteristics locally. This feedback mechanism provides information on the polarity of neighboring cells thus promoting co-alignment of polarities for individual epithelial neighbors.

Planar Organization of Polarized Cells

We have seen how individual cells may be polarized such that an apical and basal side of the cell is identified. We also noted how this polarization may be communicated to some extent to neighboring cells as they adhere to one another. This polarization in one dimension (apex to base) is only part of the key to planar cell polarization. PCP requires polarized organization in two dimensions (i.e., two axes oriented perpendicular to each other). That is, we must identify not only a top and bottom but also a distinct front and back (or left and right) sides (Figure 2–16). In cross section, epithelial sheets have a clear top (apex) and bottom (base). But if we look at the sheet from above, so that we see

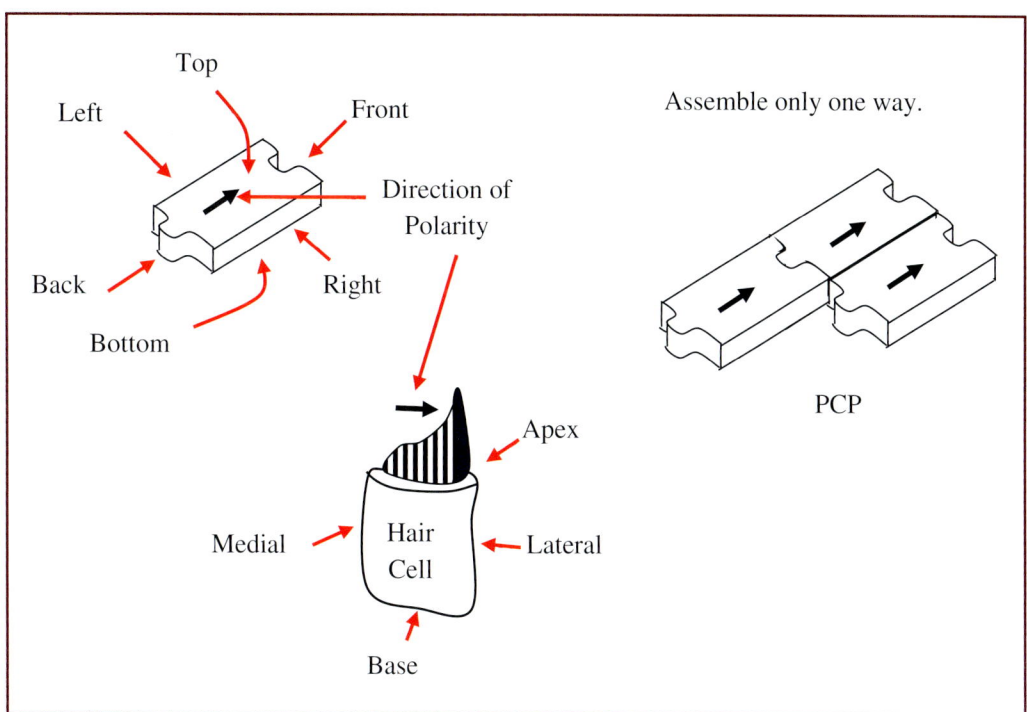

Figure 2–16. Planar cell polarity.

only the apical surfaces of cells, then we will discover a second axis of polarization. For the sake of illustration here, we will adopt the names "front" and "back" sides for the faces of our second cellular axis and symbolize the direction of the axis with an arrow, where the arrow point is directed toward the front face (see Figure 2–16). We now can speak of how the cell is pointed in the front-to-back axis and note the direction of the polarization in relation to all other cells in the epithelium.

PCP is produced with the systematic linkage of each polarized cell to complementary faces. In our general example, we will impose the following rules: (1) cell junctions will be made at all cell faces but only between regions at the top of cells and (2) for each cell there must be a junction between its front face and the back face of another cell. Given these rules, the cells can be assembled in only one way, that is, they can only form a sheet with all arrows pointing in the same direction. This planar organization requires Wnt signaling, and it is clear from this example that the key rules are: (1) there must be an intrinsic cell polarity distinguishing apex and base and (2) a lateral polarization distinguishing front and back, left and right. These asymmetries must define the rules for forming cell junctions during tissue assembly. The signaling pathways are complex however, and both partitioning (*Par*) genes and Wnt-PCP signaling are involved.

There are several examples of PCP, where individual cells organize to produce highly ordered spatial patterns in a tissue plane. One familiar example is the pattern of hair follicles on the dorsal surface of the hand or forearm. The individual hairs generally are not pointed randomly with respect to each other, but rather appear to lie in organized collective parallel patterns, which may change directions collectively over the surface. One of the most striking examples of PCP is seen in the arrangement of hair cells and supporting cells in the sensory epithelia of maculae, cristae, and the organ of Corti (see Figures 1–7 and 1–8). Many other examples are far more subtle and difficult to appreciate without special consideration. For example, it is not obvious that PCP is critical for proper execution of convergent extension (see below) during gastrulation and neural tube closure in the embryo (see Chapter 4). We know this is the case because there are genes that are critical for PCP, and mutations in these genes lead to birth defects that are manifested during these developmental periods including failure of neural tube closure (neural tube defects such as spina bifida). Several core genes important for convergent extension were first identified in *Drosophila*; however, some vertebrate Wnt-PCP genes include frizzled homolog (*Fzd3*, *Fzd6*), flamingo homolog (*Celsr1*), van Gogh-like (*Vangl2*), disheveled (*Dvl1*, *Dvl2*), scribbled homolog (*Scrib1*), and intraflagellar transport 88 homolog (*Ift88*). We will consider several of these in Chapter 6 during discussion of inner ear development.

CONVERGENT EXTENSION

What is convergent extension and why is PCP critical during the process? Embryonic morphogenesis is the most obvious feature observed during development. Primordial embryonic tissues begin as nearly amorphous sheets and during development ultimately become molded to form coherent shapes resembling the adult organism. How this is accomplished is of course the central question of the embryologist. One contributing process that occurs at various stages of development and one that has been preserved across all vertebrates and invertebrates alike is the process of convergent extension.

Convergent extension serves to elongate a structure along one dimension while shortening it in another dimension (Figure 2–17). The process results in an extension of a tissue element along one or two spatial axes in concert with a convergence and intermingling (intercalation) of cells along the third axis. In any case, cells actually move relative to one another at right angles to the axis of extension. There are many ways that convergent extension can be manifest. For example, it can occur radially in a shell of tissue (converge and intercalate along the radial axis and extend along the anteroposterior axis) or it may occur in a rectangular planar sheet or tissue block (e.g., tissue converges and

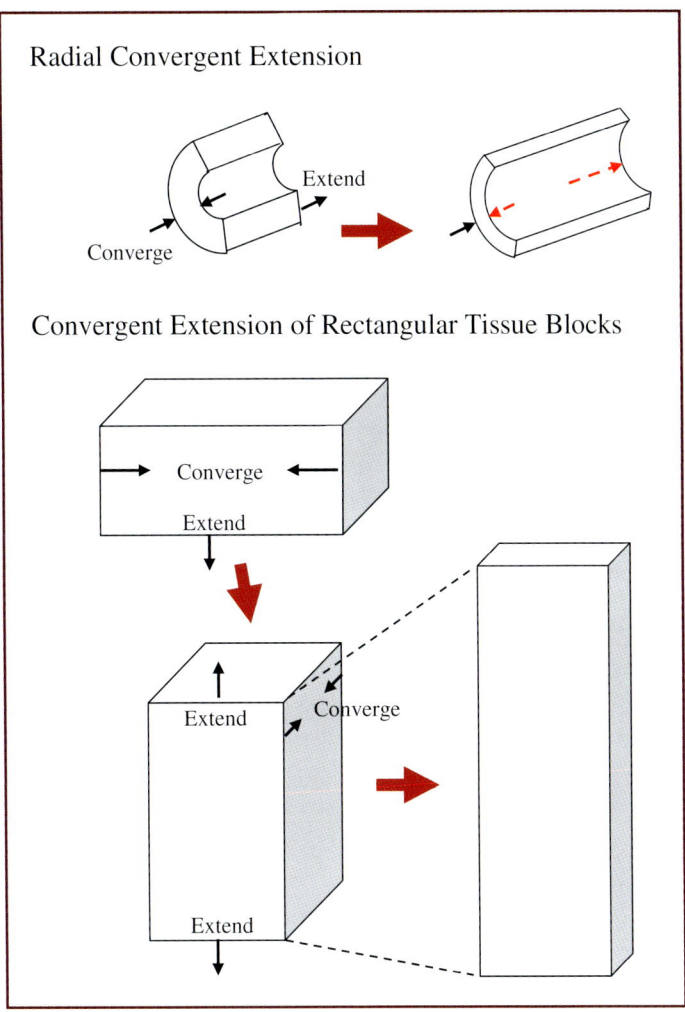

Figure 2–17. General features of convergent extension.

intercalates along the mediolateral axis and extends along either the dorsoventral or the anteroposterior axis or both). The mode of convergent extension depends on existing regional tissue forces and boundaries, on forces developed by individual cells, and on the intrinsic polarity configuration of individual cells. As you will see, normal intercalation of cells during convergence requires that the cells be individually polarized in at least two dimensions. The process does not require, but may include, concurrent cellular proliferation.

The precise mechanisms and cellular movements underlying morphogenesis and convergent extension in all its forms are not clear. However, some cases have been well studied and a modified schematic example will be considered here as it illustrates the basic processes and general requirements for all convergent extension tissue movements (Keller, 2002; Keller et al., 2002; Keller, Shook, & Skoglund, 2008). Consider a layer of mesothelial cells from the gastrula forming a mediolateral anteroposterior plane (Figure 2–18). Mediolateral convergence results in a reduction in mediolateral width with an anteroposterior lengthening of the tissue. During mediolateral convergence, it has been shown that individual cells form large tractive protrusions (lamellipodia) at their mediolateral faces, whereas along the axis of extension, the anteroposterior cell faces form transient intercellular adhesions (filiform contacts). The filiform contacts are dynamic, attaching and detaching in cycles. Presumably, these transient attachments allow cells to slide by each other in the mediolateral direction under constant traction yet maintain mechanical linkage across cells in the anteroposterior direction during the recycling of attachments. The force driving the cells to slide by one another in the mediolateral direction is the traction forces developed by the lamellipodia, which pull on neighboring cells thus developing tension that draws cells together mediolaterally. As cells pull together and converge mediolatrerally, the cells stretch under tension and slide by one another as they intercalate. The newly inserted cells force the tissue to elongate in the anteroposterior dimension. Thus, cells pull themselves along the mediolateral axis but they are pushed along the anteroposterior axis as cells wedge between other cells during intercalation.

This process is possible only because each cell is polarized in the same way (e.g., filiform adherence proteins on anterior and posterior faces, and lamellipodia on both lateral faces). More importantly, these polarized cells must organize together in very specific ways in order to converge and extend. When these cells aggregate to form the tissue planes, they do so by lining up anterior face membranes to posterior face membranes and lateral faces with medial faces. This produces a highly ordered repeating cellular spatial pattern in the plane of the tissue and this is a clear example of PCP. This organization is required for convergent extension as knockouts or mutations in the core PCP genes, or inhibition of their products and signaling cascades disables convergent extension and leads to birth defects. When PCP is blocked, cells are not properly polarized and cannot organize themselves appropriately. Interference with PCP often produces neural tube defects, because neural tube closure depends critically on PCP and convergent extension. Other effects are seen in the cochlea including improper alignment of hair cells accompanied by shorter cochlear lengths. Shorter cochlear lengths are thought to be due to interference with convergent extension.

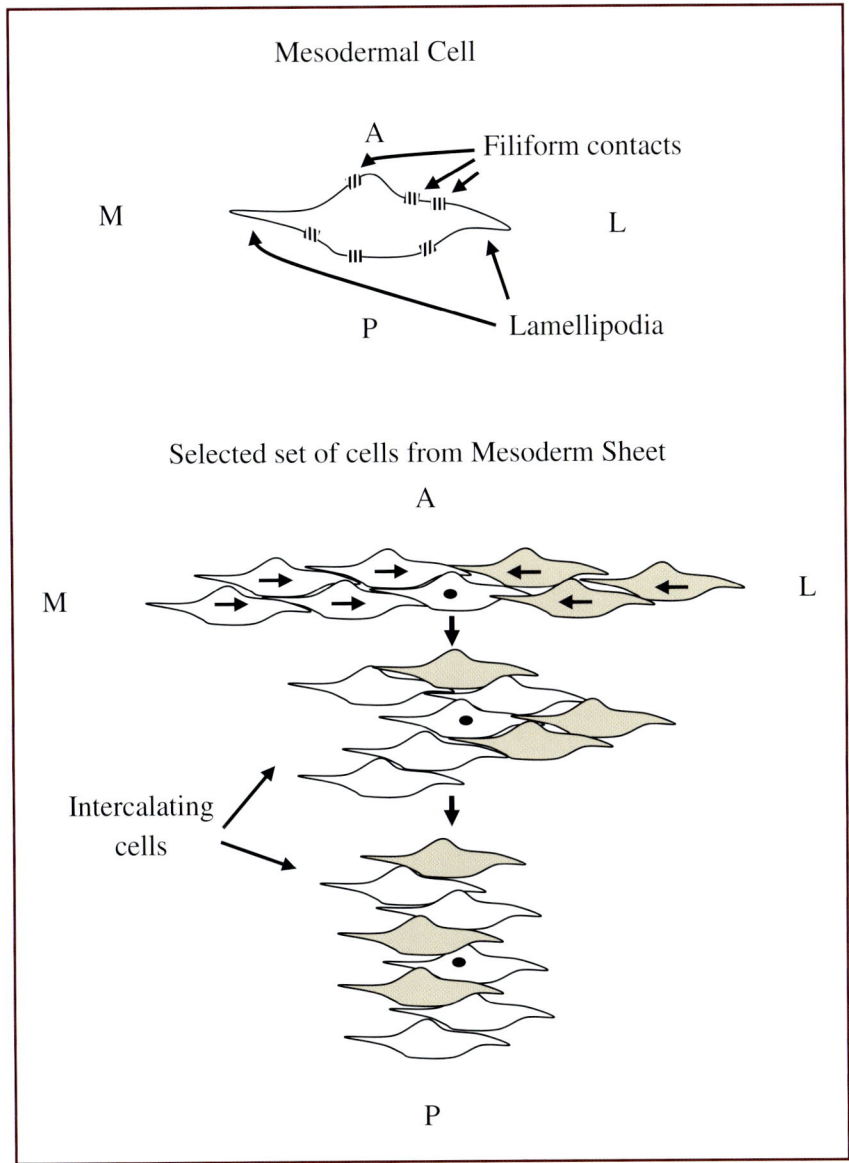

Figure 2–18. Cellular processes for convergent extension are depicted. A, anterior; P, posterior; M, medial; L, lateral.

CONCLUSION

As we see in Chapter 4, tissues and organ systems arise from a single fertilized egg, the zygote. The basic processes discussed in this chapter along with additional mechanisms are implemented repeatedly during the formation of the adult organism. The plan for directing these processes is contained in the organism's genome; therefore, we first consider some basic concepts in genetics, then move on to general embryological development.

GENERAL RESOURCES

Alberts, B., Bray, D., Hopkin, K., Johnson, A., Lewis, J., Raff, M., . . . Walter, P. (2004). *Essential cell biology* (2nd ed.) New York, NY: Garland Science.

Alberts, B., Johnson, A., Lewis, J., Raff, M., Roberts, K., & Walter, P. (2008). *Molecular biology of the cell* (5th ed.). New York, NY: Garland Science.

LITERATURE CITED

Bryant, J., Goodyear, R. J., & Richardon, G. P. (2002). Sensory organ development in the inner ear: Molecular and cellular mechanisms. *British Medical Bulletin, 63,* 39–57.

Cosgrove, D., Samuelson, G., & Pinnt, J. (1996). Immunohistochemical localization of basement membrane collagens and associated proteins in the murine cochlea. *Hearing Research, 97,* 54–65.

Dechesne, C. J. (1992). The development of vestibular sensory organs in human. In Romand, R. (Ed.), *Development of auditory and vestibular systems 2* (pp. 419–447). New York, NY: Elsevier Science Publishers BV.

Driver, E. C., Pryor, S. P., Hill, P., Turner, J., Ruther, U., Biesecker, L. G., . . . Kelley, M.W. (2008). Hedgehog signaling regulates sensory cell formation and auditory function in mice and humans. *Journal of Neuroscience, 28*(29), 7350–7358.

Gubb, D., & Garcia-Bellido, A. (1982). A genetic analysis of the determination of cuticular polarity during development in *Drosophila melanogaster*. *Journal of Embryology and Experimental Morphology, 68,* 37–57.

Hooper, J. E., & Scott, M. P. (2005). Communicating with hedgehogs. *Nature Reviews Molecular Cell Biology, 6*(4), 306–317.

Kang, S., Graham, J. M., Jr., Olney, A. H., & Biesecker, L. G. (1997). GLI3 frameshift mutations cause autosomal dominant Pallister-Hall syndrome. *Nature Genetics, 15,* 266–268.

Keller, R. (2002). Shaping the vertebrate body plan by polarized embryonic cell movements. *Science, 298,* 1950–1954.

Keller, R., Davidson, L., Edlund, A., Elul, T., Ezin, M., Shook, D., & Skoglund, P. (2000). Mechanisms of convergence and extension by cell intercalation. *Philosophical Transactions of the Royal Society of London, 355,* 897–922.

Keller, R., Shock, D., & Skoglund, P. (1998). The forces that shape embryos: Physical aspects of convergent extension by cell intercalation. *Physical Biology, 5,* doi.org/10.1088/1478-3975/5/1/15007.

Kelley, M. W. (2007). Cellular commitment and differentiation in the organ of Corti. *International Journal of Developmental Biology, 51,* 571–583.

Kelly, M. C., & Chen, P. (2009). Development of form and function in the mammalian cochlea. *Current Opinion in Neurobiology, 19,* 395–401.

Khetarpal, U., Robertson, N.G., Yoo, T. J., & Morton, C. C. (1994). Expression and localization of COL2A1 mRNA and type II collagen in human fetal cochlea. *Hearing Research, 79,* 59–73.

Klaus, A., & Birchmeier, W. (2008). Wnt signaling and its impact on development and cancer. *Nature Reviews Cancer, 8*(5), 387–398.

Kros, C. J., Ruppersberg, J. P., & Rusch, A. (1998). Expression of a potassium current in inner hair cells during development of hearing in mice. *Nature, 394,* 281–284.

Lanford, P. J., Lan, Y., Jiang, R., Lindsell, C., Weinmaster, G., Gridley, T., & Kelley, M. W. (1999). Notch signaling pathway mediates hair cell development in mammalian cochlea. *Nature Genetics, 21,* 289–292.

Logan, C. Y., & Nusse, R. (2004). The Wnt signaling pathway in development and disease. *Annual Review of Cell Biology, 20,* 781–810.

Montcouquiol, M., Crenshaw III, E. B., & Kelley, M. W. (2006). Noncanonical Wnt signaling and neural polarity. *Annual Review of Neuroscience, 29,* 363–386.

Nehrlich, A. G. (1995). Collagen types in the middle ear mucosa. *European Archives of Otorhinolaryngology, 252,* 443–446.

Nusse, R. (2005). Wnt signaling in disease and in development. *Cell Research, 15*(1), 28–32.

O'Brien, L. E., Jou, T. S., Pollack, A. L., Zhang, Q., Hansen, S. H., Yurchenco, P., & Mostov, K.

E. (2001). Rac1 orientates epithelial apical polarity through effects on basolateral laminin assembly. *Nature Cell Biology*, *3*, 831–838.

Radtke, F., Schweisguth, F., & Pear, W. (2005). The NOTCH "gospel": Workshop on Notch signaling in development and cancer. *EMBO Reports*, *6*(12), 1120–1125.

Robertson, N. G., Resendes, B. L., Lin, J. S., Lee, C., Aster, J. C., Adams, J. C., & Morton, C. C. (2001). Inner ear localization of mRNA and protein products of COCH, mutated in the sensorineural deafness and vestibular disorder, DFNA9. *Human Molecular Genetics*, *10*(22), 2493–2500.

Basic Concepts in Genetics

> **KEY POINTS**
>
> 1. The nucleus of the cell holds the chromosomes. Normally, humans have 46 chromosomes, half of which (i.e., 23) are inherited from your mother and half from your father.
> 2. Chromosomes contain deoxyribonucleic acid (DNA), which has approximately 3 billion base pairs that code for 25,000 to 30,000 genes in humans.
> 3. Genes are the fundamental units of heredity. A gene contains the necessary information for the cell to generate a functional product (i.e., protein).
> 4. Genes segregate and randomly distribute (Mendel's Principles of Segregation and Random Assortment) during gamete formation so that each union of haploid gametes (egg and sperm) produces a unique individual.
> 5. Genetic disorders arise when altered DNA is passed on to the offspring. Genetic disorders include chromosomal abnormalities, single gene disorders, nontraditional (or non-Mendelian) inheritance, and multifactorial inheritance.
> 6. Mendelian inheritance patterns may be autosomal or sex-linked and they are further distinguished as either dominant or recessive. Mendelian inheritance explains many genetic disorders including syndromic and nonsyndromic hearing loss.
> 7. Non-Mendelian inheritance patterns (e.g., mitochondrial, digenic, and multifactorial) can also explain genetic hearing loss as well as other genetic disorders.

CHROMOSOMES

Cytogenetics is the study of chromosomes (see review by Garcia-Sagredo, 2008). Within each nucleus of the cell one finds the chromosomes, which contain the genetic material (i.e., deoxyribonucleic acid or DNA) that makes us who we are. Ordinarily, the DNA packaged with other proteins (e.g., histones) is distributed throughout the nucleus in strands known as chromatin; however, during part of the cell cycle, the chromatin condenses or packs tightly and the individual chromosomes can be observed with the light microscope (Figure 3–1). Human somatic cells are diploid cells containing 46 chromosomes ($2n$ where $n = 23$), 22 pairs of autosomes, and 2 sex chromosomes. Gametes, or germ cells (i.e., egg and sperm), are known as haploid cells as they contain 23 chromosomes (22 autosomes and 1 sex chromosome). The complement of chromosomes is known as the karyo-

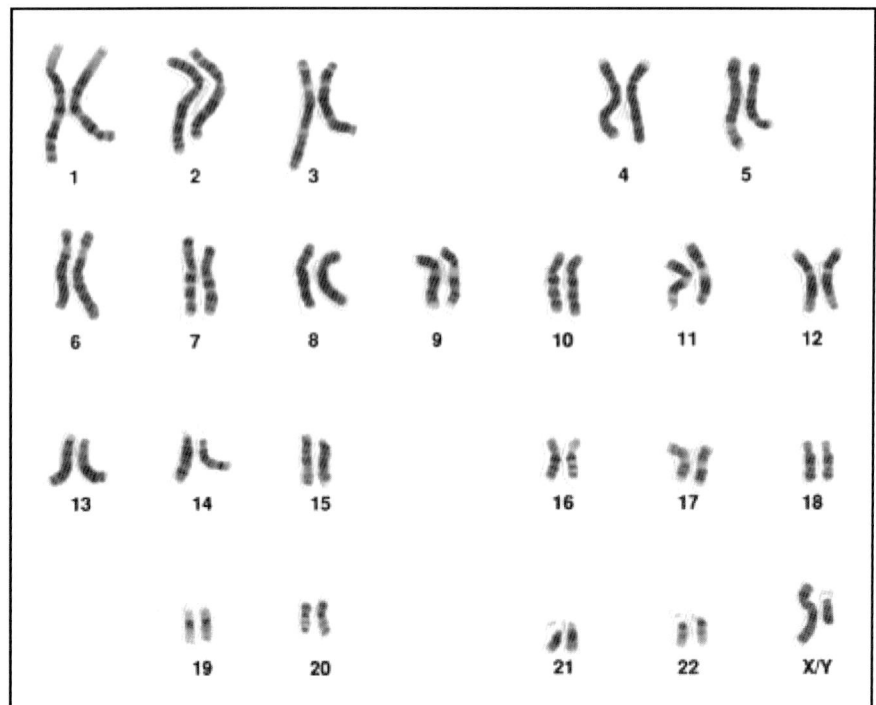

Figure 3–1. Chromosomes displayed in the karyotype. Note that a pair of homologous chromosomes are present for all autosomes and in this case distinct X and Y chromosomes make up the sex chromosomes for a male. Two homologous X chromosomes would be present in a female. One chromosome of each pair is derived from the mother and one from the father. Paternal and maternal genes located on homologous chromosomes may be the same or may be distinct alleles. The Y sex chromosome can be derived only from the father. Distinct light and dark bands can be observed in the stained chromosomes, which are better represented in Figure 3–2. Figure courtesy of the National Human Genome Research Institute (Artist: Darryl Leja).

type (Figure 3–2). The normal karyotype is 46, XX (44 autosomes and two X sex chromosomes indicating a female) or 46, XY (44 autosomes, one X, and one Y chromosome indicating a male). The autosomes are numbered from the largest pair (1) to the smallest (22), with the sex chromosomes making up the final pair. Chromosomes viewed in the karyotype typically have replicated in preparation for cell division; therefore, each chromosome in the karyotype is composed of two sister chromatids (Spurbeck, Adams, Stupca, & Dewald, 2004). Chromatids are identical copies of the original chromosome that are tethered together at the centromere. The chromosomes also are commonly stained such that they display alternating light and dark regions or colored bands along their length. Several staining techniques can be used; however, one of the most common stains is Giemsa and the staining technique is known as "G-banding" ("G" for Giemsa). Fluorescence in situ hybridization (FISH) is another common banding technique. The bands produced by the staining technique are numbered according to an international standard (Schafer, Slovak, & Campbell, 2009).

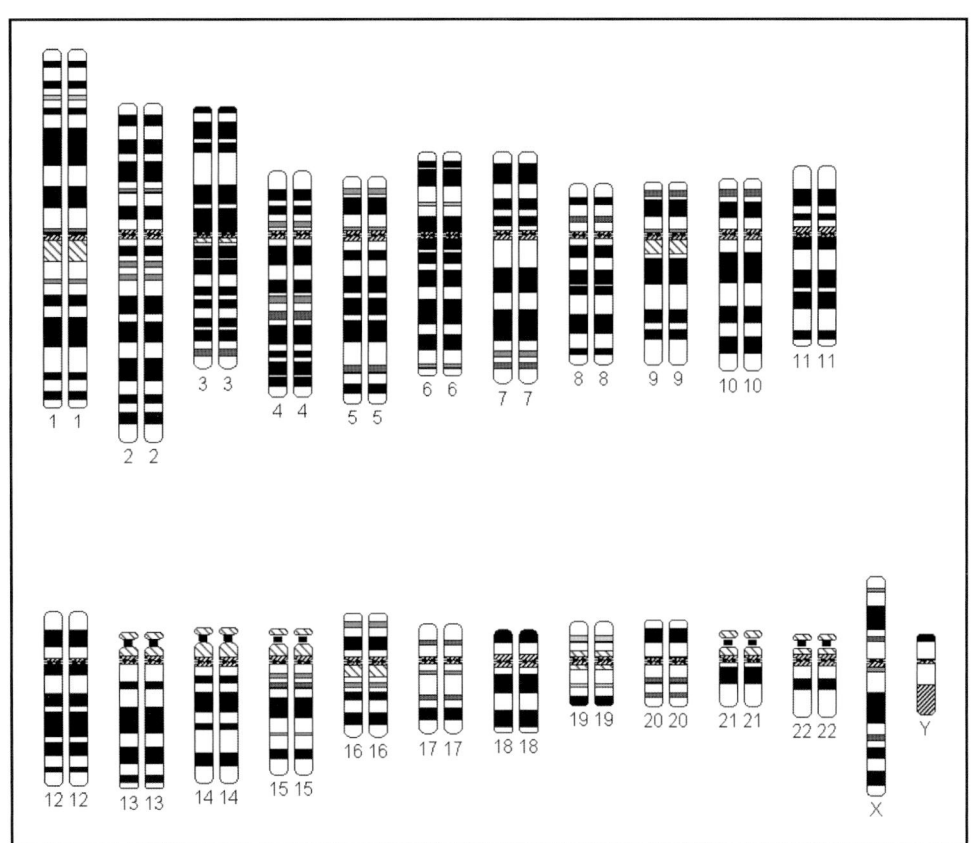

Figure 3–2. Karyogram (schematic representation of the chromosomes) depicting the human karyotype for a male, 46, XY. Drawn using Hiller, Bradtke, Balz, and Rieder (2004).

In addition to the distinct bands produced by stains, several other structural features are visible when viewing the chromosomes in the karyotype (Spurbeck, et al., 2004). Each chromosome has a narrowed region, or constriction, known as the centromere while the ends are called the telomeres. From the centromere, each chromosome has two arms labeled "p" and "q" (Figure 3–3). The p-arm is shorter

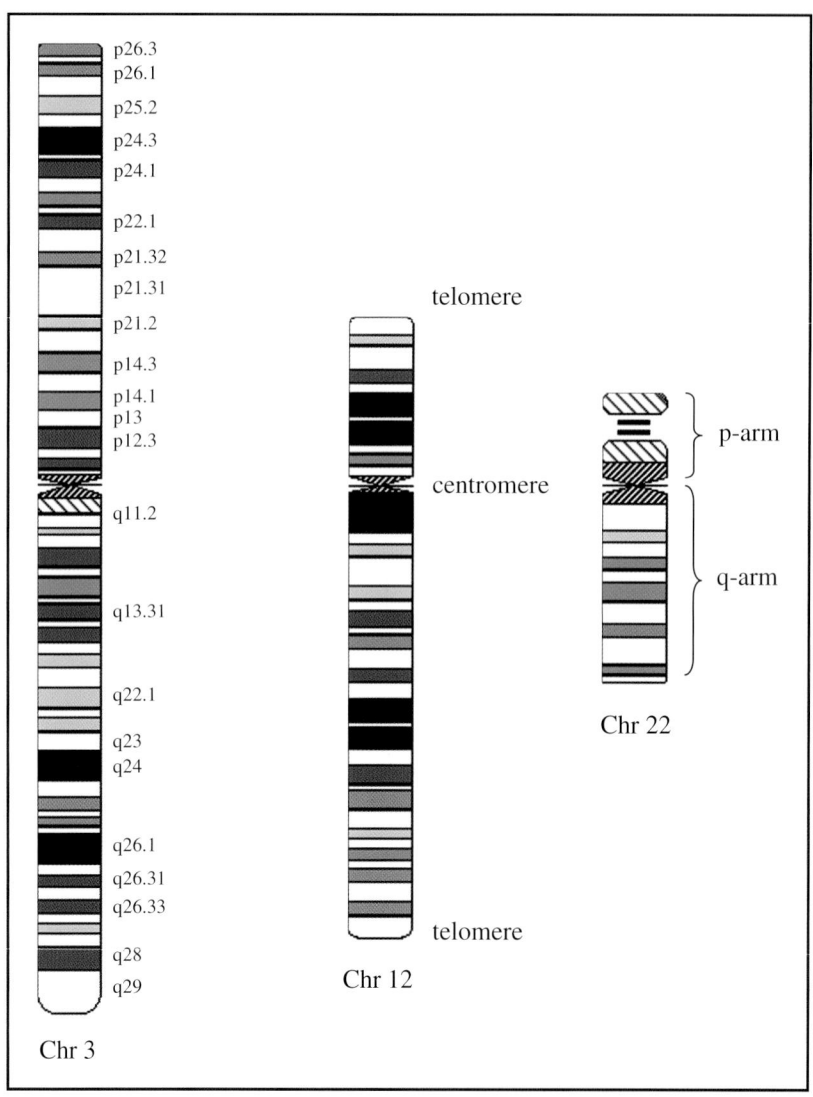

Figure 3–3. Ideograms for three chromosomes (3, 12, and 22) showing the telomeres and centromere as well as demonstrating metacentric (chromosome 3), submetacentric (chromosome 12), and acrocentric (chromosome 22) locations for the centromere. Regions, bands, and subbands are numbered (numbered regions and bands are shown for chromosome 3) so that mapped positions on the chromosome (i.e., loci) can be identified. Ideograms drawn using Hiller et al. (2004).

than the q-arm. The location of the centromere can be described as metacentric (essentially centered in the middle of the p- and q-arms), submetacentric (closer to the telomere of the p-arm), or acrocentric (very near the telomere of the p-arm). The numbering standard begins at the centromere and increases as one moves to the telomeres. Therefore, the lowest numbered regions are found near the centromere.

The chromosome number, arm, numbered regions, and bands are used to identify locations or mapped positions (loci) on the chromosomes. For example, the location at region 1 on the short arm of chromosome 3 would be identified as 3p1. Xq2 identifies region 2 on the long arm of the X chromosome. The regions often are further divided into numbered bands and subbands, which allows one to identify more specific positions on the chromosome. The nomenclature 3p12.1 (read "three p one two point one") identifies the location at subband 1 within band 2 within region 1 on the short arm of chromosome 3. Location 3p12.1 identifies a more precise position than 3p1.

Chromosomes contain DNA, the alphabet of an organism's genome (Figure 3–4). DNA is composed of nucleotides; a nucleotide consists of deoxyribose sugar, phosphate, and one of four bases: adenine (A), thymine (T), guanine (G), or cytosine (C). The structure of DNA is a double helix, which resembles a twisting ladder where the sides of the ladder are composed of the sugar and phosphate molecules and each rung of the ladder contains a pair of complementary bases (A exclusively pairing with T and C exclusively pairing with G). If one splits the helix lengthwise, the DNA is separated into two strands, each strand containing one half of the nucleotide pairs. The structure of the sugar and phosphate backbone further identifies one end of each DNA strand as 5' (five prime, referring to the fifth carbon of the ribose molecule) and the other end as 3' (three prime, the third carbon of ribose). During transcription, which is the process of reading the DNA, the nucleotide sequence is read in the 5' to 3' direction much like you are reading each line of this text in a left to right direction.

The human genome consists of 3 billion base pairs (3 billion letters on each strand of DNA), which code for 25,000 to 30,000 genes. Each gene is located at a specific place on a chromosome that can be identified by the locus. Broadly defined, a locus is simply a mapped position on a chromosome. Genes are the fundamental units of heredity and contain the instructions or code (sequence of letters) cells use to make proteins or various ribonucleic acids (RNAs) as well as the regulatory sequences that influence gene expression. Some genes are a few kilobases in length whereas others are several hundred or thousands of kilobases (1 kilobase = 1,000 base pairs).

Proteins are made up of amino acids held together by peptide bonds. Amino acids contain an amino group (-NH2) and a carboxyl group (-COOH). The amino group of one amino acid can be combined with the carboxyl group of another amino acid to produce a covalent peptide bond and water (NHCO + H_2O). Twenty amino acids are available to make up a wide array of proteins that are necessary for cell function (Table 3–1). A string of amino acids linked together by peptide bonds is called a polypeptide chain. One or more polypeptide chains are linked by peptide bonds to produce proteins. The chemical structure and sequence of the amino acids within the peptide chains is the primary determinant of the three-dimensional structure (conformation) of the protein.

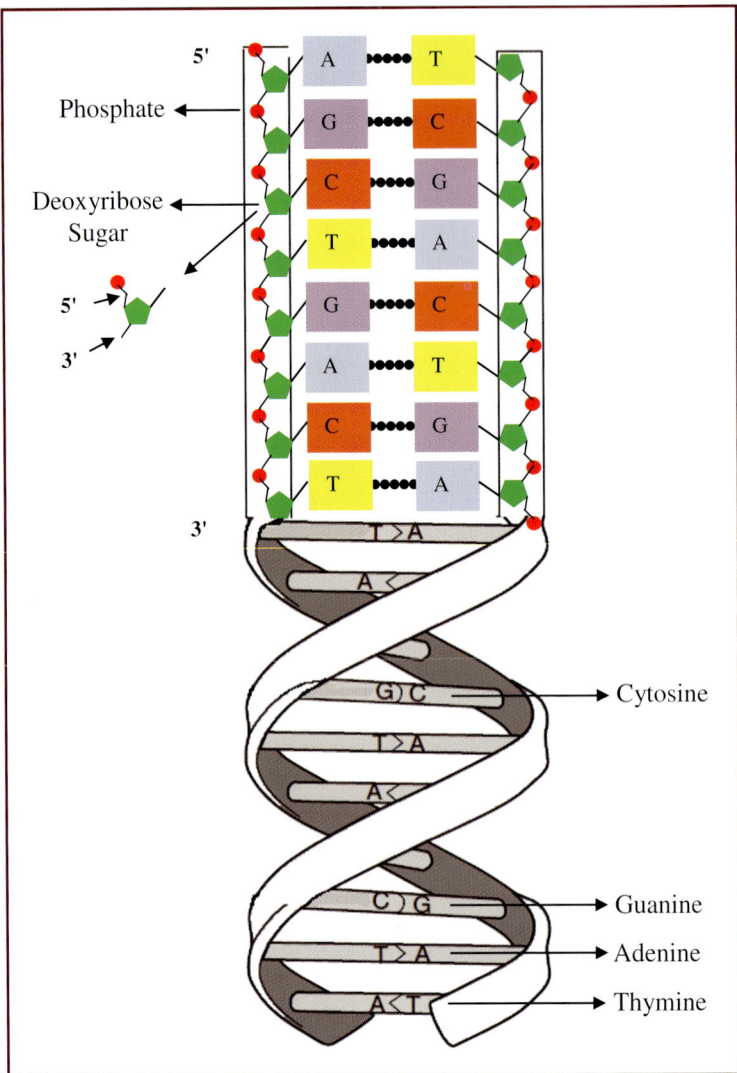

Figure 3–4. *Deoxyribonucleic acid (DNA). Modified from a figure courtesy of the National Human Genome Research Institute (Artist: Darryl Leja). The sugar element of the DNA chain is deoxyribose (green pentagon) having five carbon atoms. The fifth carbon is bound to phosphate (red dot) thus representing the 5' (read "five prime") end of deoxyribose. The 5' phosphate group covalently bonds to the 3' carbon of an adjacent deoxyribose molecule to form a chain of sugars each binding a particular nucleotide A (adenine), T (thymine), C (cytosine), or G (guanine). The 5'–3' bonds form the chain linking the nucleotide coding sequence.*

The amino acid sequence creates specific longitudinal patterns of polypeptide alignment or complex folding. Such patterns include α helices, β sheets, and coiled coils for example. Proteins can have a very simple structure or be quite complex; however, four levels of organization can be described for a protein. The primary

Table 3–1. List of Amino Acids that Combine to Form Proteins. Three-letter and single-letter abbreviations are also given.*

Amino Acid	Three-letter Code	Single-letter Code
Alanine	Ala	A
Arginine	Arg	R
Asparagine	Asn	N
Aspartic Acid	Asp	D
Cysteine	Cys	C
Glutamic Acid	Glu	E
Glutamine	Gln	Q
Glycine	Gly	G
Histidine	His	H
Isoleucine	Ile	I
Leucine	Leu	L
Lysine	Lys	K
Methionine	Met	M
Phenylalanine	Phe	F
Proline	Pro	P
Serine	Ser	S
Threonine	Thr	T
Tryptophan	Trp	W
Tyrosine	Tyr	Y
Valine	Val	V

*Abbreviations are a type of shorthand for listing the amino acid chain that makes up a protein of interest (GJB2 amino acid chain is shown in Figure 3–10, see *B*).

structure is the amino acid sequence (see Figure 3–10 for example). Secondary structure is the polypeptide sequences that form α helices and β sheets. Tertiary structure is the complete three-dimensional folded conformation of the protein. Quaternary structure signifies that the protein is created by two or more polypeptide chains (chains that are not linked by peptide bonding). As an aside, approximately 18% of the human body is protein (whereas it is 60% fluid).

RNA is an important component of the genetic machinery. Like DNA, RNA consists of nucleotides composed of a sugar, a phosphate, and one of four bases

(Figure 3–5). Three bases, guanine (G), cytosine (C), and adenine (A), are used by both RNA and DNA. RNA differs from DNA in that RNA is single-stranded, the sugar is ribose, and uracil replaces thymine as one of the bases. RNA also can leave the nucleus, whereas DNA remains inside the nucleus. Table 3–2 compares DNA and RNA.

CELL CYCLE

Somatic cells cycle through two main stages: interphase and cell division (Figure 3–6). During interphase, the cell is preparing to divide by growing and undergoing chromosome replication. When the chromosomes replicate, an exact copy of each chromosome is produced. Cell division consists of mitosis and cytokinesis. Mitosis is the process of cell division in which the replicated chromosomes condense or pack tightly (prophase), line up on the midline of the cell (metaphase), separate (anaphase), and move to opposite sides of the cell (telophase). Cytokinesis then occurs and the one cell becomes two cells each with 46 chromosomes (Figure 3–7). Examination of the chromosomes for karyotype analysis typically is completed on metaphase chromosomes.

Germ cells or gametes (ovum and sperm) are formed in the reproductive organs by meiosis, in which DNA replication is followed by two cell divisions to

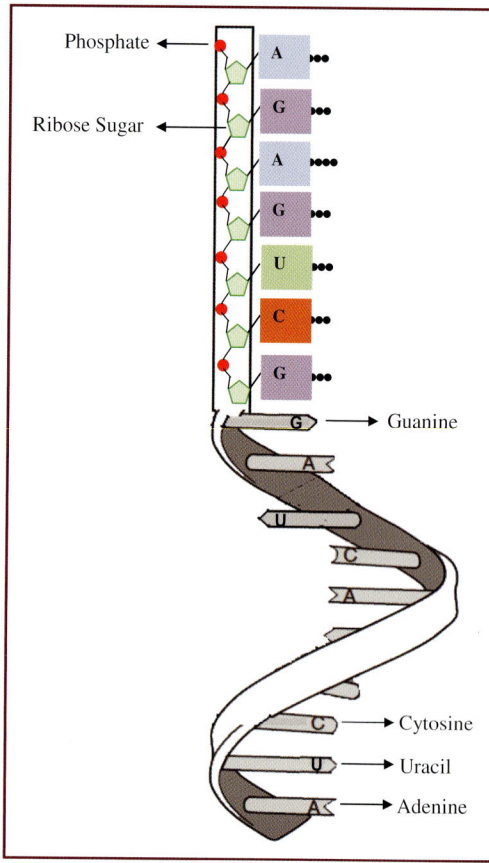

Figure 3–5. Ribonucleic acid (RNA). Modified from a figure courtesy of the National Human Genome Research Institute (Artist: Darryl Leja).

Table 3–2. Comparison of DNA and RNA

DNA	RNA
Remains in the nucleus of the cell.	May leave the cell nucleus.
Composed of sugar and phosphate backbone.	Composed of sugar and phosphate backbone.
Sugar: Deoxyribose	Sugar: Ribose
Double-stranded	Single-stranded
Bases: A, C, G, T	Bases: A, C, G, U

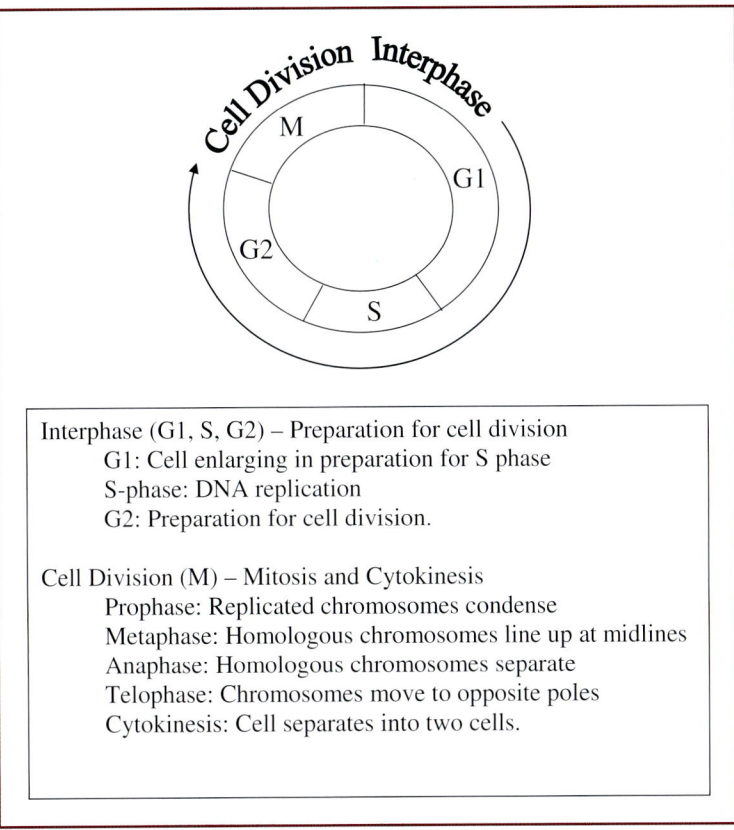

Figure 3–6. *Phases in the cell cycle. Cells cycle through two major stages, interphase and cell division, each of which consists of several phases.*

Figure 3–7. *Steps in mitosis, which produces two daughter cells from one parent cell. During the interphase of the cell cycle, DNA synthesis occurs ("S phase," Figure 3–6) as each of the 46 chromosomes replicates itself. The original and its replica are called chromatids and these become visible during the prophase when the two identical chromatids are attached at the centromere. The chromatids are separated and distributed to each of two daughter cells during the anaphase and telophase of mitosis.*

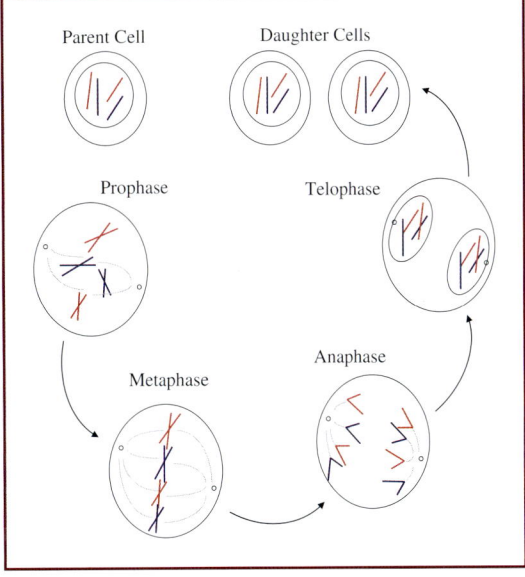

61

produce haploid instead of diploid cells (Figure 3–8). Four haploid cells are produced during spermatogenesis whereas oogenesis results in one haploid cell and two polar bodies. During meiosis, segments of homologous (i.e., matching) chromosomes cross over and exchange genetic material, and when the chromosomes separate each haploid cell contains a unique set of genetic alleles (an allele is simply one form or version of a gene). This explains why we share similar traits with our relatives, but we are not exact copies. What about multiple births? Fraternal (diyzygotic) twins or multiples also each have a unique set of genetic alleles even though they develop simultaneously in the uterus. Fraternal twins or multiples result when two or more eggs are fertilized and implant in the uterus. Identical (monozygotic) twins or multiples, however, share the same genes, as they are the result of a single fertilized egg separating very early in development and proceeding to become two or more individuals.

The segregation and separation of the genetic material into haploid cells normally follows Mendel's principles of segregation and random assortment. Johann Gregor Mendel was an ordained priest in a Moravian monastery who developed and carried out an eloquent set of experiments to investigate the concept of inheritance (see reviews by Chudley, 1998; Jay, 2001). He studied pea plants and carefully documented specific characteristics (traits) of the plants and seeds across several generations. His studies took many years to complete and he replicated his findings to be certain of his conclusions. Indeed, the design and conduct of Mendel's experiments are an excellent example of the scientific process. Cummings (2003) provides an engaging summary of Mendel's work.

Mendel's studies led to several conclusions. First, genes exist in pairs and the parents contribute equally to the offspring's genetic makeup. Second, genes are responsible for the visible characteristics of the offspring and some traits are visible (expressed) while others are hidden. Mendel appropriately recognized that offspring could have identical traits but carry different gene pairs. Therefore, he coined the terms dominant and recessive to characterize traits. In addition, it became important to distinguish between the visible expression of a trait (phenotype) and the individual's genetic makeup (genotype).

Mendel's Principle of Segregation

For every trait, each parent has a gene pair to contribute. During gamete formation, the parental gene pairs separate so that each gamete contains one member of the pair. The offspring then receives only one member of the mother's gene pair and one member of the father's gene pair. This is Mendel's principle of segregation, which maintains the total number of genes and chromosomes that are inherited. The principle of segregation also explains inheritance of single traits and allows one to predict the possible genotypes for each offspring when the parental genotypes are known. The visible traits in the offspring together with the knowledge that the trait is dominant or recessive often will further refine the list of possible genotypes.

Let us use an example from genetic studies of cerumen to illustrate the concepts just presented. As defined previously, an allele is simply a form of a gene, one allele inherited from the mother and one from the father. Therefore, an individual has two alleles for every gene in his or her genome. If a person has identical alleles for a gene, he is homozygous (or a homozygote) for that allele. If the person

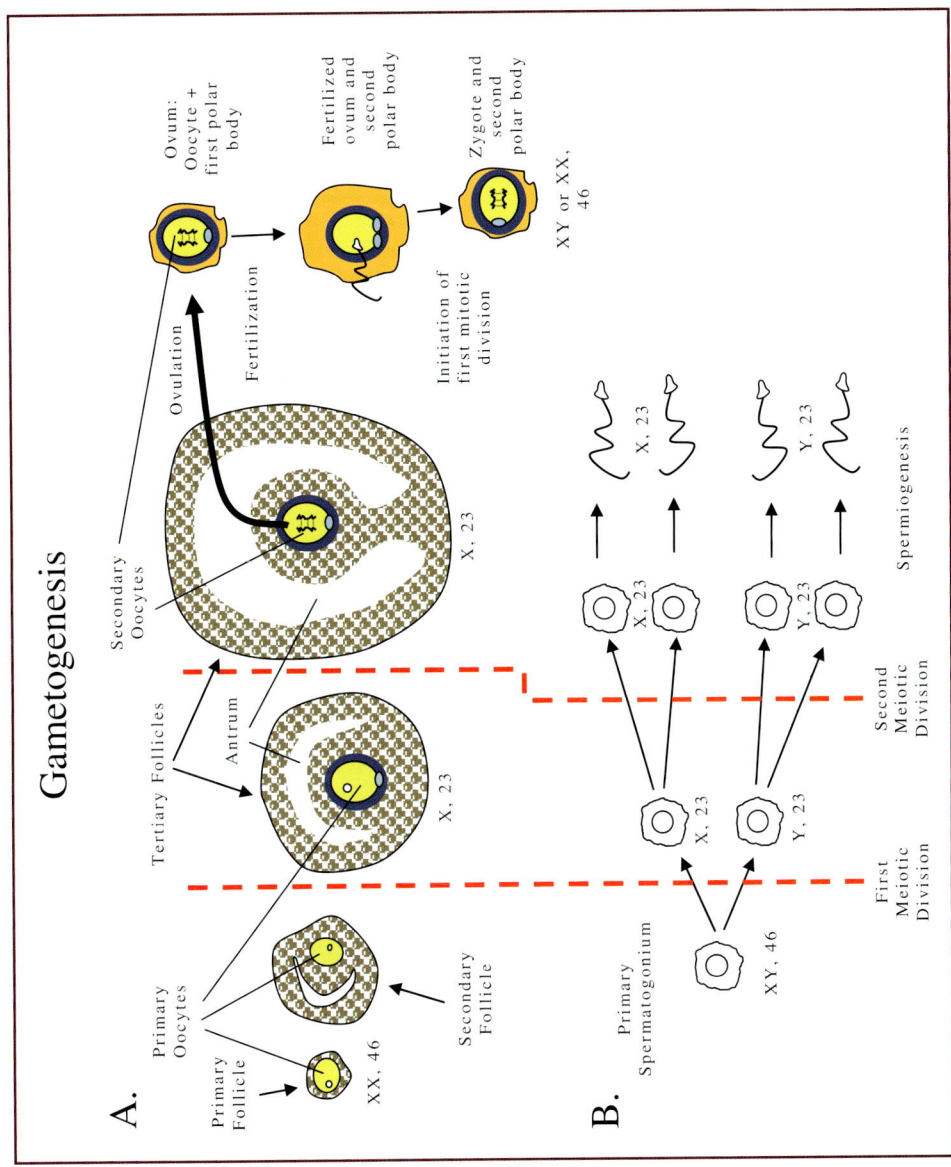

Figure 3–8. Gametes are haploid cells formed during meiosis in the gonads. A maternal and paternal gamete will join during fertilization to produce a new, unique human being.

has different alleles, he is considered to be heterozygous (or a heterozygote). The homozygous or heterozygous nature of an individual's alleles defines the genotype (or genetic makeup). The outcome or observable expression (traits) of the genotype is the phenotype. Alleles for the gene ABCC11 at 16q12 determine whether a person will have wet or dry earwax (Yoshiura et al., 2006). There are two known earwax alleles for this gene where a single nucleotide within the gene may be adenine (we'll call this the *A* allele producing dry earwax) or guanine, the *G* allele, producing wet earwax. If one has two *A* alleles, he is said to be homozygous; the genotype would be described as *AA* (or some set of symbols that identifies two identical alleles) and the phenotype is dry earwax. If the person is homozygous for the *G* allele (i.e., *GG*), then he will have a phenotype of wet earwax. An individual who has one *G* allele and one *A* allele is heterozygous (*GA*) and will display a phenotype of wet earwax. The fact that the heterozygote displays the wet earwax phenotype illustrates the concept of dominant versus recessive. Dominant alleles produce the associated phenotype when the genotype contains one or both copies of the allele. In the present example, *GG* or *GA* genotypes produce the wet earwax phenotype. Recessive alleles will produce only the associated phenotype when both recessive alleles are present in the genotype (in this case, *AA*). We will return to these concepts later when we discuss Mendelian inheritance patterns.

Mendel's Principle of Random Assortment

In addition to the segregation of genes, Mendel discovered another important phenomenon that takes place during gamete formation. The principle of random or independent assortment explains inheritance of two or more traits. When the chromosomes are segregating during gamete formation, they do so randomly and independently for each gene pair. The separation of alleles for one gene pair into haploid cells has no effect on the segregation of the alleles for another gene pair.

The principles of segregation and random assortment produce new combinations of genes during meiosis. One additional way that new gene combinations are formed is through recombination. Recall that, in preparation for cell division, the chromosomes replicate and homologous chromosomes pair up. During this pairing process, nonsister chromatids of the homologous pair overlap (cross over each other) at random sites and physically exchange chromosomal segments. This recombination of genetic material creates new complements of alleles on each chromosome and increases genetic variability among the gametes. Random assortment and recombination increase the number of different genetic combinations that can be produced by a set of parents to an estimated 8×10^{23} (Cummings, 2003), which explains why we share visible traits with our family members but are not identical to our family members. We each have our own unique combination of our parents' genes.

GENE NOMENCLATURE

Previously, we described how one identifies or maps a position on a chromosome using a standardized numbering scheme, but what about the gene or genes that reside at that location? We highlighted the ABCC11 gene. Who produced this nomenclature and what does it mean? Given the fact that the human genome contains perhaps 30,000 genes and some organisms may have as many as 90,000 to 100,000 genes, it is extremely important to have a unified strategy to give each gene a unique name and symbol. Furthermore, some genes are conserved, in other words found in many species, or are orthologs (genes in different species that derive from a common ancestor), and ideally such genes would share the same nomenclature.

The Human Genome Organisation (HUGO) is an international organization of scientists who study human genetics (http://www.hugo-international.org/). HUGO has committees on ethics, public awareness, intellectual property, and gene nomenclature. The HUGO Gene Nomenclature Committee (HGNC, http://www.genenames.org/) approves a name and symbol for every known human gene. They work with scientists, journals, database organizers, and other gene-naming committees (e.g., International Committee on Standardized Nomenclature for Mice) to ensure unique names for all genes and agreed upon nomenclature guidelines. Human gene nomenclature guidelines state that human genes will be designated with capitalized Latin letters and Arabic numerals (Wain et al., 2002); hence, the gene ABCC11 designates a human gene. Genes from other species (mouse, rat) also have guidelines regarding nomenclature (Maltais et al., 2002), which stipulate that gene symbols be written in italics with the first letter capitalized and subsequent letters in lower case. Therefore, if the ABCC11 gene had a homologous gene identified in the mouse, its symbol would be written *Abcc11*. Human and nonhuman gene nomenclature are used throughout the text. For simplicity, we will use human and mouse nomenclature. Therefore, uppercase letters indicate a human gene whereas italicized uppercase and lowercase letters indicate a nonhuman gene. Genes critical for structural and functional development of the ear have been initially identified in species other than human or mouse, including zebrafish, Drosophila (fruit fly), and birds (chicken).

Gene symbols represent a shortened version of the gene name, which also follows guidelines for naming. The gene name should begin with the same letter as the gene symbol; it should be as short as possible yet convey some meaningful information about the gene function. ABC stands for ATP-binding cassette. GJ signifies gap junction (recall from Chapter 1 that the inner ear has gap junctions that are critical for endolymph homeostasis).

Gene families usually have several members and letters or numerals are used in gene names to designate classes, families, or subfamilies of genes. The ABC gene family has seven subfamilies labeled A through G and each subfamily has several members identified by numerals. Therefore, ABCC11 stands for ATP-binding cassette, subfamily C, member 11. The gap junction (GJ) gene family also has several members (alpha, beta, delta, gamma, epsilon) identified by appropriate Latin alphabet letters (A, B, C, D, E, respectively) with numbers identifying different members of each group. Therefore, GJB2 stands for gap junction beta 2. GJA3 is the symbol for gap junction alpha 3.

DNA TO RNA TO PROTEIN

How does a gene create a functional product that the cell can use? In order for a gene to generate a protein, the DNA must be transcribed to produce ribonucleic acid (RNA, specifically messenger RNA or mRNA) and then the mRNA must be translated in the cytoplasm to construct the correct amino acid sequence that makes up the protein. The process from DNA to RNA to protein is necessary because the DNA must remain in the nucleus of the cell and proteins are constructed outside of the nucleus in the cell's cytoplasm at the ribosomes. Therefore, an intermediary (RNA) must bring the appropriate code from the nucleus to the ribosome.

Genes are composed of several distinct elements. Every gene includes coding sequences for the amino acids that will make up the protein and adjacent nucleotide sequences for proper gene expression. Proper gene expression is necessary for the production of RNA in the correct amount, correct place, and at the correct time during development or during the cell cycle. Figure 3–9A shows a schematic of the identifiable regions that make up a gene. Beginning at the 5' end, one finds the promoter region. The nucleotide sequence in the promoter region is responsible for the initiation of transcrip-

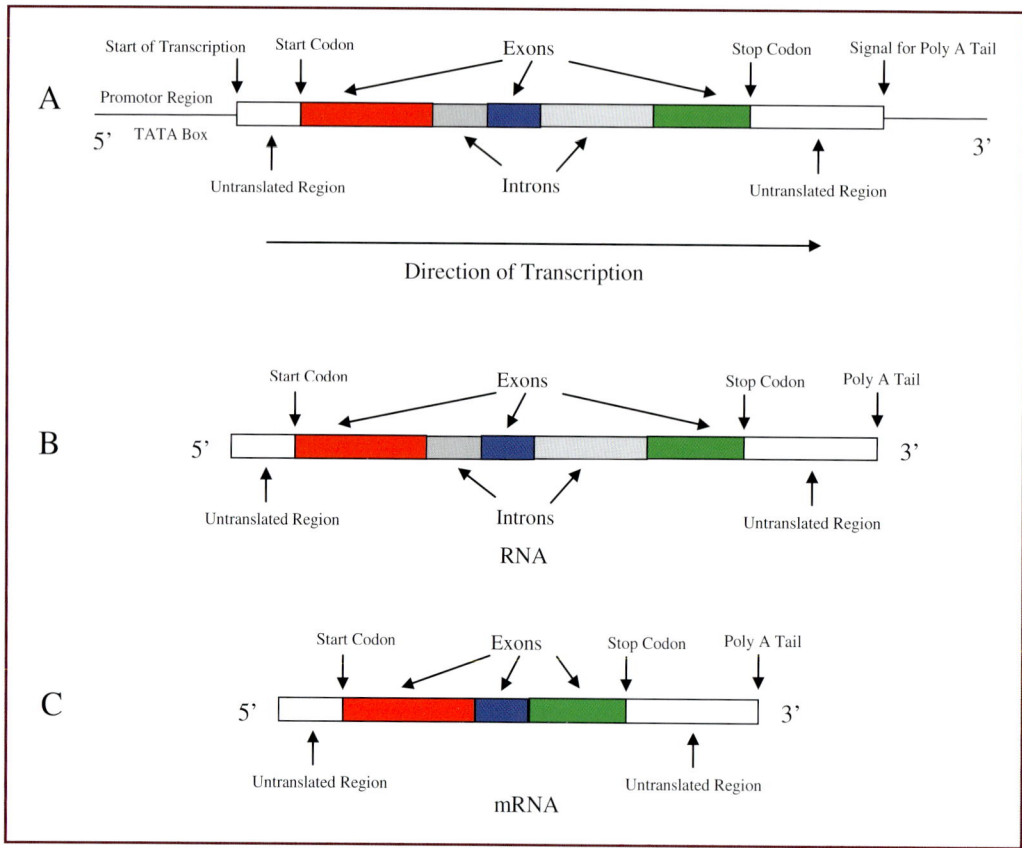

Figure 3–9. Genes consist of several regions. Transcription begins at the TATA box and extends to the poly A tail (**A**). The RNA produced by transcription (**B**) undergoes further modification (introns are spliced out) to become mRNA before leaving the nucleus of the cell (**C**). The mRNA contains all of the exons, here colored red, blue, and green, which contain the coding sequence for the protein's amino acids.

tion. The initiator codon specifies where translation will begin. A codon is simply a sequence of three adjacent nucleotides (e.g., ATG, UGA, GGG). After the initiator codon, there are several alternating regions of exons and introns. Exons contain the nucleotide sequences that code for specific amino acids, which will eventually be translated to construct a protein. The intervening sequences, or introns, are removed before the mRNA leaves the nucleus. Genes typically contain several exons and introns. The termination codon stops translation and thus ends protein synthesis and at the 3' end of the gene is the polyadenylation tail or Poly A tail. Every portion of the gene is likely important for proper gene function; however, gene expression is complex and specific function(s) for some segments (e.g., introns) are not yet completely understood.

Transcription begins at the 5' end of the gene at a location specified by the TATA box, where the nucleotide sequence "TATA" or some variation of T's and A's occurs (see Figure 3–9A). An enzyme (RNA polymerase) in concert with appropriate transcription factors binds to the TATA box and unwinds the DNA helix so that the nucleotide sequence can be read. RNA then pairs with the DNA from the transcription start site forward until the 3' end is reached. A sequence at the 3' end leads to the end of transcription and the addition of a poly A tail (a sequence of several adenine nucleotides) to the RNA.

Once the RNA is synthesized, it is further modified before it leaves the nucleus. RNA ribozymes splice out the intron sequences leaving the exon coding sequences intact in the resultant single-stranded mRNA (see Figure 3–9C). After splicing, the mRNA is ready to be transported from the nucleus into the cytosol.

At the ribosomes in the cytosol, the mRNA is translated and the chain of amino acids is constructed that will form the eventual protein. The key to translation and assembling the proper order of amino acids is that sequences of three adjacent nucleotides in the mRNA code for specific amino acids. Recall that each sequence of three nucleotides is called a codon. Table 3–3 identifies the various codons and their respective amino acids. Except for methionine, amino acids are represented by two or more codons. At the ribosome, the mRNA is read until the codon AUG is encountered, which codes for the amino acid methionine. When AUG is encountered tRNA brings methionine to the ribosome and the construction of the amino acid chain begins. From the start codon, each subsequent codon (also known as a reading frame) is read and the corresponding amino acid is added to the chain (Figure 3–10). Translation continues until one of three possible stop codons is read: UGA, UAA, or UAG. Once a stop codon is encountered, no additional amino acids are added to the chain. The polypeptide chain now undergoes folding and perhaps other structural modifications to become the functional product that will be used by cells or tissues.

In summary, the process of going from DNA to RNA to protein involves several steps. The first step is transcription where the DNA is unwound and read letter by letter to produce RNA. Second, RNA splicing removes the introns to produce mRNA. Third, mRNA is transported from the nucleus. Fourth, in the presence of ribosomes, translation of mRNA occurs and the proper amino acid sequence is constructed. Finally, protein assembly produces the final functional product needed for the normal function of the cell(s) or tissue(s). This multistep process is gene expression.

Table 3–3. Codons for the 20 Amino Acids Available to Make Proteins*

UUU	Phe	UCU	Ser	UAU	Tyr	UGU	Cys
UUC		UCC		UAC		UGC	
UUA	Leu	UCA		UAA	Stop	UGA	Stop
UUG		UCG		UAG		UGG	Trp
CUU	Leu	CCU	Pro	CAU	His	CGU	Arg
CUC		CCC		CAC		CGC	
CUA		CCA		CAA	Gln	CGA	
CUG		CCG		CAG		CGG	
AUU	Ile	ACU	Thr	AAU	Asn	AGU	Ser
AUC		ACC		AAC		AGC	
AUA		ACA		AAA	Lys	AGA	Arg
AUG	Met	ACG		AAG		AGG	
GUU	Val	GCU	Ala	GAU	Asp	GGU	Gly
GUC		GCC		GAC		GGC	
GUA		GCA		GAA	Glu	GGA	
GUG		GCG		GAG		GGG	

*Each grouping of three bases codes for a specific amino acid. Methionine (Met) is the start codon for translation. UAA, UAG, and UGA are stop codons that end translation.

Source: Information from HHMI (2001, p. 27).

CONTROL OF TRANSCRIPTION AND TRANSLATION

Regulation of gene expression can occur at one or more steps in the process described above (Figure 3–11). Regulatory mechanisms at any of the steps of transcription and translation can alter the protein product. Specific mechanisms operating during transcription and splicing can lead to the production of multiple protein isoforms (different versions of the same protein) from a single gene (see review by Breitbart, Andreadis, & Nadal-Ginard, 1987).

One site for regulation is at the DNA where transcription occurs. If transcription is turned on or off, enhanced or repressed, then the types of proteins

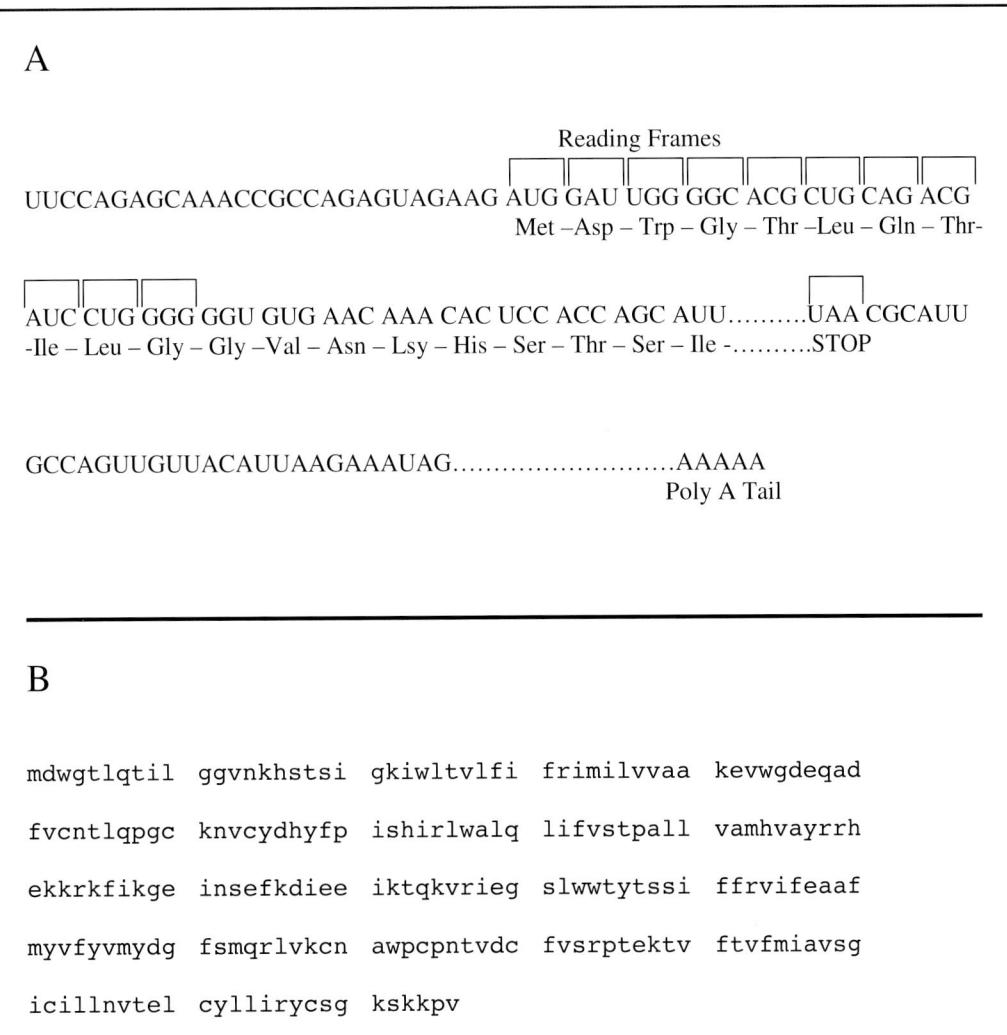

Figure 3–10. A. Translation of GJB2 mRNA. At the ribosome, the mRNA is read until the codon for methionine signifies the start of the amino acid chain. Each reading frame thereafter signals another amino acid to the chain (three letter codes are presented) until one of the stop codons is read (in this case UAA). After the stop codon, no additional amino acids will be added to the polypeptide chain. At the end of the mRNA is the poly A tail. B. The complete amino acid sequence for GJB2 is shown. Each letter represents one of the amino acids in the 226 amino acid peptide chain. Amino acid sequence from GenBank (Benson, Karsch-Mizrachi, Lipman, Ostell, & Sayers, 2008) database for the sequence originally published by Najmabadi et al. (2005), GenBank accession number AY953441. Amino acids are grouped in sections of 10 for easier reading. After the peptide chain is complete, the protein will undergo folding to achieve its final conformation.

produced as well as the amount of protein synthesized by the cell will certainly be affected.

Other sites where regulation can occur are during RNA splicing, transport of the mRNA from the nucleus into the

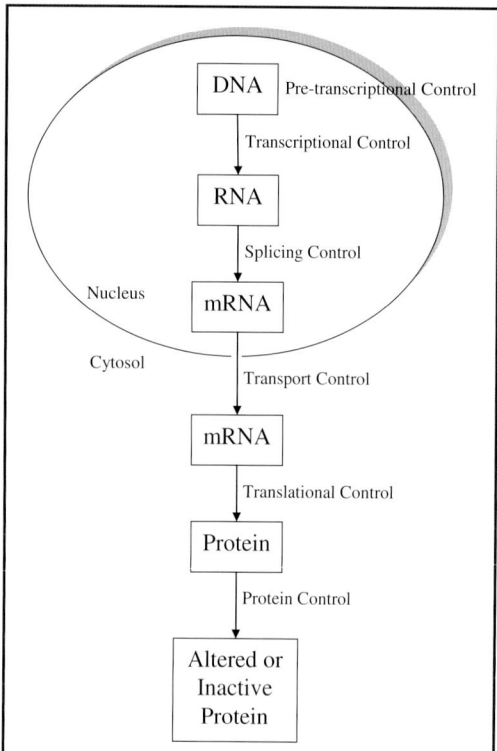

Figure 3–11. Gene regulation can occur at any of several steps from DNA to RNA to protein. One of the most effective places for gene regulation is transcriptional control; however, regulation even at later steps can inactivate the protein product. Pretranscriptional control shown here is the idea that the DNA can be packaged such that it is inaccessible to transcription (epigenetics).

cytosol, translation, and protein folding. Even at these latter stages, regulation can result in protein inactivation and thus altered gene expression. One other important control of gene expression is epigenetics, which are biological processes that control DNA packaging (recall that DNA is packaged with other proteins to produce chromatin). When the chromatin is tightly packed, the DNA in those packed regions is inaccessible to the transcription machinery (see review by Provenzanno & Domann, 2007), and therefore, the genes in the packed region are silenced. One could consider this type of control to be pretranscriptional control. That said, controlling transcription at the DNA is one of the most effective and well-studied ways to regulate or influence gene expression.

Transcription Factors

Transcription factors are proteins that act directly or indirectly to control transcription and the amount of mRNA that a target gene produces. Transcription factors are themselves produced by genes and, once synthesized or activated, they act to influence other genes by binding to DNA or other regulators of transcription (e.g., RNA polymerase). Several outcomes of transcription factor activity include switching genes on and off as well as repression or enhancement of gene expression. Transcription factors are categorized into families on the basis of the structure of the protein, which allows the protein to bind to regions of DNA and control gene expression.

General transcription factors (TFIIA, TFIIB, etc.) are necessary for activating transcription. When RNA polymerase binds to the TATA box, it is unable to initiate transcription unless general transcription factors bind to it or near it in the promoter region of the gene. These transcription factors work with or influence RNA polymerase in order for the enzyme to function appropriately and transcribe the DNA into RNA.

Basic Helix-Loop-Helix (bHLH) transcription factors have α protein structure with two α helices separated by a loop. Several bHLH transcription factors are important in the development of skeletal muscle; however, they are also important for ear development. Examples of bHLH transcription factors that are important for the ear include Neurogenin 1 (Ngn1) and Neurogenic Differentiation 1 (NeuroD).

Zinc finger transcription factor proteins have a structure such that an α helix and a β sheet are held together by zinc. This structure produces fingerlike protrusions that bind to DNA. Gata binding protein 3 (Gata3) is one example of a zinc finger transcription factor that is critical for differentiation of auditory neurons and may distinguish auditory from vestibular neurons during development (Jones & Warchol, 2009; Lawoko-Karali et al., 2004).

Helix-turn-helix transcription factors are a large family that contains several members including homeodomain, POU domain (POU), paired (Pax), and forkhead/winged helix to name a few. Homeodomain proteins contain a highly conserved sequence of amino acids called the homeodomain (60 amino acids), and most are critical for embryogenesis and organogenesis. One large family of genes that produce homeodomain proteins are homeobox (HOX) genes. HOX genes are critical for early development of the embryo including segmentation of the central nervous system along the rostral-caudal axis (see Chapter 4). In humans, there are 39 HOX genes arranged in clusters on four different chromosomes with each cluster of genes arranged according to a spatiotemporal expression pattern during development (Favier & Dolle, 1997). The POU gene family codes for proteins with both a homeobox region and a bHLH region. Fifteen POU genes are grouped into six classes. Examples of POU domain transcription factors that are critical for inner ear development are found in class III and class IV. Pou3f4 (POU domain class III factor 4, also called Brn4) is a transcription factor that arises from mesenchyme surrounding the otic vesicle. It is required for normal coiling of the cochlea and other morphological features. Pou4f1 (Brn3a or POU domain, class 4, transcription factor 1) plays a role in growth and migration of sensory neurons and axon guidance in inner ear neurogenesis. Brn3c (Pou4f3) is essential for hair cell differentiation and survival. Pax proteins may or may not have a homeodomain, but all have a paired domain that can bind directly to DNA (Lang, Powell, Plummer, Young, & Ruggeri, 2007), and some of these proteins are critical for inner ear development—particularly PAX3, in which mutations result in Waardenburg syndrome type I (dystopia canthorum, hypopigmentation such as white forelock and/or heterochromia, hearing loss).

GENETIC DISORDERS

A mutation is a permanent change in the nucleotide sequence of DNA. Mutations occur spontaneously, or they can be induced by exposure to mutagens such as radiation. Alterations in the DNA do not always have to be detrimental to the organism. Sometimes a permanent change in the DNA may actually be beneficial and allow an organism to survive or thrive in its environment. Although it is important to recognize this fact, we focus here on genetic disorders.

Genetic disorders have a net gain or loss of DNA or alteration of the DNA sequence, which will affect protein synthesis (the functional product of genes). Disorders are inherited when the altered DNA is passed on to the offspring. Genetic disorders account for approximately 30 to 45% of birth defects and may account for some of the 50% of birth defects with an unknown cause. Genetic disorders include chromosomal abnormalities, single gene disorders, nontraditional (or non-Mendelian) inheritance, and multifactorial inheritance.

Chromosome Abnormalities

Chromosome abnormalities are either numerical or structural, are visible with the light microscope or other specialized techniques (e.g., fluorescence in situ hybridization), and generally result in clinical syndromes (of which hearing loss may be one of the identifiable features). Large segments of DNA and potentially multiple genes are affected because portions of chromosomes, entire chromosomes, or the entire karyotype may be altered. Overall, chromosome abnormalities occur in 5 to 7 per 1000 live births, 2% of pregnancies among women age 35 and older, and are responsible for 50 to 70% of spontaneous abortions (miscarriages) that occur in the first trimester of pregnancy (Brent, 2004; McFadden & Friedman, 1997).

Numerical Abnormalities

Numerical abnormalities affect the number of chromosomes. Recall that the normal number of chromosomes is 46 (or 2*n* where *n* = 23) with a karyotype of 46, XX or 46, XY. This is known as euploidy. Abnormalities that produce an extra copy of the entire set of chromosomes are called polyploidy. For example, three copies of the chromosomes (3n or triploidy) would have a karyotype of 69, XXX or 69, XXY or 69, XYY. Four copies (4n) is termed tetraploidy. Triploidy is the most common form of polyploidy in humans; however, triploidy as well as other multiple copies of the entire set of chromosomes are lethal in humans.

A loss or gain of only one chromosome is known as aneuploidy. Aneuploidy is the most common chromosome abnormality. A loss of one chromosome is termed monosomy whereas an increase of one chromosome is termed trisomy. Perhaps the most well-known aneuploidy disorder is trisomy 21 or Down syndrome (Hernandez & Fisher, 1996; Wiseman, Alford, Tybulewicz, & Fisher, 2009), which occurs in 1 in 700 live births although incidence varies across reports. The karyotype for someone with Down syndrome is 47, XX +21 for a female or 47, XY +21 for a male demonstrating a gain (or trisomy) of chromosome 21. A disorder is characterized as a syndrome when two or more clinically identifiable symptoms are present in the phenotype. The phenotype for Down syndrome includes short stature, short fingers, mental retardation, craniofacial features such as flat face and protruding tongue, and hearing loss. One example of monosomy is Turner syndrome (45, X or monosomy X), which occurs in 1 in 2500 live births and has a phenotype that includes small stature, short or webbed neck, broad chest, renal anomalies, and hearing loss (Morgan, 2007).

Numerical abnormalities can occur when the chromosomes do not separate properly during meiosis and the resulting gamete contains the incorrect number of chromosomes. This situation is termed nondisjunction. If the gamete formed by nondisjunction contributes to fertilization, then the resulting zygote (i.e., fertilized egg) will have polyploidy or aneuploidy. Another cause for polyploidy is dispermy. Normally, only one sperm fertilizes the egg; however, in dispermy two sperm enter the ovum during fertilization resulting in a zygote with an additional set of paternal chromosomes and overall 3n total chromosomes.

Structural Abnormalities

Structural chromosome abnormalities involve portions of one or more chromosomes. Four types of structural abnormalities include deletions, duplications, inversions, and translocations (Figure 3–12). During meiosis, structural abnormalities can occur when the chromosomes are crossing over and recombining. For example, a portion or portions of a chromosome undergoing recombination may not recombine and be lost, may invert before recombining, or may recombine on the wrong chromosome. Structural abnormalities also can occur during early embryogenesis when cells are undergoing prolific mitosis.

As the term implies, a deletion involves a loss of part of a chromosome and hence, a net loss of genetic material. The nomenclature 46 XX (5p-) indicates deletion of the short arm of chromosome 5 in a female, which is the chromosome abnormality in Cri du Chat syndrome (Mainardi, 2006). 22q11 deletion (or 22q11-) causes several syndromes including Shprintzen, diGeorge, and velocardiofacial syndromes (Goldmuntz, 2005; Shprintzen, 2008).

Duplications produce two or more copies of a DNA segment on the same chromosome resulting in a net gain of genetic material. The nomenclature dup(1q) indicates a duplication of the long arm of chromosome 1.

An inversion results when a portion of one chromosome reverses direction. Paracentric inversions involve only one arm of the chromosome (e.g., inv(4) (q12.21;12.31)) whereas pericentric inversions involve both arms.

Translocations are structural abnormalities that involve more than one chromosome. In this case, a portion of one chromosome breaks off and attaches to another chromosome. If one chromosome loses genetic material and the other chromosome gains genetic material such that there is no net loss, the translocation is termed balanced. An unbalanced translocation occurs when there is an overall net loss of genetic material.

Single Gene Disorders

Single gene or monogenic disorders are caused by alterations (mutations) in the DNA for a specific gene and account for a significant portion of inherited deficits including hearing loss (Toriello, Reardon, & Gorlin, 2004). Mendel's principles apply here and these disorders follow Mendelian inheritance patterns, which include characterization of the inherited allele as dominant or recessive and as autosomal or sex-linked. Single gene disorders also can occur in the mitochondrial DNA (discussed later).

Recall that a single gene consists of hundreds or thousands of nucleotides and various types of mutations can occur potentially anywhere along the length of the gene's DNA code. Mutations in the exons will directly affect the amino acid sequence and protein synthesis; however, mutations in the other regions of the gene (introns, promoter region, etc.) can be equally detrimental to transcription or translation and ultimately protein synthesis. For example, mutations in the intron nucleotide sequences can lead to splicing errors, which will alter the mRNA sequence.

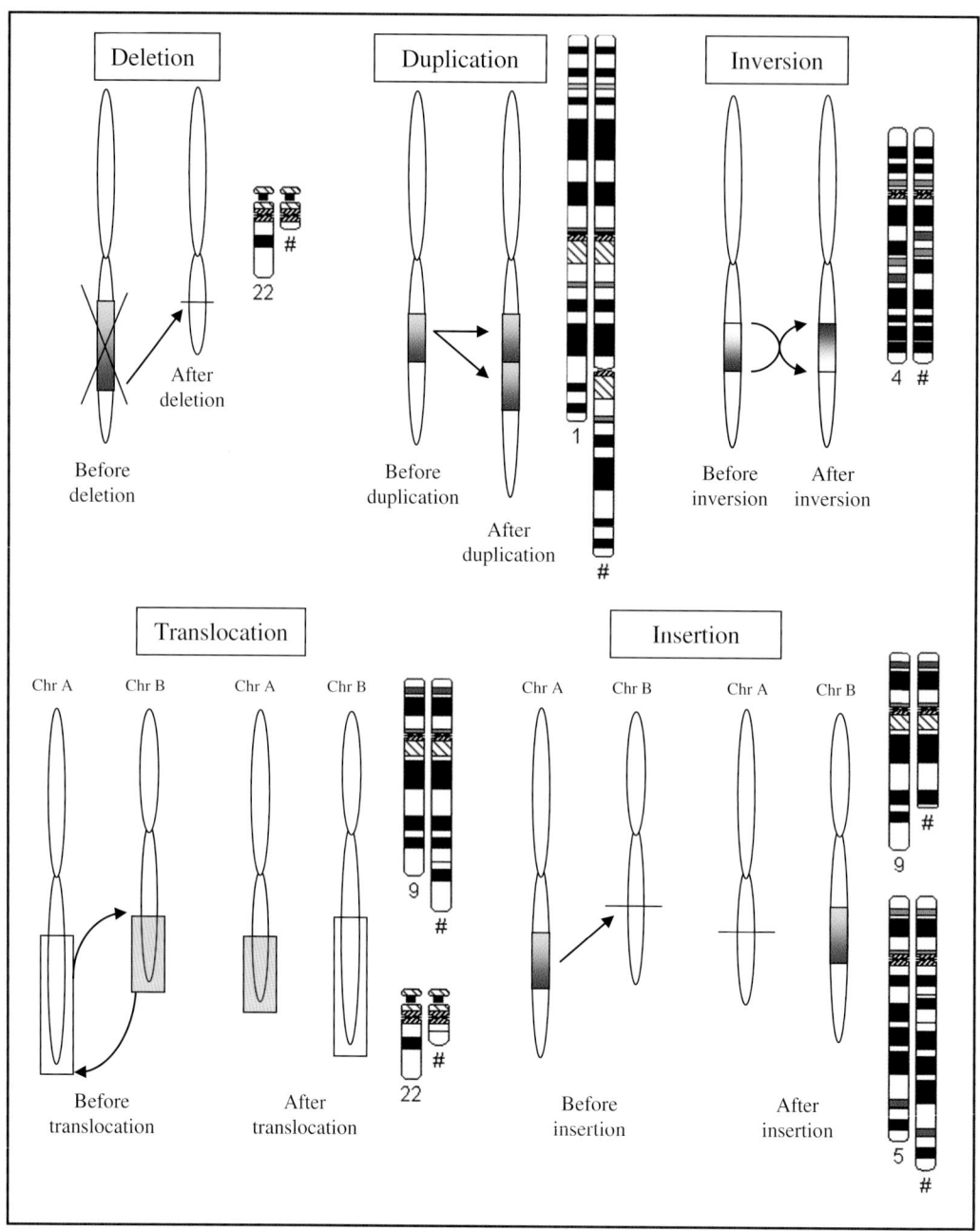

Figure 3–12. Structural chromosomal abnormalities have several forms. Deletions, duplications, and inversions occur on a single chromosome. Translocations and insertions involve two chromosomes. Cartoons depict the structural abnormality in general while the ideograms represent actual stained chromosomes. Ideograms drawn using Hiller et al. (2004).

Mutations in the promoter region may affect the onset of transcription or may prohibit access to the DNA altogether.

What types of mutations can occur in our genes? A point mutation is a single nucleotide change where one letter in the DNA is substituted by another but the total number of nucleotides remains unaltered. Insertions and deletions are mutations where nucleotides are added or removed from the DNA, respectively. If the mutation ultimately affects the mRNA reading frames such that one or more codons are altered, then amino acids may be altered. When a mutation results in an amino acid change such that the wrong amino acid is placed in the polypeptide chain, this is called a missense mutation. A nonsense mutation occurs when the mutation produces a stop codon. A stop codon encountered prior to the normal stop codon will truncate or end the amino acid sequence prematurely and the protein will be altered. Single nucleotide changes or point mutations often result in missense or nonsense mutations. Insertions or deletions of nucleotides in multiples of three will often result in the insertion or deletion of one or more amino acids or could also cause a premature stop codon. Insertions or deletions of nucleotides in multiples other than three will shift the reading frames in the mRNA, thus altering codons and the amino acid sequence of the protein. This situation is termed a frameshift mutation.

Sometimes a single nucleotide is altered but the amino acid sequence and the protein are not affected. This polymorphism is possible because some amino acids are coded by two or more different codons (see Table 3–3) so a point mutation may change one letter in the codon but the same amino acid gets placed in the polypeptide change and the overall amino acid sequence remains intact. Single nucleotide polymorphisms (or SNPs) are an interesting area of research. Scientists have identified about 1.4 million places on the genome where SNPs occur (The International Hapmap Consortium, 2005). Research in this area may lead to a better understanding of complex diseases, susceptibility for disease, and insights into human history.

Mutations can occur anywhere along the length of a gene and a given gene may exhibit several types of mutations in a population. Therefore, different mutations in the same gene may result in different forms of hearing loss (syndromic, nonsyndromic, dominant, or recessive). For example, different mutations in GJB2 cause DFNB1, DFNA3, palmoplantar keratoderma with deafness, keratitis-ichthyosis-deafness syndrome, or Vohlwinkel syndrome depending on the location and type of mutation and the ultimate effect of the mutation on the protein product. Pandya et al. (2003) reported that more than 50 different mutations have been identified in the GJB2 gene.

Single gene disorders can result in clinical syndromes or may result in hearing loss alone. Hearing loss associated with a syndrome is called syndromic hearing loss, whereas the inheritance of hearing loss as an isolated phenotype is called nonsyndromic hearing loss. A specific nomenclature is used to identify nonsyndromic hearing loss: The letters DFN are followed by a letter and a number. The abbreviation DFN indicates the nonsyndromic nature of the hearing loss. The next letter identifies the inheritance pattern as autosomal dominant, autosomal recessive or sex-linked (A for dominant, B for recessive, Y for the male sex chromosome, X indicates the X chromosome). The number at the end of this nomenclature symbolizes a unique locus (a number identifies each unique locus). For example, DFNB1 indicates the first locus identified for an autosomal recessive form of nonsyndromic hearing loss. DFNA9 indicates the ninth locus identified for nonsyndromic hearing loss with an autosomal dominant mode of inheritance. DFNX1 signifies an X-linked nonsydromic locus. This nomenclature does not tell us the specific chromosome locus, the specific gene that is affected, the protein product, or the type of mutation that has occurred in the gene, but we do know that it is a nonsydnromic hearing loss, dominantly or recessively inherited, and the mutation resides on one of the autosomes in the first two examples and on the X chromosome in the third example. Table 3–4 shows some of the loci for nonsyndromic forms of hearing impairment. Once a locus is identified and given a unique label, additional studies will eventually reveal

Table 3–4. Examples of Nonsyndromic Hearing Loss Loci Along With the Genes and Proteins Produced Where Such Information Is Known.

Inheritance	Name	Locus	Gene	Protein
Dominant	DFNA1	5q31	DIAPH1	Diaphanous
	DFNA9	14q12-q13	COCH	Cochlin
	DFNA11	11q12.3-q21	MYO7A	Myosin 7A
	DFNA12	11q22-q24	TECTA	α tectorin
	DFNA15	5q31	POU4F3	POU domain TF
	DFNA48	12q13-q14	MYO1A	Myosin 1A
Recessive	DFNB1	13q12	GJB2	Connexin 26
	DFNB9	2p22-p23	OTOF	Otoferlin
	DFNB16	15q21-q22	STRC	Stereocilin
	DFNB17	7q31	Unknown	Unknown
	DFNB19	18p11	Unknown	Unknown
	DFNB29	21q22	CLDN14	Claudin 14
X-linked	DFNX3	Xq21.1	POU3F4	POU Domain TF
	DFNX4	Xp21.2	Unknown	Unknown
	DFNX6	Xp22	Unknown	Unknown
Y-linked	DFNY1	Y	Unknown	Unknown

TF = transcription factor.

Source: All information is from Van Camp and Smith, Hereditary Hearing Loss Homepage, http://hereditaryhearingloss.org/. Retrieved July 17, 2009.

the specific gene and protein that are affected. Such information is already known for many nonsyndromic loci (Van Camp & Smith, Hereditary Hearing Loss Homepage. http://hereditaryhearingloss.org), some of which are included in Table 3–4. For example, the DFNB1 locus is 13q12, where the gene gap junction beta 2 (GJB2) encoding the protein connexin 26 is affected. DFNA9 locus on 14q11 contains mutations in the gene COCH that encodes the protein cochlin. As of July 2010, the Hereditary Hearing Loss Homepage showed 60 autosomal dominant (DFNA) loci with 24 genes identified, 85 autosomal recessive (DFNB) loci with 35 genes identified, 5 X-linked (DFNX) loci with 2 known genes, and 1 Y-linked locus where the gene is still unknown. In addition, 2 modifier gene loci have been identified.

As you can see from the above discussion, hearing loss is genetically heterogeneous. No single gene is the root cause for all hearing loss; rather, many genes cause hearing loss. This is one of several reasons why finding genes for hearing loss in humans is challenging. A second reason is that genetics is not the only cause for hearing loss. Several environmental factors (e.g., noise, drugs) also cause hearing loss. The environmental factor mimics a genetic condition resulting in the same phenotype. This is known as phenocopy. Noise exposure and taking ototoxic medications are two examples of environmental factors that cause hearing loss and would be considered phenocopies. Third, finding genes in humans requires large families with a number of affected individuals in order to have appropriate samples for analyses. Large, consanguineous families have been instrumental in the discovery of many of the currently known genes. Even though challenges exist, new genes are being discovered regularly in humans and other animal models.

Mendelian Inheritance Patterns

Mendelian inheritance patterns explain heredity of many traits (or phenotypes), both normal and abnormal. Here we focus on genetic disorders. Single gene disorders can be classified as autosomal (mutation resides on one of chromosomes 1 to 22) or sex-linked (mutation resides on the X or Y chromosome). Furthermore, single gene disorders may be inherited via a dominant or recessive mode of transmission. Mendelian inheritance patterns include autosomal dominant, autosomal recessive, and sex-linked.

The inheritance pattern is often discerned from the case history and a pedigree. The pedigree is a symbolic representation of an individual's family history (Figure 3–13). Each row represents a generation; symbols designate males (squares) and females (circles), matings, offspring, the proband (identified by an arrow), and identify those individuals in the family who are affected and those who are not (filled and unfilled symbols, respectively). Various hypothetical pedigrees will help us understand the characteristics of different Mendelian inheritance patterns. We use hearing loss as the phenotype of interest; however, Mendelian inheritance patterns apply to numerous phenotypes.

Autosomal dominant inheritance patterns (Figure 3–14) show that males and females are equally likely to have the phenotype. Affected individuals can be seen in every generation (called vertical transmission), and an affected parent may likely have an affected child. Recall that an autosomal dominant mode of inheritance requires that the individual have

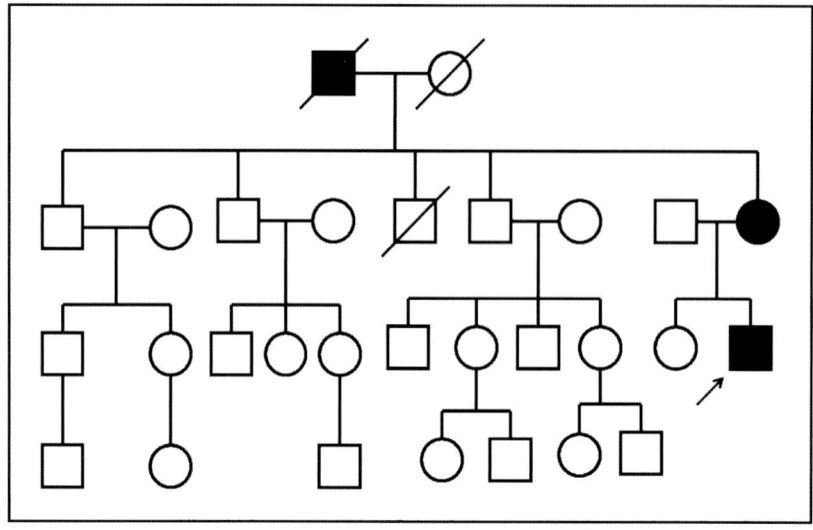

Figure 3–13. *The pedigree is a symbolic representation of the family history showing individuals for each generation identifying those who have a phenotype of interest and those who do not. Various symbols are used to designate gender, mating (including consanguinity), offspring, deaths, twins, and so on.*

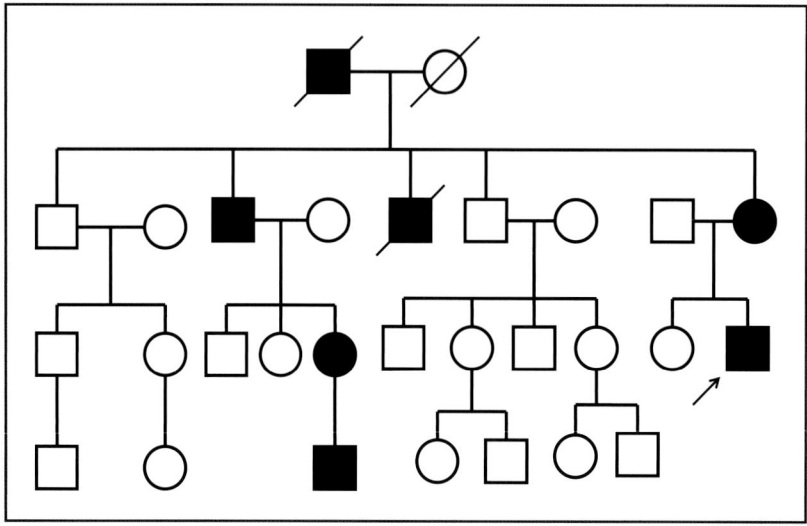

Figure 3–14. *Autosomal dominant pedigree.*

only one allele to express the phenotype. Therefore, all heterozygotes (+/− genotype) will display the phenotype; in this case, hearing loss in an autosomal dominant inheritance pattern. For every pregnancy that occurs between an affected heterozygote and unaffected person, there is a 50% risk that the offspring will inherit the allele for hearing loss. The 50% risk is due to the fact that the offspring will always inherit a normal allele from the unaffected parent but has a 50% chance of inheriting the hearing loss allele from the affected heterozygote parent. If the affected parent happens to be homozygous for the dominant hearing loss allele (−/− genotype), then all of the offspring from this individual will inherit the allele (100% risk) and all will be affected. If we know the genotypes for the parents, we can use the Punnett square to demonstrate the risk for each offspring. Figure 3–15 shows the Punnett square for the proband and his partner, which demonstrates that there is a 50% risk to pass on the hearing loss allele each time this couple becomes pregnant.

Autosomal recessive inheritance patterns (Figure 3–16) show that males and females can both be affected; however, one does not see affected individuals in every generation. Recessive inheritance patterns require that the individual inherit two recessive alleles (homozygous genotype) in order to show the phenotype. Heterozygotes (+/− genotype) have one of the recessive alleles, but will not show the phenotype. Therefore, the affected

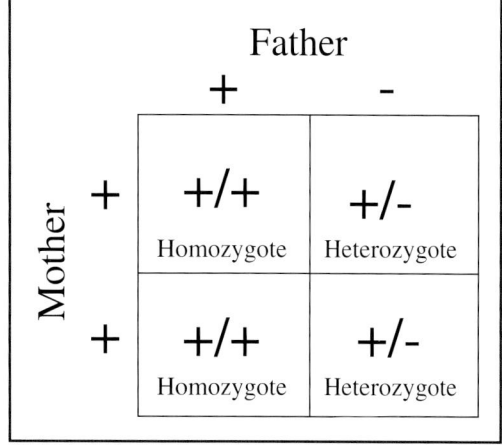

Figure 3–15. The Punnett square reveals the possible genotypes for offspring given the known genotypes of the parents. In this case, one parent is +/− (carries a dominant allele for hearing loss) and the other parent is normal (+/+). The risk for each pregnancy to inherit the dominant disorder is 50% (i.e., 50% of the possible genotypes have the dominant mutation).

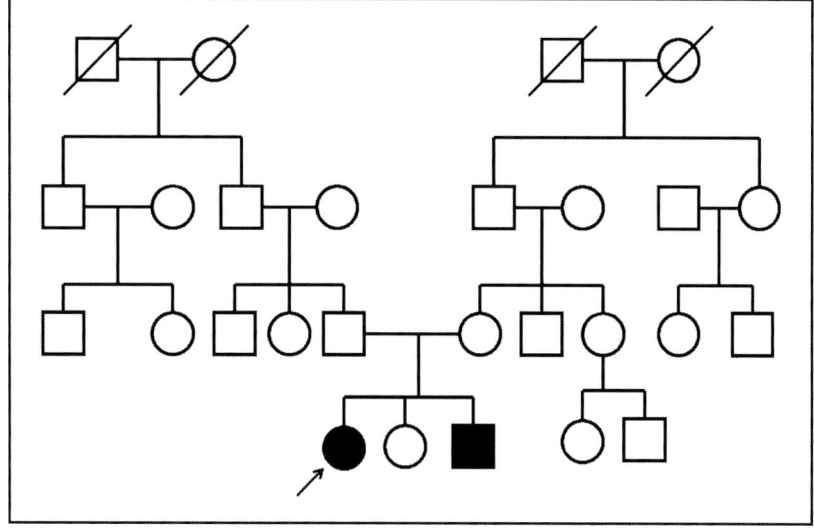

Figure 3–16. Autsomal recessive pedigree.

individuals in the autosomal recessive pedigree must be homozygous for the hearing loss gene (–/– genotype) and their unaffected parents must be heterozygote carriers (+/– genotype). Other unaffected persons in the pedigree may be heterozygous (+/– genotype) or may be homozygous for the normal (or wildtype) alleles (+/+ genotype). The Punnett squares for autosomal recessive genotypes (Figure 3–17) demonstrate different risks for each pregnancy depending on the genotypes of the parents:

Homozygote (–/–) X Homozygote (–/–): 100% affected

Homozygote (–/–) X Heterozygote (+/–): 50% affected, 50% carrier

Heterozygote (+/–) X Heterozygote (+/–): 25% affected, 50% carrier, 25% normal

Wildtype (+/+) X Homozygote (–/–): 100% carrier

Wildtype (+/+) X Heterozygote (+/–): 50% carrier, 50% normal

Sex-linked genetic disorders are X-linked if the mutation is located on the X chromosome or Y-linked if the mutation is on the Y chromosome. The female (XX) parent only provides an X chromosome to the offspring and the male (XY) can provide an X or Y chromosome. If the father passes on an X chromosome, the offspring will be a female. If a Y chromosome, the

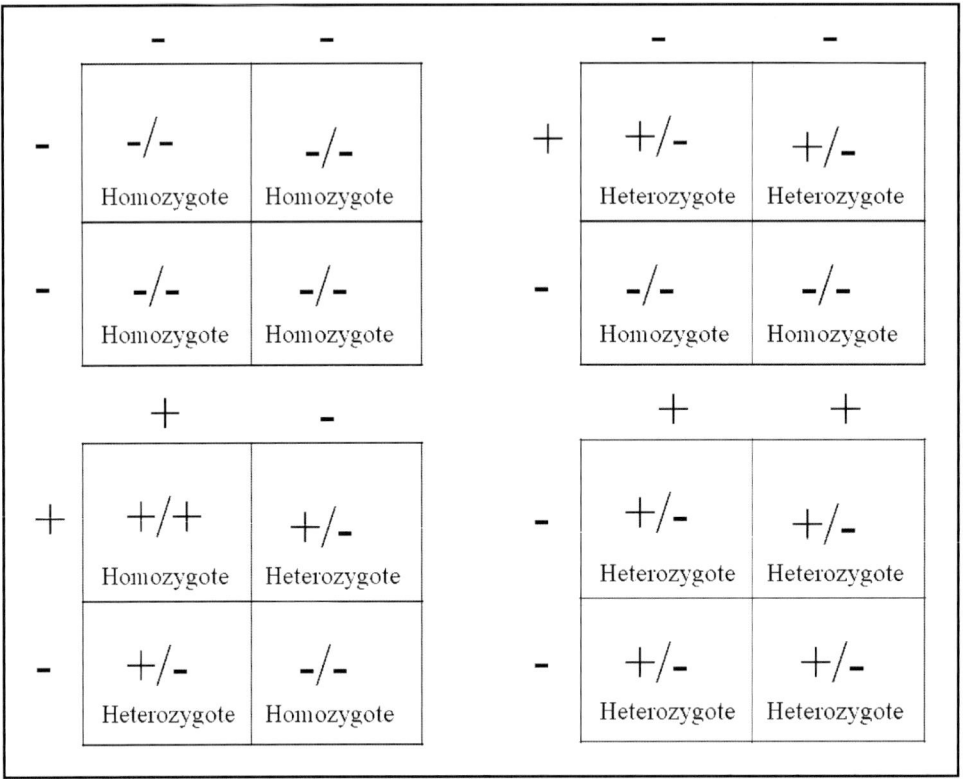

Figure 3–17. Punnett squares for different matings of various genotypes. Risks for the offspring to inherit the recessive disorder vary depending on the parental genotypes.

offspring will be male. Therefore, sex-linked inheritance patterns show characteristics indicative of the sex chromosome that contains the mutation. Y-linked inheritance patterns show, as expected, male-to-male transmission and all male offspring of affected males are affected. The Y chromosome is small and contains few genes, but there is one report of Y-linked nonsyndromic hearing loss in the scientific literature (Wang et al., 2004).

X-linked inheritance patterns show no male-to-male transmission; however, males are affected if they inherit a mutated X chromosome from the mother. X-linked dominant and recessive modes of transmission are also possible (Figure 3–18). In X-linked dominant patterns, affected males will have all affected daughters and no affected sons. A heterozygous female may have affected sons and daughters (50% chance for each pregnancy). Both females and males are affected in the X-linked dominant pattern. X-linked recessive patterns show more affected males than females since males with one mutated X chromosome will be affected. Affected males (who must get the mutant allele from their mother) have unaffected sons and heterozygous obligate carrier daughters. Sons of heterozygous females have a 50% chance of inheriting the recessive gene and being affected, whereas daughters have a 50% chance of inheriting the recessive gene and being a carrier.

Some additional terms should be defined that characterize some features present in Mendelian inheritance patterns including expressivity and penetrance. *Penetrance* indicates the percentage of individuals with the genotype that actually present with the clinical phenotype. For example, dystopia canthorum (wide set eyes) is 100% penetrant in Waardenburg syndrome type I (WSI), which means that every individual with WSI will have wide set eyes. Hearing loss in WSI, however, has a penetrance of 25 to 75% (reviewed by

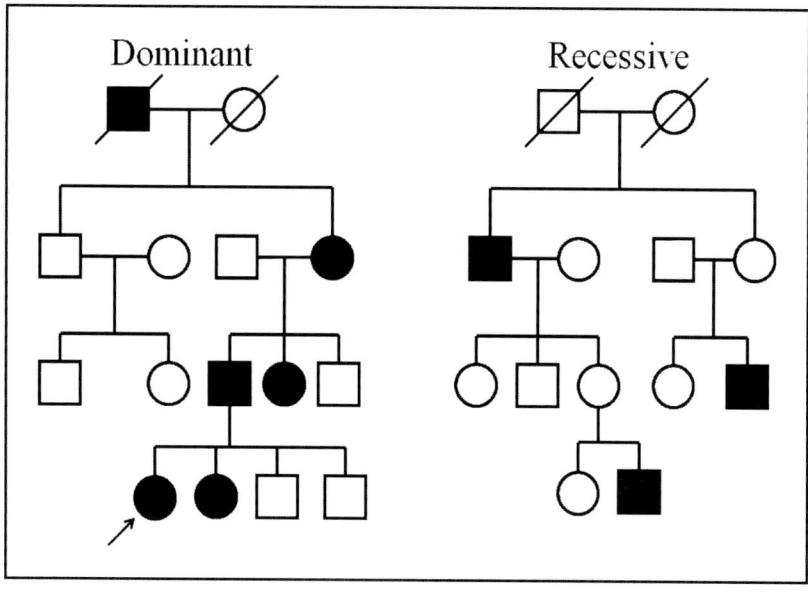

Figure 3–18. *X-linked dominant* (left) *and recessive* (right) *pedigrees.*

Read & Newton, 1997), which means that as few as one quarter (25%) to as many as three quarters (75%) of individuals with WSI will have hearing loss, and the remaining individuals with the genotype will show no hearing loss. *Expressivity* indicates that the severity of the phenotype varies among affected individuals. Hearing loss in WSI shows variable expressivity such that hearing loss may range from mild (particularly for the low frequencies) to profound hearing loss at all frequencies. In addition, some individuals may show unilateral hearing loss even though bilateral hearing loss is most common.

Non-Mendelian Inheritance

Non-Mendelian inheritance, or nontraditional inheritance patterns do not follow the characteristics described above for Mendelian inheritance. Several forms of non-Mendelian inheritance also cause hearing loss or may explain phenotypic variations. A few examples are discussed below.

Mitochondrial

Recall, that mitochondria are organelles inside the cell that produce energy molecules (e.g., ATP) for normal cell function. Mitochondria are inherited entirely from the mother. Each cell in the body can contain thousands of mitochondria, each of which contains its own DNA. The mitochondrial DNA is circular in shape and contains 16,500 base pairs that code 37 genes. Thirteen of these genes code for proteins associated with the production of ATP (i.e., oxidative phosphorylation), 2 genes produce rRNA, and 13 genes produce tRNA. Therefore, the mitochondria have the genetic machinery in place to synthesize some of their own proteins.

Mutations in the mitochondrial DNA also cause genetic disorders. Because each cell contains many mitochondria, it is possible that only some of the mitochondrial DNA will harbor a mutation, therefore, two conditions may exist. *Homoplasmy* indicates that all of the mitochondria in a cell have the genetic mutation, and *heteroplasmy* means that only some of the mitochondria have a mutation. Both sexes can be affected with mitochondrial mutations, but recall that only the mother passes mitochrondria onto the offspring. Therefore, the characteristics of mitochondrial inheritance are such that all offspring of an affected mother will be affected and there is no transmission of the mutation through an affected father (Figure 3–19). Variable expressivity, reduced penetrance, and pleiotropy (the same genetic mutation produces multiple phenotypes) are common with mitochondrial inheritance. A number of mitochondrial syndromes have associated hearing loss such as mitochondrial myopathy, encephalopathy, lactic acidosis, strokelike episodes (MELAS), and myoclonus epilepsy associated with ragged-red fibers (MERRF). Nonsydromic hearing loss also can occur with mitochondrial mutations (Fischel-Ghodsian, 1999; Kokotas, Petersen, & Willems, 2007).

Digenic Inheritance

Digenic inheritance involves genes at two different loci interacting to cause a phenotype. We return to gap junction genes to illustrate this form of inheritance. As previously stated, DFNB1 (locus 13q12) is a recessive form of nonsyndromic hear-

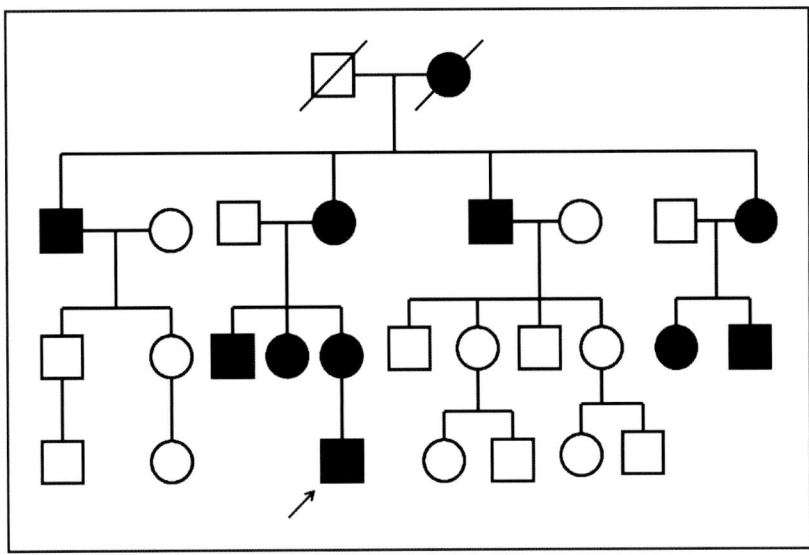

Figure 3–19. Mitochondrial inheritance.

ing loss that is caused by inheriting two mutated alleles for the GJB2 gene. We know from the previous discussion that parents who are heterozygous for mutations in GJB2 have normal hearing, but their offspring will have a 25% risk to inherit the recessive hearing loss alleles. However, a proportion of individuals who are heterozygous for GJB2 have a hearing loss phenotype. This seems strange because we would expect such a genotype to have normal hearing; however, these heterozygotes for GJB2 also have inherited a mutation in a completely different gene, GJB6. Such individuals are heterozygous for two different genes, GJB2 and GJB6, both of which are critical for normal inner ear function, and as a result have hearing loss due to digenic inheritance (del Castillo et al., 2002). An example pedigree is shown in Figure 3–20. As it turns out, the GJB6 gene resides close to GJB2 at the 13q12 locus; therefore, DFNB1 can be caused by homozygous mutations for GJB2 (most common), digenic inheritance involving both GJB2 and GJB6 (rare), and perhaps homozygous mutations for GJB6 (very rare). Digenic inheritance for GJB2 and GJB3 has also been shown to account for nonsyndromic hearing loss in Chinese families (Liu et al., 2009).

One should not confuse digenic inheritance with compound heterozygosity, which can also cause hereditary hearing and vestibular impairment. Although digenic inheritance involves two genes, compound heterozygosity involves a single gene. An individual with compound heterozygosity has inherited two different mutations in the same gene, one mutation from the mother and the other mutation from the father. The individual is considered a heterozygote because he or she has two different alleles; however the two mutant alleles together produce the disordered phenotype. One example of compound heterozygosity occurs for Jervell and Lange-Nielsen syndrome (JLNS), tra-

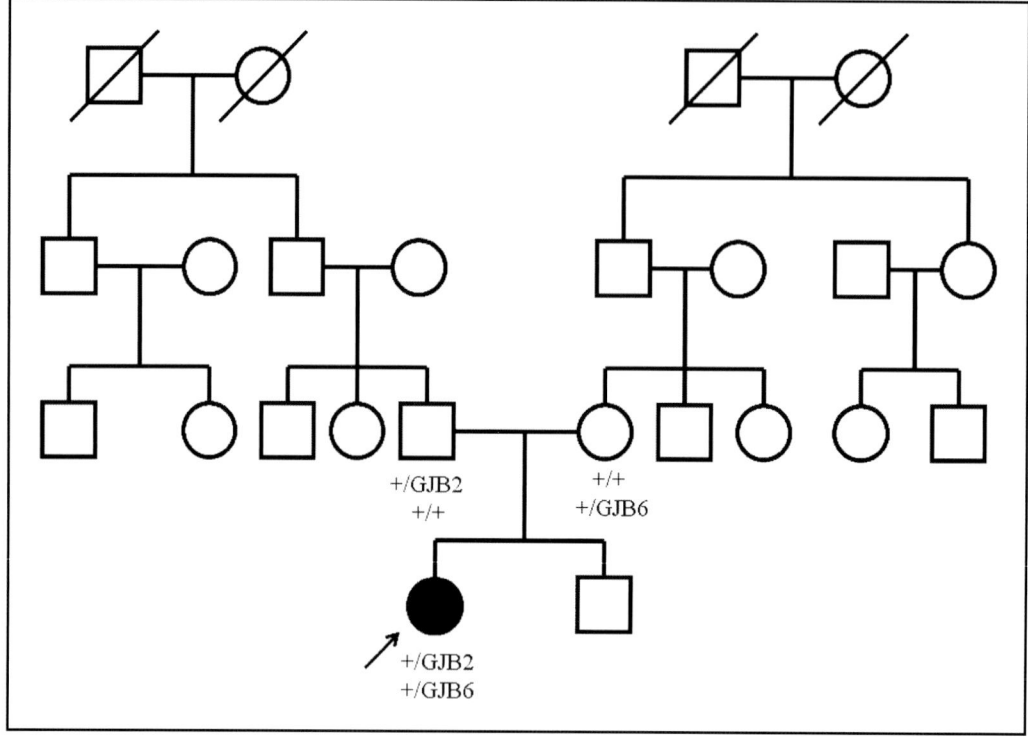

Figure 3–20. Pedigree showing digenic inheritance for DFNB1. The proband inherited one mutation in GJB2 from the father and a mutated GJB6 allele from the mother resulting in a phenotype of hearing loss.

ditionally thought to be an autosomal recessive disorder caused by homozygous mutations in KCNE1 or KCNQ1 genes (Splawski et al., 2000; Tyson et al., 1997). Recall from Chapter 1 that KCNE1 and KCNQ1 encode potassium channel proteins important for ionic homeostasis in the inner ear endolymph. JLNS has a phenotype that includes sensorineural hearing loss and heart problems as the affected potassium channel genes also contribute to normal timing of heart contractions. Wang et al. (2002) reported on two families with JLNS where the mother and father were heterozygous carriers, each carrying a different mutation in the KCNQ1 gene. In one example, the mother had a missense mutation in exon 6 of the gene while the father had a missense mutation in exon 3. One of their children inherited both mutations and had JLNS due to the compound heterozygosity.

Modifier Genes

Genes do not act in isolation. Genetic background can profoundly influence the phenotype expressed by a given gene. Variable expression and variable penetrance in any genetic disorder, therefore, could be due to the presence of modifier genes in the genetic background of an individual. Modifier genes are independent genes that alter the phenotype produced by another gene. They can enhance or suppress the phenotype, thereby making one more susceptible or less susceptible, respectively, to disease caused by other alleles (see review by Nadeau, 2003).

Many modifier genes have been discovered in animal models, including those that modify hearing loss or dizziness (Friedman et al., 2000; McHugh & Friedman, 2006; Nadeau, 2003). In humans, the protective effect of a modifier gene was found in a large Pakistani family with recessive hearing loss DFNB26 (Riazuddin et al., 2000). Individuals homozygous for DFNB26 (locus 4q31) had the expected phenotype: congenital profound sensorineural hearing loss. However, several family members were homozygous for DFNB26 but had normal hearing. Riazuddin et al. (2000) discovered that the normal hearing homozygotes had a dominantly inherited modifier gene at locus 1q24 that suppressed the DFNB26 phenotype. This modifier gene, now called DFNM1 (deafness modifier 1), prevented the hearing loss that would have been caused by the mutant DFNB26 alleles. Further research eventually will identify the specific genes for DFNB26 and DFNM1 as well as the mechanism for the protective effect. A second modifier gene (DFNM2) has been suggested that may modify hearing associated with a mitochondrial mutation (Bykhovskaya et al., 2000).

Multifactorial Inheritance

The interaction between genes and the environment is multifactorial inheritance. An individual may have a genetic susceptibility for a disease or disorder, but may not develop a phenotype unless exposed to the environmental factor. Neural tube defects (e.g., spina bifida) can result from gene-environment interactions and the risk for such can also be reduced by environmental factors (e.g., taking folic acid before and during pregnancy). Age and environmental factors (e.g., noise, toxins) are well known causes of hearing loss. In animal models, genes are being identified that cause susceptibility to noise-induced hearing loss or age-related hearing loss (see reviews by Davis, Kozel, & Erway, 2003; Johnson, Zheng, & Noben-Trauth, 2006; Konings, Van Laer, & Van Camp, 2009; Ohlemiller, 2006). The implication is that such genetic mutations may also exist in humans. One mitochondrial mutation identified in humans is the A1555G mutation that makes one genetically susceptible to aminoglycoside ototoxicity (Guan, Fischel-Ghodsian, & Attardi, 2000; Prezant et al., 1993; Pandya et al., 1997). Individuals with this mutation may not have hearing loss due to the mutation alone, but will lose hearing quickly when exposed to aminoglycosides.

CONCLUSIONS

The DNA in our genome contains all of the instructions necessary to produce the complex, multicellular organism of a human being. Genes control development, which we discussed briefly in the preceding chapters and will see more of in the next chapter. Understanding genetics has expanded our knowledge base in cellular function, and in understanding dysfunction from genetic mutations. Inheritance patterns allow one to understand genotype-phenotype relationships as well as make predictions regarding risk for inheriting genetic disorders. Hearing loss can be caused by genetic mutations. Indeed, genetic causes account for 50 to 60% of all hearing loss and may contribute to a significant portion of age-related and noise-induced hearing loss.

GENERAL RESOURCES

Alberts, B., Bray, D., Hopkin, K., & Johnson, A. (2004). *Essential cell biology* (2nd ed.). New York, NY: Garland Science.

Alberts, B., Johnson, A., Lewis, J., Raff, M., Roberts, K., & Walter, P. (2002). *Molecular biology of the cell* (5th ed.). New York, NY: Garland Science.

Carlson, B. M. (2004). *Human embryology and developmental biology* (3rd ed.). Philadelphia, PA: Mosby.

Cummings, M. R. (2003). *Human heredity: Principles and issues* (6th ed.). Pacific Grove, CA: Thompson Learning.

Department of Energy Human Genome Program. (1992). *Primer on molecular genetics.* Washington, DC: Department of Energy.

Keats, B. J. B., Popper, A. N., & Fay, R. R. (2002). *Genetics and auditory disorders.* Berlin, Germany: Springer-Verlag.

Micklos, D. A., Freyer, G. A., & Crotty, D. A. (2003). *DNA science: A first course* (2nd ed.). New York, NY: Cold Spring Harbor Laboratory Press.

Nelson, D. L., & Cox, M. M. (2005). *Lehninger principles of biochemistry* (4th ed.). New York, NY: W. H. Freeman and Company.

LITERATURE CITED

Benson, D. A., Karsch-Mizrachi, I., Lipman, D. J., Ostell, J., & Sayers, E. W. (2009). GenBank. *Nucleic Acids Research, 37*(Database Issue), D26–D31.

Breitbart, R. E., Andreadis, A., & Nadal-Ginard, B. (1987). Alternative splicing: A ubiquitous mechanisms for the generation of multiple protein isoforms from single genes. *Annual Review of Biochemistry, 56*, 467–495.

Brent, R. L. (2004). Environmental causes of human congenital malformations: The pediatrician's role in dealing with these complex clinical problems caused by a multiplicity of environmental and genetic factors. *Pediatrics, 113*, 957–968.

Bykhovskaya, Y., Estivill, X., Taylor, K., Hang, T., Hamon, M., Casano, R. A. M. S., Yang, H., Rotter, J. I., . . . Fischel-Ghodsian, N. (2000). Candidate locus for a nuclear modifier gene for maternally inherited deafness. *American Journal of Human Genetics, 66*, 1905–1910.

Chudley, A. E., (1998). Genetic landmarks through philately—Gregor Johann Mendel (1822–1884). *Clinical Genetics, 54*, 121–123.

Davis, R. R., Kozel, P., & Erway, L. C. (2003). Genetic influences in individual susceptibility to noise: A review. *Noise and Health, 5*(20), 19–28.

del Castillo, I., Villamar, M., Moreno-Pelayo, M. A., del Castillo, F.J., Alvarez, A., Telleria, D., . . . Moreno, F. (2002). A deletion involving the connexin 30 gene in nonsydromic hearing impairment. *New England Journal of Medicine, 346*, 243–249.

Favier, B., & Dolle, P. (1997). Developmental function of mammalian *Hox* genes. *Molecular Human Reproduction, 3*(2), 115–131.

Fischel-Ghodsian, N. (1999). Mitochondrial deafness mutations reviewed. *Human Mutation, 13*, 261–270.

Friedman, T., Battey, J., Kachar, B., Riazuddin, S., Noben-Trauth, K., Griffith, A., & Wilcox, E. (2000). Modifier genes of hereditary hearing loss. *Current Opinion in Neurobiology, 10*, 487–493.

Garcia-Sagredo, J. M. (2008). Fifty years of cytogenetics: A parallel view of the evolution of cytogenetics and genotoxicology. *Biochimica et Biophysica Acta, 1779*, 363–375.

Goldmuntz, E., (2005). DiGeorge syndrome: New insights. *Clinical Perinatology, 32*(4), 963–978.

Guan, M. X., Fischel-Ghodsian, N., & Attardi, G. (2000). A biochemical basis for the inherited susceptibility to aminoglycoside ototoxicity. *Human Molecular Genetics, 9*, 1787–1793.

Hernandez, D., & Fisher, E. M. (1996). Down syndrome genetics: Unravelling a multifactorial disorder. *Human Molecular Genetics, 5*(Suppl. 1), 1411–1416.

Hiller, B., Bradtke, J., Balz, H., & Rieder, H. (2004). CyDAS Online Analysis Site. Retrieved from http://www.cydas.org/OnlineAnalysis/

The International Hapmap Consortium. (2005). A haplotype map of the human genome. *Nature, 437*, 1299–1320.

Jay, V., (2001). A portrait in history: Gregor Johann Mendel. *Archives of Pathology and Laboratory Medicine, 125,* 320–321.

Johnson, K. R., Zheng, Q. Y., & Noben-Trauth, K. (2006). Strain background effects and genetic modifiers of hearing in mice. *Brain Research, 1091,* 79–88.

Jones, J. M., & Warchol, M. E. (2009). Expression of the GATA3 transcription factor in the acoustic ganglion of the developing avian inner ear. *Journal of Comparative Neurology, 516,* 507–518.

Kokotas, H., Petersen, M. B., & Willems, P. J. (2007). Mitochondrial deafness. *Clinical Genetics, 71,* 379–391.

Konings, A., Van Laer, L., & Van Camp, G. (2009). Genetic studies on noise-induced hearing loss: A review. *Ear and Hearing, 30*(2), 151–159.

Lang, D., Powell, S. K., Plummer, R. S., Young, K. P., & Ruggeri, B. A. (2007). PAX genes: Roles in development, pathophysiology, and cancer. *Biochemistry and Pharmacology, 73,* 1–14.

Lawoko-Kerali, G., Rivolta, M. N., Lawlor, P., Cacciabue-Rivolta, D. I., Langton-Hewer, C., van Doorninck, J. H., & Holley, M. C. (2004). GATA3 and NeuroD distinguish auditory and vestibular neurons during development of the mammalian inner ear. *Mechanisms of Development, 121,* 287–299.

Liu, X. Z., Yuan, Y., Yan, D., Ding, E. H., Ouyang, X. M., Fei, Y., . . . Dai, P. (2009). Digenic inheritance of non-syndromic deafness caused by mutations at the gap junction proteins Cx26 and Cx31. *Human Genetics, 125,* 53–62.

Mainardi, P. C. (2006). Cri du Chat syndrome. *Orphanet Journal of Rare Diseases, 1,* 33, doi:10.1186/1750-1172-1-33.

Maltais, L. J., Blake, J. A., Chu, T., Lutz, C. M., Eppig, J. T., & Jackson, I. (2002). Rules and guidelines for mouse gene, allele, and mutation nomenclature: A condensed version. *Genomics, 79*(4), 471–474.

McFadden, D. E., & Friedman, J. M. (1997). Chromosome abnormalities in human beings. *Mutation Research, 396,* 129–140.

McHugh, R. K., & Friedman, R. A. (2006). Genetics of hearing loss: Allelism and modifier genes produce a phenotypic continuum. *Anatomical Record Part A, 288A,* 370–381.

Morgan, T. (2007). Turner syndrome: Diagnosis and management. *American Family Physician, 76*(3), 405–410.

Nadeau, J. (2003). Modifier genes and protective alleles in humans and mice. *Current Opinion in Genetics and Development, 13,* 290–295.

Najmabadi, H., Nishimura, C., Kahrizi, K., Riazalhosseini, Y., Malekpour, M., Daneshi, A., . . . Smith, R. J. (2005). GJB2 mutations: Passage through Iran. *American Journal of Medical Genetics A, 133A*(2), 132–137.

Ohlemiller, K. K. (2006). Contributions of moue models to understanding of age-and noise-related hearing loss. *Brain Research, 1091,* 89–102.

Pandya, A., Arnos, K. S., Xia, X. J., Welch, K. O., Blanton, S. H., Friedman, T. B., . . . Nance, W. E. (2003). Frequency and distribution of GJB2 (connexin 26) and GJB6 (connexin 30) mutations in a large North American repository of deaf probands. *Genetic Medicine, 54*(4), 259–260.

Pandya, A., Xia, X., Radnaabazar, J., Batsuuri, J., Dangaansuren, B., Fischel-Ghodsian, N., & Nance, W. E. (1997). Mutation in the mitochondrial 12S ribosomal RNA gene in two families from Mongolia with matrilineal aminoglycoside ototoxicity. *Journal of Medical Genetics, 34,* 169–172.

Prezant, T. R., Agapian, J. V., Bohlman, M. C., Bu, X., Oztas, S., Qiu, W. Q., Arnos, K. S., . . . Fischel-Ghodsian, N. (1993). Mitochondrial ribosomal RNA mutation associated with both antibiotic-induced and non-syndromic deafness. *Nature Genetics, 4,* 289–294.

Provenzano, M. J., & Domann, F. E. (2007). A role for epigenetics in hearing: Establishment and maintenance of auditory specific gene expression patterns. *Hearing Research, 233,* 1–13.

Read, A. P., & Newton, V. E. (1997). Waardenburg syndrome. *Journal of Medical Genetics, 34,* 656–665.

Riazuddin, S., Castelein, C. M., Ahmed, Z. M., Lalwani, A. K., Mastroianni, M. A., Naz,

S., . . . Wilcox, E. R. (2000). Dominant modifier DFNM1 suppresses recessive deafness DFNB26. *Nature Genetics*, 26, 431–434.

Shaffer, L. G., Slovak, & M. L., & Campbell, J. L. (Eds.). (2009). *ISCN 2009: An international system for human cytogenetic nomenclature.* Basel, Switzerland: Karger.

Shprintzen, R. J. (2008). Velo-cardio-facial syndrome: 30 years of study. *Developmental Disabilities Research Reviews*, 14(1), 3–10.

Splawski, I., Shen, J., Timothy, K. W., Lehmann, M. H., Priori, S., Robinson, J. L., Moss, A. J., . . . Keating, M. T. (2000). Spectrum of mutations in long-QT syndrome genes. KVLQT1, HERG, SCN5A, KCNE1, and KCNE2. *Circulation, 102*, 1178–1185.

Spurbeck, J. L., Adams, S. A., Stupca, P. J., & Dewald, G. W. (2004). Primer on medical genomics Part XI: Visualizing human chromosomes. *Mayo Clinic Proceedings*, 79(1), 58–75.

Toriello, H. V., Reardon, W., & Gorlin, R. J. (2004). *Hereditary hearing loss and its syndromes.* New York, NY: Oxford University Press.

Tyson, J., Tranebjarg, L., Bellman, S., Wren, C., Taylor, J. F. N., Bathen, J., Aslaksen, B., . . . Bitner-Glindzicz, M. (1997). IsK and KvLQT1: Mutation in either of the two subunits of the slow component of the delayed rectifier potassium channel can cause Jervell and Lange-Nielsen syndrome. *Human Molecular Genetics*, 6, 2179–2185.

Van Camp, G., & Smith, R. J. H. Hereditary Hearing Loss Homepage. URL: http://hereditaryhearingloss.org

Wain, H. M., Bruford, E. A., Lovering, R. C., Lush, M. J., Wright, M. W., & Povey, S. (2002). Guidelines for human gene nomenclature. *Genomics*, 70(4), 464–470.

Wang, Q. J., Lu, C. Y., Li, N., Rao, S. Q., Chi, Y. B., Han, D. Y., Li, X., . . . Shen, Y. (2004). Y-linked inheritance of nonsyndromic hearing loss in a large Chinese family. *Journal of Medical Genetics*, 41:e80. Retrieved May 12, 2008 from http://www.jmedgenet.com/cgi/content/full/41/6/e80

Wang, Z., Li, H., Moss, A. J., Robinson, J., Zareba, W., Knilans, T., . . . Towbin, J. A. (2002). Compound heterozygous mutations in KvLQT1 cause Jervell and Lange-Nielsen syndrome. *Molecular Genetics and Metabolism*, 75, 308–311.

Wiseman, F, K., Alford, K. A., Tybulewicz, V. L. J., & Fisher, E. M. C. (2009). Down syndrome—recent progress and future prospects. *Human Molecular Genetics*, 18(R1), R75–R83.

Yoshiura, K., Kinoshita, A., Ishida, T., Ninokata, A., Ishikawa, T., Kaname, T., . . . Niikawa, N. (2006). A SNP in the ABCC11 gene is the determinant of human earwax type. *Nature Genetics*, 38(3), 324–330.

Basic Concepts in Embryology

> **KEY POINTS**
>
> 1. Embryology is the study of the embryo. Embryologists are experts in development.
> 2. Human development begins with fertilization of the egg by sperm.
> 3. Cell fate (what a cell will become) is influenced by several mechanisms including cell lineage, restriction, and induction.
> 4. The embryonic period of development begins at fertilization and lasts until the end of the eighth week after fertilization.
> 5. Many events occur during the embryonic period leading to the morphogenesis of all major organ systems.
> 6. Important embryonic developmental events critical for the head and neck include gastrulation, neurulation, segmentation, regionalization of the nervous system, appearance of the cranial placodes, and development of the pharyngeal apparatus.
> 7. The fetal period extends from nine weeks postfertilization until birth. Differentiation of organ systems continues during this time period as well as tremendous growth of the fetus.
> 8. Development is influenced by both genetic and environmental factors.

FERTILIZATION AND CLEAVAGE

Development of a new mammalian organism begins in the ampulla of the fallopian tube with the fertilization of the ovum by sperm (ovum or oocyte: the gamete [egg] from the female; sperm or spermatozoa: the gamete [seed] from the male). Fertilization initiates the development of an entirely new organism. At the completion of fertilization, a new single cell called the zygote is formed. This single cell represents the fusion of genetic material from the male and female and will give rise to all tissues and structures in the new individual. The newly formed genome is now at work orchestrating a complex sequence of events that involves cell proliferation, cell migration, cell and tissue interactions, and even programmed cell death. Normal development depends on precise, coordinated interactions between genetic and environmental factors.

At fertilization and indeed during early embryogenesis, cells are totipotent, which means that they have the capacity to follow any developmental pathway and become any cell type or tissue in the body. In other words, cells can make lots of choices when they are totipotent. As development proceeds, cells become more restricted in the choices they can make and ultimately they acquire specialized functions.

How do cells make choices about what they will become? One mechanism that guides developmental change is found within the cells themselves. Molecules produced by the cell can act on the cell itself to impart change. A second mechanism that determines the choices available for the cell is cell lineage. Persistent changes may occur in each generation of descendant cells during proliferation. This imparts a kind of "memory" to each cell that reflects its molecular history and can limit its final fate. The pedigree of each cell traces its history back to a specific progenitor cell or cell type where the process of restriction sent it down a particular developmental pathway for a specialized function. For example, gastrulation produces three embryonic germ layers (ectoderm, mesoderm, and endoderm) from which specific organs and tissues develop. Neurons develop from ectoderm and muscles develop from mesoderm, not vice versa. The cell lineage determines and restricts the choices for ectodermal, mesodermal, and endodermal cells. A third way that cells make choices is through induction. Induction is an interaction between cells and tissues such that cells respond to cues in their immediate environment. Interactions occur between the inducer cell and reactor cell through diffusible molecules (or signaling molecules), extracellular matrix, or physical contact. The induction of the neural plate by signaling molecules from the notochord is one example. In order for the reacting cell to respond to the inducing molecule, the reacting cell must possess: (1) the appropriate receptors to receive the signaling molecule or interpret the physical contact; (2) the necessary components of the signal transduction pathway; and (3) the appropriate transcription factors to control gene expression. Three signaling pathways, Notch, Sonic hedgehog (Shh), and Wnt, were discussed in some detail in Chapter 2. There are many others, including signaling molecules that belong to a large family of proteins known as growth factors, which are secreted by the inductor. These growth factors bind to receptors on the reacting cell and initiate a cascade of intracellular reactions ultimately affecting transcription factors that control gene expression in the reacting cell. Importantly, the ability of the reacting cell to respond to the inductive signal often is influenced by the cell's previous interactions. Examples of these mechanisms can be found throughout the following discussion.

The early zygote is encapsulated by the corona radiata and zona pellucida (Figure 4–1) and remains for some time in the fallopian tube. The end of fertilization is signaled by the duplication of all genetic material and cleavage of the zygote into

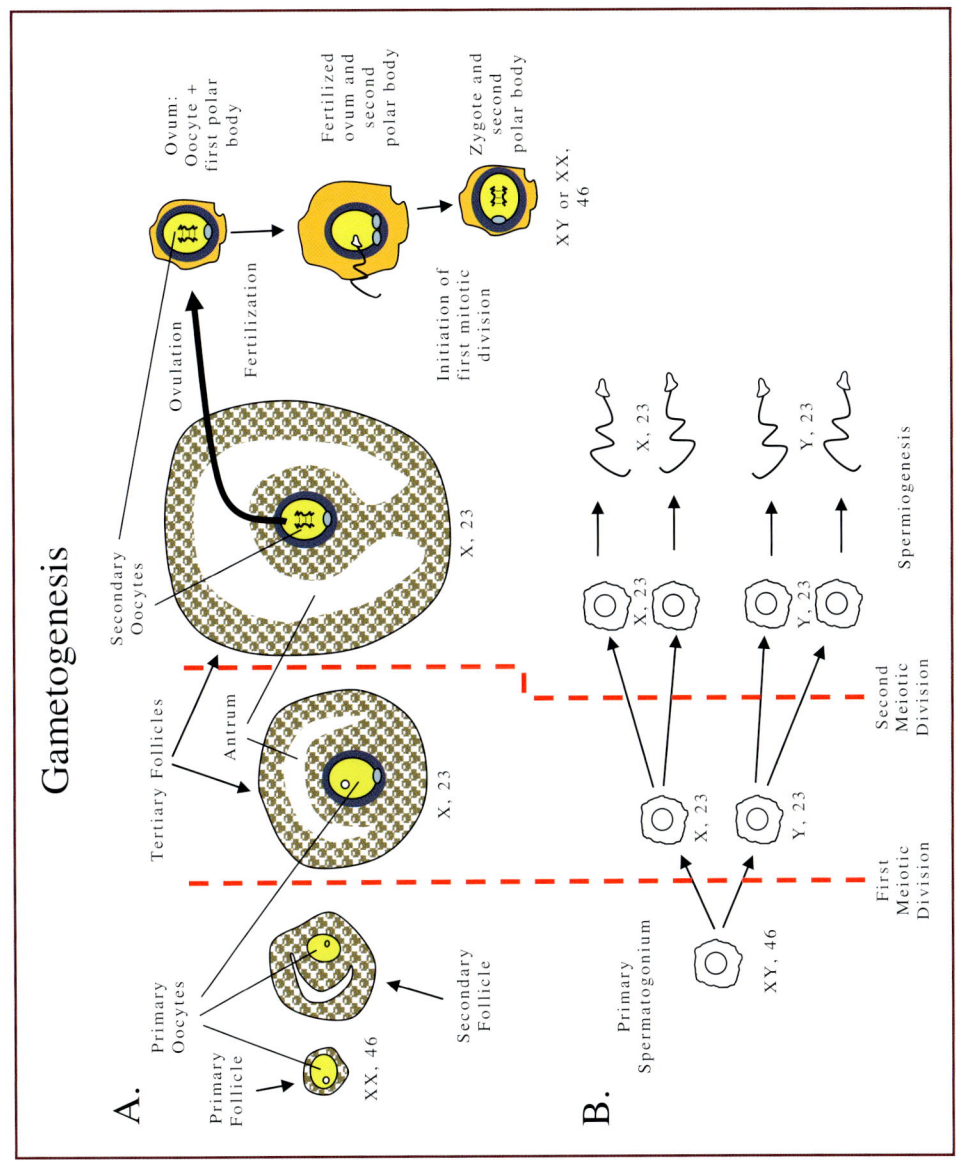

Figure 4–1. *Gametogenesis for the egg (**A**) and sperm (**B**). Spermatogenesis includes meiosis and spermiogenesis. Spermiogenesis represents the differentiation of spermatids into mature sperm.*

two smaller cells. Cleavage represents the first appearance of mitosis. As the zygote divides, the duplication of all genetic material and formation of two new cells occurs in a manner similar to normal cell division. However, it is not exactly like normal cell division and growth. The difference is that the daughter cells of cleavage, termed blastomeres, are not the same size as the parent cell. Instead, it appears as though the original volume of the zygote is divided among each of the new cells and, as a result, the cells become progressively smaller with each mitotic division. In mammals, subsequent mitosis does not necessarily occur at the same time in each blastomere. Cleavage continues over the course of several days as the embryo is swept along the fallopian tube toward the uterus. The movement of the embryo is caused by peristaltic contractions of the fallopian tube walls. This is likely a carefully controlled process that depends on molecular signals exchanged between the embryo and uterine structures. The cleavage process continues slowly over the course of days. Over the same period, the cells change their shape and begin to form a compact ball. This process is called "compaction" and is aided by the appearance of cell surface adhesion molecules.

FORMATION OF THE BLASTOCYST

The number of blastomeres increases and ranges between 12 and 32 as the embryo approaches the entrance to the uterus (approximately 3 days after fertilization). The corona radiata is lost within the fallopian tube but the zona pellucida persists until the embryo reaches the uterus. At this stage, the embryo is termed a morula and the blastomeres form a tight ball of cells that are held encapsulated by the hardened zona pellucida (Figure 4–2).

Initially, the morula floats freely as it enters the uterine fluids. Soon thereafter, individual blastomeres occupying the outer cell layer of the morula begin to form tight junctions and gap junctions between cells at their apical surfaces. These junctions are composed of cell adhesion molecules (E-cadherins) and the outer sheet of tightly held blastomeres form a diffusion barrier surrounding the proliferating interior blastomeres. The outer sheet of cells ultimately becomes the trophoblast. The presence of Na^+-K^+ ATPase membrane pumps causes the net movement of Na^+ from uterine fluids into the intercellular spaces of the morula interior. Water follows the resulting osmotic gradient, thus creating an interior fluid space that steadily increases in size. This process is called cavitation and the interior fluid space is called the blastocoele. Once the blastocoele is formed, the embryo is called a blastocyst or blastula and consists of several parts: (1) an outer zona pellucida; (2) the outer cell layer (trophoblast); (3) the inner cell mass (embryoblast); and (4) the blastocoele. The embryoblast gives rise to future cells of the embryo itself whereas the trophoblast develops into nutrient exchange tissues between the embryo and the mother (including the embryonic component of the placenta).

The morula and early blastocyst are immersed in uterine fluids for a period of days. Enzymes begin to degrade the zona pellucida such that the zona barrier weakens and ultimately fails so that the trophoblast emerges as the outermost layer of cells. The blastocyst is said to hatch from the encasement of the zona pellucida and then floats naked in uterine fluids ultimately landing against the endometrial wall of the uterus. At this stage, the

Basic Concepts in Embryology   93

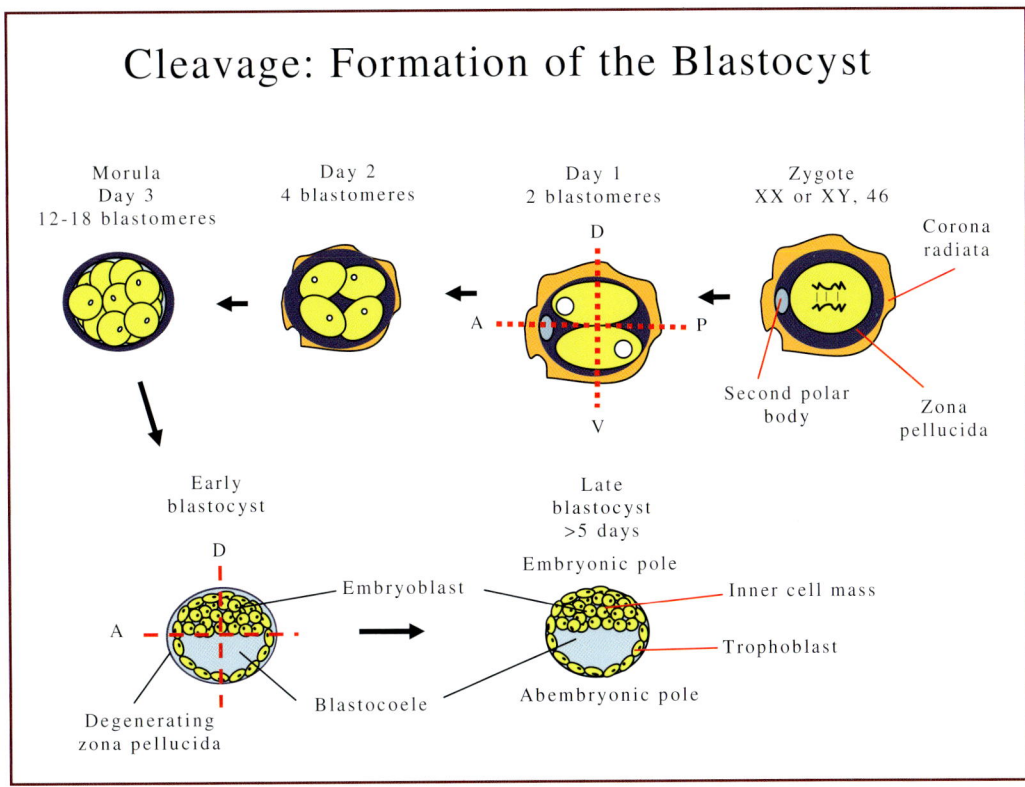

Figure 4–2. *Cleavage and formation of the blastocyst.*

embryo is considered to be a late blastocyst or blastula, which marks the end of the cleavage stage of development.

IMPLANTATION

The endometrium of the uterus has been prepared by hormones over the preceding month and is ready to accept the blastocyst once it emerges from the zona pellucida. The late blastocyst attaches to the endometrium at a position near its embryonic pole (Figure 4–3A). After attaching, the trophoblast (outer cell layer) begins to grow and develop extensively within the endometrium. By the end of the first week, the blastocyst is superficially attached to the endometrium (Figure 4–3B) and receives nourishment from it. Over the following week, the blastocyst embeds itself completely within the endometrium, begins to secrete human chorionic gonadotrophine hormone (hCG) into the surrounding endometrial tissue, and forms an elaborate exchange apparatus with the maternal tissue and blood supply (Figure 4–3C). Ultimately, the endometrium completely surrounds the embedded blastocyst and the trophoblast joins with other extraembryonic cells to form the chorion and large chorionic cavity (Figure 4–4). During the final phases of implantation, the blastocyst forms the amnion, the bilaminar (two-cell layers) embryonic disc and the yolk sac (Figures 4–3C and 4–4). The embryonic disc separates the amnionic and yolk sac fluid chambers.

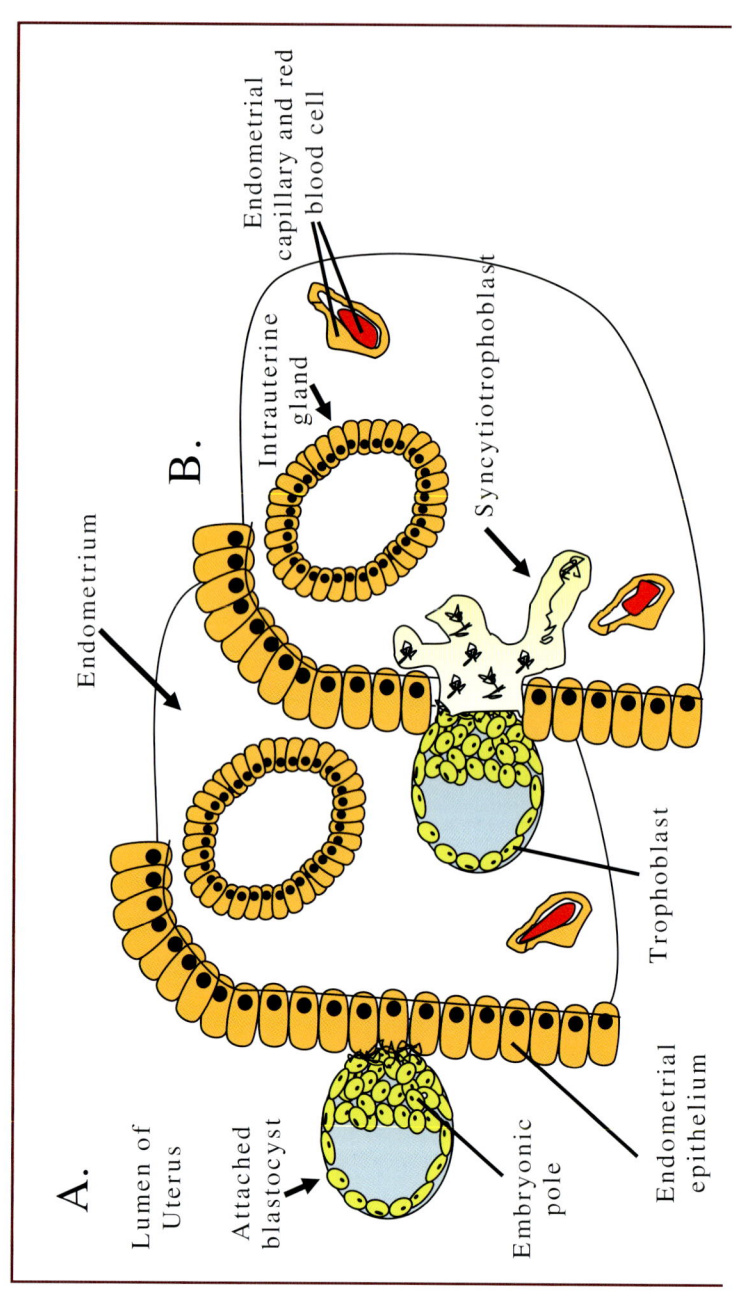

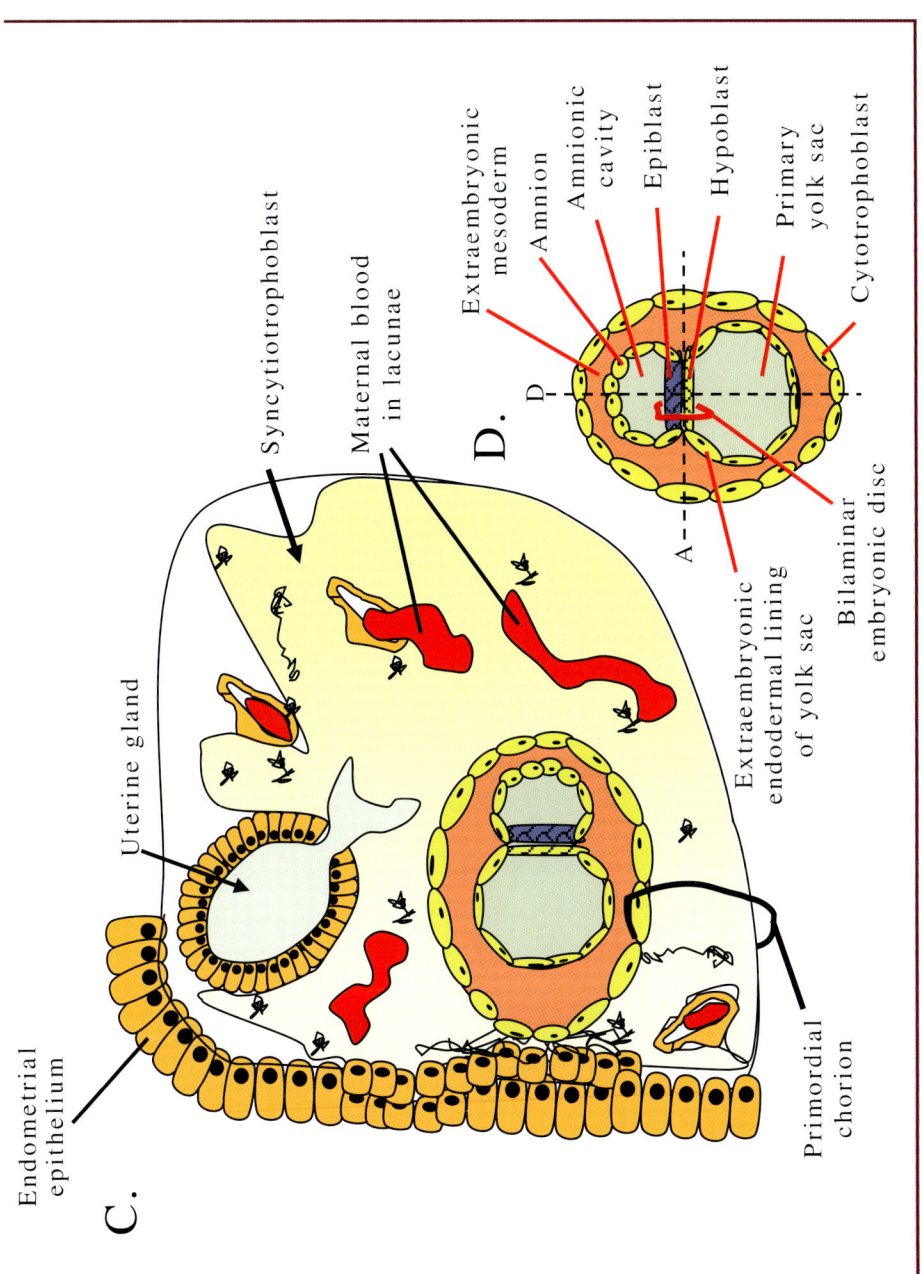

Figure 4-3. Attachment and implantation of the blastocyst into the uterine wall. **A.** Initial attachment days 5 to 6. **B.** Blastocyst at days 6 to 7, where the syncytiotrophoblast expands into endometrial tissue. **C.** Blastocyst at days 9 to 10. **D.** Illustration of the embryonic anterior-posterior and the dorsal-ventral axes and anatomical compartments in the 9- to 10-day-old blastocyst.

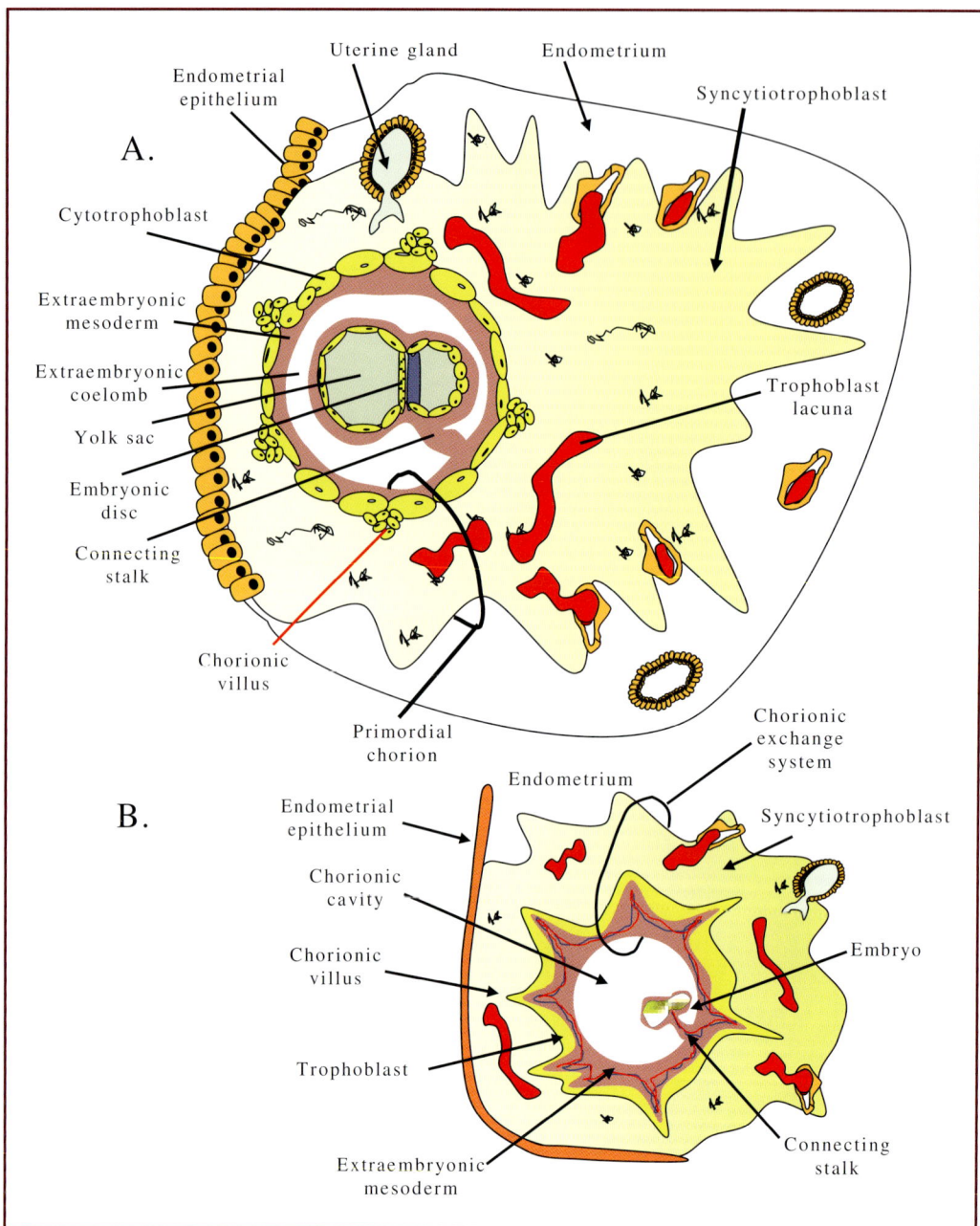

Figure 4–4. Structures associated with the implanted blastocyst. **A.** Day 12. **B.** Day 14.

The bilaminar embryonic disc is composed of one layer of epiblast cells (on the amnion, dorsal side) and a lower layer of hypoblast cells on the yolk sac side of the disc (ventral side, see Figure 4–3D). The epiblast is reminiscent of an epithelial sheet and ultimately develops a basal lamina on its inner surface. Implantation is complete by about 10 to 12 days in the human. Ultimately, the amnion, embryonic disc, and yolk sac assembly remain suspended in the chorionic cavity fluids by a short tether (connecting stalk). By the end of the second week, implantation is complete and the chorionic, amnionic, and yolk cavities are formed and serve along with associated structures to support the developing two-layered embryonic disc. Nutrients are supplied and metabolic products ultimately are removed by an elaborate exchange apparatus between the embryo and endometrial capillaries. This marks the completion of blastogenesis and the embryo is now poised to begin the morphogenic period of gastrulation. During the coming third week, the embryo will begin the formation of germ layers as well as early tissue and organ differentiation.

ANATOMICAL LANDMARKS ON THE EMBRYONIC DISC

Figure 4–5 illustrates the anatomical planes and axes used to describe spatial organization in the developing embryo. The embryonic disc lies in the horizontal plane whereas the dorsoventral axis extends at a right angle to the horizontal plane and measures distances between the top and bottom surfaces of the disc. The epiblast lies on the dorsal side and the hypoblast on the ventral side of the embryonic disc. Similarly, the amnion lies dorsally and the yolk sac is ventral to the embryonic disc (see Figure 4–3D). The horizontal plane is formed by the lateral (left and right) and the anterior-posterior (AP, or cranial-caudal) axes (see Figure 4–5C). The AP and dorsoventral (DV) axes are perpendicular to one another and they form the median or sagittal plane of the embryo at its midline (see Figure 4–5B). Points laying to the left or right of the median plane are said to be lateral to it. The lateral axis is perpendicular to the median plane and it identifies lateral distances from the midline. The lateral axis and the dorsoventral axis form the transverse plane (see Figure 4–5A). A transverse section through the embryonic disc reveals structures in cross section showing the left and right lateral extent of the tissue. Sagittal sections also produce cross sections but the view reveals the anterior-posterior extent of the tissue. The caudal pole of the embryonic disc is easily identified as the site of attachment for the connecting stalk (see Figure 4–5E).

GASTRULATION

Gastrulation marks the beginning of morphogenesis. Embryonic cells and tissues are highly dynamic during gastrulation. Cells of the embryonic disc proliferate extensively and migrate in systematic ways to form new shapes, tissue layers, and specialized regions of cellular organization. The simple bilaminar construction of the embryo is transformed into a trilaminar disc. The cells of the epiblast will give rise to all of the cells of the final adult. Cells of the hypoblast will form extraembryonic structures and the hypoblast will be replaced by mesendoderm cells, that is, cells that will differentiate into mesoderm and endoderm.

The first frank anatomical evidence that gastrulation has begun is the appearance of the primitive streak on the caudal dorsal surface of the epiblast (Figure 4–6). The primitive streak is a visible tissue stripe that can be seen in the epiblast layer of the embryonic disc near its caudal edge (near the stalk attachment). The primitive streak is an important early source for patterning signals such as chordin (Chrd), nodal, and other members of the transforming growth factor beta (TGF-β) super family, (Carlson, 2004). The primitive

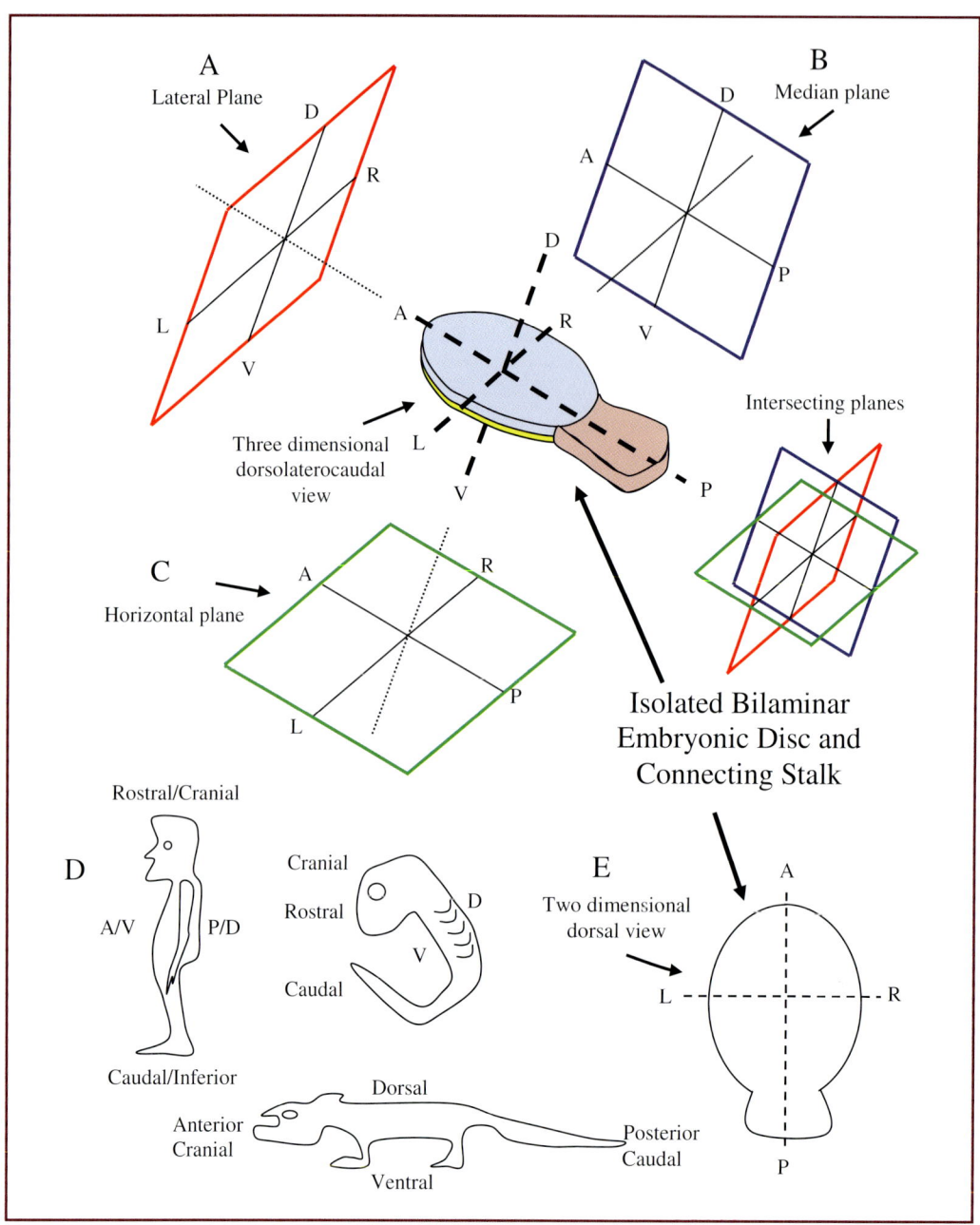

Figure 4–5. *Anatomical axes and planes in the developing embryo. **A.** The lateral plane contains the left-right and dorsal-ventral axes. **B.** The medial plane contains the anterior-posterior and dorsal-ventral axes. **C.** The horizontal plane contains the anterior-posterior and left-right axes. **D.** Anatomical directions in the adult human, embryo, and one amphibian species. **E.** Isolated bilaminar embryonic disk as seen from the dorsal view. Anterior-posterior and left-right axes are shown as dashed lines.*

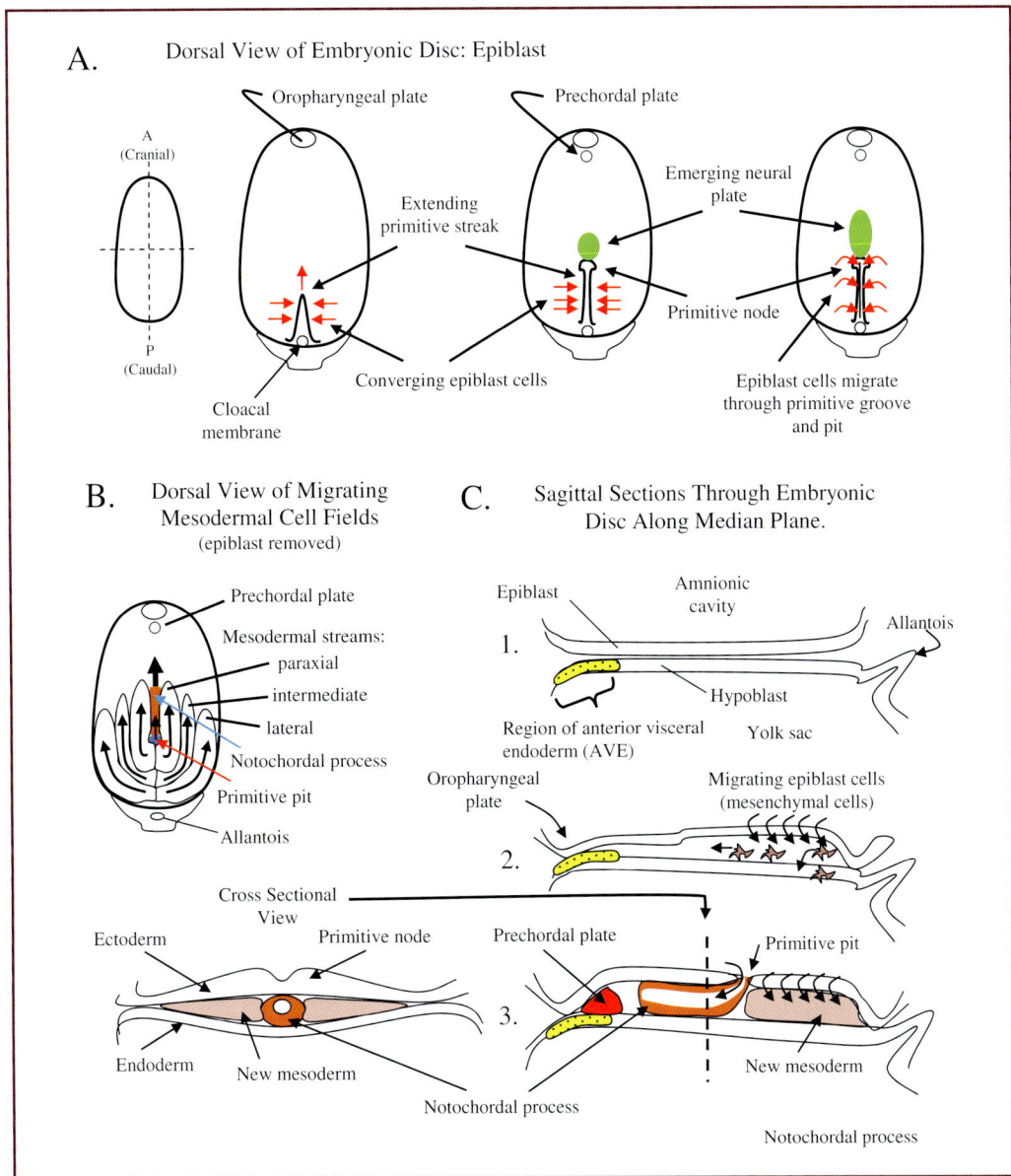

Figure 4–6. Schematics represent the bird model although similar processes occur in other species including the mammal.

streak initially forms as a raised thickened region (ridge) along the median line of the embryonic disc. New cells are continually produced in the surrounding epiblast and the addition of new cells causes the epiblast sheet to migrate toward the median plane (i.e., toward the streak). The streak emerges with the insertion of new cells from surrounding regions. The thickened streak elongates cranially in a narrow band as proliferation continues and cells converge at the streak. The elongation of the streak is an example of convergent-extension (discussed in Chapter 2). The cranial (anterior) edge of the elongating streak forms a round elevated primitive node containing a central depression or primitive pit. The primitive node is termed Hensen's node in birds, which corresponds to Spemanns' organizer in amphibians, and is often referred to as the "primary organizer." At the same time, the midline of the streak begins to fold downward (ventralward) to form a sulcus called the primitive groove running along the midline of the primitive streak (see Figure 4–6). The formation of the primitive groove results from cells migrating downward through the epiblast layer, a process called ingression. The primitive groove extends cranially to the primitive node where it merges with the primitive pit at the center of the primitive node. After diving downward, the migrating cells lose their epithelial characteristics, become mesenchymal-like, and move laterally and rostrally beneath the surface of the epiblast and its basal lamina (see Figures 4–6B and 4–6C).

Even before the proliferation and migration of epiblast cells, cells of the hypoblast are highly mobile and gene markers demonstrate specific cell migrations from the middle of the hypoblast anterior-posterior axis to the anterior pole of the hypoblast. This aggregation of marked hypoblast cells in anterior regions occurs before the appearance of the primitive streak and it forms the anterior visceral endoderm (AVE; Beddington & Robertson, 1998). Thus, the AVE forms one of the first markers of the anteroposterior axis. Epiblast cells that migrate through the primitive groove and node to form new embryonic mesoderm and endoderm are referred to as mesendoderm cells. The earliest migrating epiblast cells passing through the primitive node move to regions adjacent to the AVE on the midline and form the axial or prechordal mesoderm (in birds a circular prechordal plate, see Figure 4–6). The AVE is specialized endoderm that is thought to contribute to the organizing influences on the formation of the head in early gastrulation in mammals. Signals produced by cells of this region, such as cerberus (Cer1) and dickkopf (Dkk1), are thought to influence the overlying ectoderm to ensure the formation of anterior structures of the head and brain. These signals are required to inhibit and prevent signals from the node and primitive streak (e.g., Wnt) from transforming the neuroepithelium into more posterior phenotypes. This nascent influence later is reinforced by overlying anterior mesendodermal cells including the anterior axial mesoderm and prechordal plate. In birds, the prechordal plate is thought to exert primary control of anterior neural patterning although regions of the anterior hypoblast (analog to the mammalian AVE) also may contribute. Properties similar to the avian anterior hypoblast have also been ascribed to the anterior endoderm in frogs (Rallu et al., 2002; Wilson & Houart 2004).

The AVE and prechordal plate are important structures in that they also mediate patterning signals that aid in subdividing forebrain and midbrain regions during neural development. These structures together are sometimes referred to as the prechordal plate complex or simply as the anterior axial mesendoderm. The complex is slightly caudal to the future oropharyngeal membrane and oral cavity (see Figure 4–6).

Based on work in animal models, some of the earliest cells migrating through the primitive groove reach the surface of the hypoblast and begin to replace the extraembryonic endoderm cells. Indeed, the cells of the hypoblast are replaced to form a new embryonic endoderm composed of these transformed migrating epiblast cells. The new embryonic endoderm becomes the dorsal cap of the yolk sac. The original extraembryonic endoderm hypoblast cells are displaced to occupy positions lining the yolk sac over regions outside the .perimeter of the embryonic disc. These displaced hypoblast cells do not form part of the embryo tissue proper.

Migrating epiblast cells not only form a new embryonic endoderm but, as noted above, they also begin to fill the spaces between the eipiblast layer and the new endoderm. These cells become the embryonic mesoderm whereas the remaining cells of the epiblast are termed embryonic ectoderm. Thus, the processes of epiblast proliferation and migration lead systematically to the three primary embryonic germ layers: (1) ectoderm, (2) mesoderm, and (3) endoderm.

Eventually, the primitive streak and groove elongate cranially to about one third of the distance to the cranial edge of the embryonic disc. From this position, epiblast ectoderm cells continue to move through the primitive groove and primitive pit to form mesodermal cell flows. The ability to separate and fill the space between the epiblast and hypoblast layers is, in part, dependent on the presence of hyaluronic acid and, more importantly, fibronectin in association with the basal lamina of the epiblast layer. The migration of mesoderm laterally and cranially from the primitive groove and node forms three major mesodermal flanks to the midline: (1) the paraxial mesoderm, (2) the intermediate and lateral mesoderm, and (3) the extraembryonic mesoderm (see Figures 4–6B, 4–7B, and 4–8). Each flank occupies successively more lateral positions from the midline where the paraxial mesoderm lies adjacent to the midline. Mesoderm cells taking positions on the midline cranial to the primitive pit (axial mesoderm) become the notochordal process, or chordamesoderm. The notochordal process elongates cranially beneath the epiblast until it reaches the circular prechordal plate. Caudal to the primitive streak is the circular cloacal membrane, which represents the most caudal limit of mesodermal migration and becomes the future anus. At about day 18 in the human, the primitive streak begins to regress caudally. As the most rostral tip of the streak moves caudally, the notochord grows caudally with it. Normally, the primitive streak disappears by the end of the fourth week. During this time, the mesoderm continues to elaborate by means of cellular migration through the primitive groove (Figures 4–6, 4–7, and 4–8).

The presence of a notochord distinguishes the cranio-caudal axis of the embryo. In fact, the notochord distinguishes animals of an entire phylum from all others in the animal kingdom. The phylum Chordata is named for animals having a notochord or remnants thereof during the course of a lifetime. This phylum includes all vertebrates inasmuch as the notochord forms part of the vertebral column in the mature animal. Depending on the phylogenetic class of an animal, the notochord (located on the dorsal midline) provides either direct structural support for body movements as a cartilaginous rod or indirect support as an integral component of the vertebral column. However, as a rostral extension and derivative of the node the notochord (chordamesoderm) secretes both transcription factors (e.g., hepatocyte nuclear factor 3 beta [HNF-3β] and signaling molecules (e.g., noggin [Nog], Chrd, Shh) that play critical roles in mediating many inductive events during development. The roles include the induction and maintenance of the neural plate and patterning of ventral structures in the nervous system.

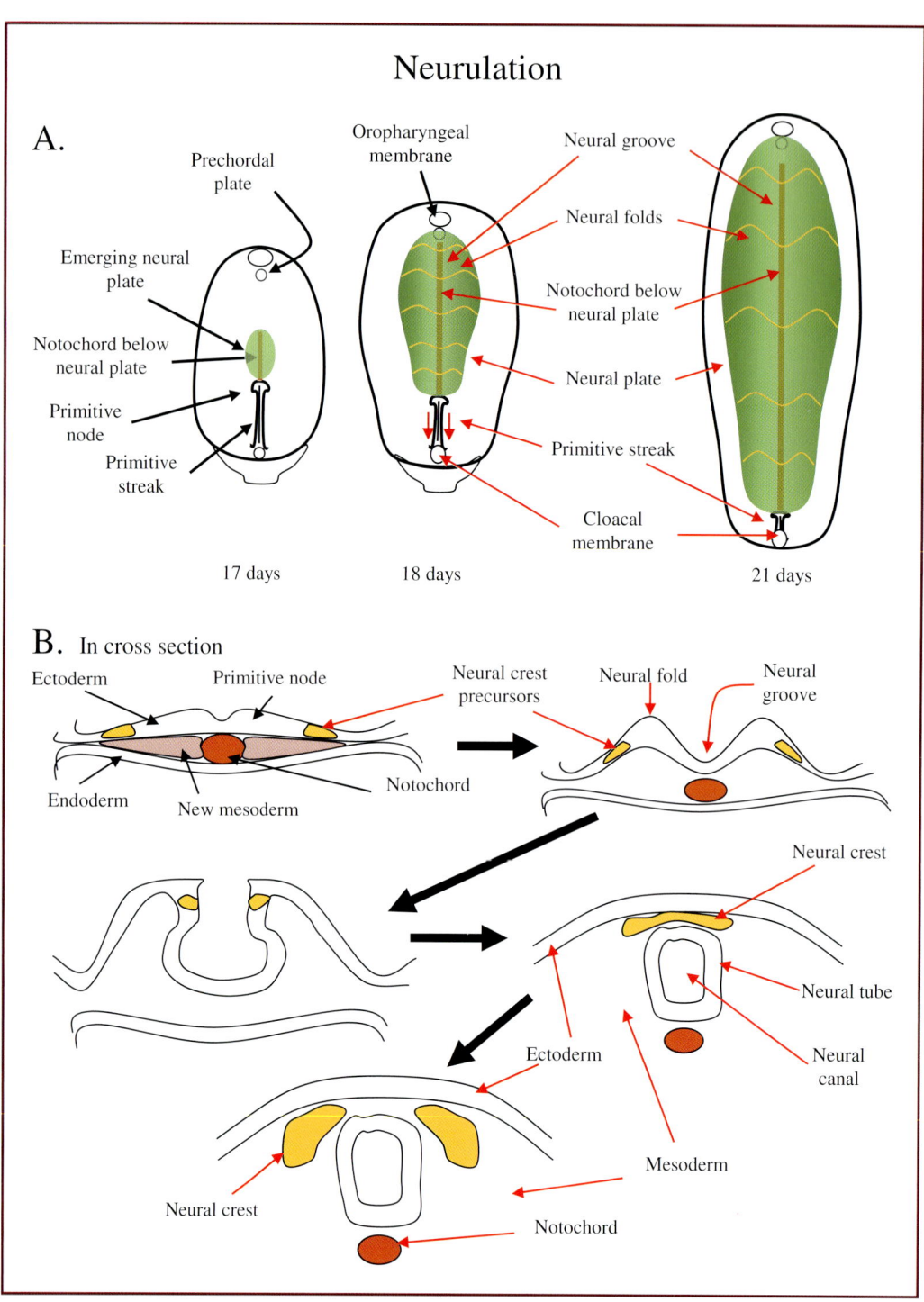

Figure 4–7. Schematics showing the process of neurulation represent the bird model. Similar processes occur in other species including the mammal. Ages shown represent days post fertilization in humans.

NEURULATION

Three germ layers, ectoderm, mesoderm, and endoderm, are formed during gastrulation by the processes of ectoderm proliferation and migration as described above. All of our tissues will develop from these germ layers (see Table 4–1 for examples). In addition, gastrulation produces the specialized regions of mesoderm including the prechordal plate and notochordal process on the midline. The first gross sign of neural induction is the formation of the neural plate, which is a thickening of ectoderm cells overlying the notochordal process (see Figures 4–6 and 4–7). The neural plate eventually extends slightly beyond the notochord to the oropharyngeal and cloacal membranes at the cranial and caudal ends, respectively. Ultimately, the central nervous system will develop from the neural plate.

In all vertebrates, induction of the neural plate by the primitive node or its' derivatives (e.g., notochord or chordamesoderm) is thought to result from the inactivation of bone morphogenetic proteins (BMPs, especially BMP4). Convincing evidence for this in mammals has recently been reported (Beddington & Robertson, 1998; Di-Gregorio et al., 2007; Levine & Brivanlou, 2007; Smukler, Runciman, Xu, & van der Kooy., 2006; Tropepe et al., 2001). BMPs are members of the TGF-β super family and are secreted by dorsal ectodermal cells. BMPs act on receptors present on the same ectodermal cells. Binding of BMPs to their cognate receptors inhibits the differentiation of ectoderm into neuroectoderm. In the absence of BMP4, the

Table 4–1. Some of the Anatomic Structures Developed from the Specific Germ Layers

Ectoderm		Mesoderm			Endoderm
	Neural Crest	*Paraxial*	*Intermediate*	*Lateral*	
Central Nervous System	Schwann Cells	Axial Skeleton	Urogenital System	Limb Skeleton	Lungs
Peripheral Nervous System	Sensory Nerves	Axial Muscles		Pleura	Trachea
Sensory Epithelia	Pigment Cells	Limb Muscles		Pericardium	Liver
Cornea	Cornea	Dermis of Skin		Peritoneum	Pancreas
Lens	Sympathetic Ganglia			Blood Cells	Digestive Tract
Hair, Nails	Skull			Endothelium of Blood Vessels	Gut
Tooth Enamel	Dentine of Teeth			Heart	Bladder
Epidermis of Skin				Lymph System	Pharynx
				Spleen	Pharyngeal Pouches
					Thyroid

Source: Information from Carlson (2004).

ectoderm forms neural tissue and it is thought that, during induction, the early gastrula organizer, the subsequent node, and the chordamesoderm (notochord) produce signals that lead to the inhibition of BMP4. Key players in this inhibitory signaling during induction are the gene products noggin, chordin, and follistatin.

As the neural plate forms (see Figure 4–7A), it also is shaped in that the cranial portion (future brain) becomes wider than the caudal region (future spinal cord). Shaping of the neural plate and initiation of neural tube closure is due to convergent extension. The neural plate folds along the midline at about day 18 to form the neural groove and neural folds on either side of the groove. Along the lateral edge of the neural folds are a group of cells that will become the neural crest (see Figure 4–7B). The transcription factor *Pax7* is required for the formation of neural crest cells (Aboitiz & Montiel, 2007; Basch & Bronner-Fraser, 2006). Neural crest cells actually are specified during gastrulation before the neural plate appears. This occurs when BMP and Wnt signals act to induce the secretion of Pax7 in prospective neural crest cells. The expression of *Pax7* induces *Slug* (a transcription factor of the zinc finger family) specifically marking prospective neural crest cells, which allows them to be observed during neurulation. The neural folds migrate toward each other until they meet at the midline and join to form the neural tube. The neural tube then separates from the surface ectoderm. The neural crest cells segregate from and gather above the neural tube. Eventually, the neural crest cells separate to form two specialized regions of cells along each side of the neural tube.

There are three sites along the length of the embryo where neural tube closure originates: (1) the hindbrain-cervical boundary, (2) forebrain-midbrain boundary, and (3) the rostal-most forebrain (Nakatsu, Uwabe, & Shiota, 2000). Neural tube closure occurs at the hindbrain-cervical boundary (near the middle of the embryonic craniocaudal axis) on about day 21 and then extends cranially and caudally from this point. The second site of closure at the forebrain-midbrain boundary (although the exact location varies) progresses rostral and caudal to meet the rostral progression of closure one and the caudal extent of the third closure site, which begins at the rostral-most forebrain. Neural tube closure should be complete by 28 days. As the neural tube closes, the surface ectoderm fuses dorsal to the neural tube and eventually will become the skin.

The genetic control of neurulation is complex and not completely understood. As previously stated, noggin and chordin may play roles in the induction of neurulation. Genes responsible for shaping the neural plate and initiating tube closure are those important in planar-cell polarity signaling. Such genes include vang-like 2 (*Vangl2*), cadherin EGF LAG seven-pass G-type receptor 1 (*Celsr1*), and homolog of *Drosophila* scribble (*Scrib*). Elevation and bending of the neural folds toward the midline requires *Shh* signaling that varies over time along the length of the neuraxis. Ephrin ligands and their receptors, specifically EphA5 and EphA7, appear to be important for midline fusion of the neural folds at least at the cranial regions. Finally, fusion and separation of the neural tube from the surface ectoderm may be due to differential expression of cadherins or other cell adhesion molecules (reviewed by Copp, Greene, & Murdoch, 2003).

The neural plate and neural tube form the central nervous system. Neural crest cells give rise to the dorsal root ganglia of the spinal cord, ganglia of the autonomic nervous system, sensory ganglia of cranial nerves V, VII, IX, and X, Schwann cells that produce the myelin around peripheral nerves, the meninges of the brain and spinal cord, and melanocytes. They also will contribute to some of the muskulosketetal structures of the head as the pharyngeal arches develop.

SOMITES

As gastrulation and neurulation are occurring, the mesoderm is undergoing developmental changes. As previously stated, ectodermal cells form the mesoderm by migrating through the primitive streak. Ultimately, this newly formed embryonic mesoderm has three regions: paraxial, intermediate, and lateral. Each region of mesoderm ultimately will form specific structures (see Table 4–1). The paraxial mesoderm on either side of the primitive streak (see Figure 4–6) becomes segmented into paired somitomeres beginning at the prechordal plate. Somitomeres are not easily discerned with the unaided eye but can be resolved using the scanning electron microscope. Somitomeres occupying positions adjacent to the prospective hindbrain ultimately contribute to the formation of muscle tissue in the pharyngeal arches. As the primitive streak regresses toward the caudal end, additional pairs of somitomeres form. At about day 20, the first pair of somites appears as bulged areas on either side of the midline (Figure 4–8). Somites are formed from the paraxial mesoderm and are readily seen on the surface of the embryo as a series of paired rectangular-shaped raised bumps on either side of the notochord beneath the neural folds (see Figure 4–8). The appearance of new somites comes with the generation of new tissue, where each additional pair segmentally extends along the length of the embryo caudally. Somites provide visible landmarks demarking regions of the nearby neural tube. For example, the spinal cord is formed from the neural tube cells caudal to the first four pairs of somites. Somites also can be used to stage embryonic development during neurulation such that an embryo is at the one- to three-somite stage when the neural groove and neural folds appear (Table 4–2). On day 28, when the caudal neural tube should be completely closed, the embryo has 30 to 35 somites. Somites eventually will develop into vertebrae, axial muscles, and skin.

SEGMENTATION

The development of the somitomeres and the appearance of somites are examples of segmentation due to a timed cycle of molecular signals. The sequential expression of genes produces the appearance of new segments in a sequential order along a specified axis (in this case the craniocaudal axis). Axial segmentation is a process occurring during development where discrete morphological regions of embryonic tissues become distinguishable from one another along the anterior-posterior axis. On the surface, these regions initially appear as a series of bumps or ridges of varying widths and shapes, many of which make their appearance in repeating temporal succession from apical to more posterior locations. These segments of tissue represent the segregation of cells with particular properties into compartments that ultimately will form distinct structures in the embryo and the adult. Anterior segments represent groups of cells that initially distinguish parts of the brain including the prosencephalon (forebrain), mesencephalon (midbrain), and rhombencephalon (hindbrain) and somewhat later more caudal groups distinguish segments of the spinal cord and

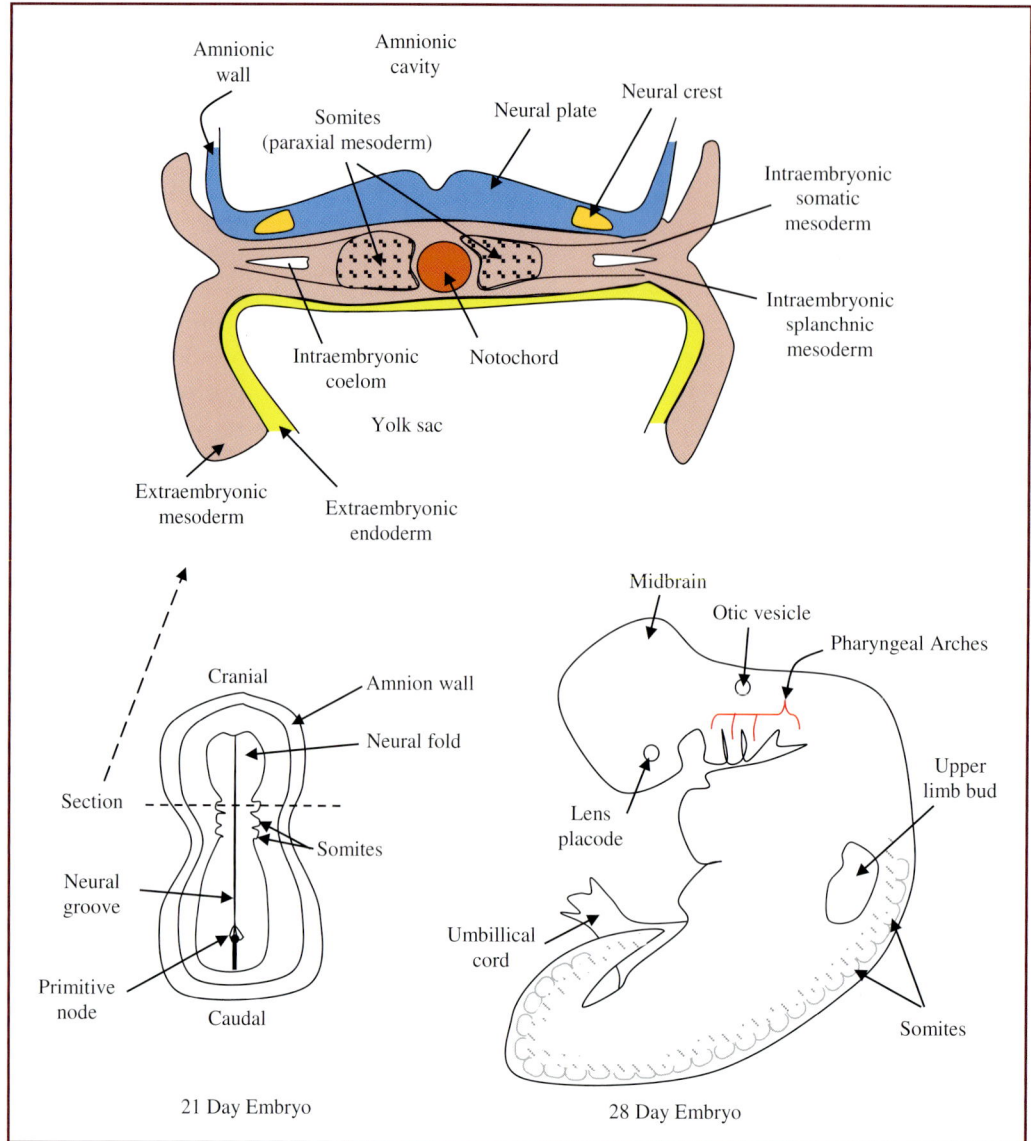

Figure 4–8. *Somites and their sequential appearance during development.* Top, *Cross section through embryo (location indicated in sketch at lower left) illustrating the anatomical location of somites in the 21 day embryo.* Lower right, *somites continue to elaborate thus forming 30 to 35 pairs, which ultimately form segmental vertebrae, axial muscles, and skin.*

vertebral column. Such segmentation extends beyond the central nervous system to include the peripheral nerves, skeleton, blood vessels, and muscles associated with each segment.

Considered together, the neural, skeletal, vascular, and muscular elements of a given segment may be referred to as a "metamere." This process of segmentation is orchestrated by a variety of molecular patterning transcription factors and their corresponding genes. The process of

Table 4–2. Timetable for Embryogenesis

Weeks	Somites	Feature
1		Cleavage Formation of morula Formation of blastocyst
2		Implantation Bilaminar embryonic disc formed Preplacodal domain appears
3		Appearance of primitive streak Gastrulation
	1	Neurulation begins
	4–12	Neural tube forming but still open rostrally and caudally First and second pharyngeal arches appear
	13–20	Cranial placodes appear
	21–29	Upper limb buds appear Rostral neuropore closed
4	30–35	Four pharyngeal arches appear Caudal neuropore closed Otic vesicle forms
5		Eye begins forming Nasal pits form Primitive mouth forms Middle ear begins forming Hands begin to form Lower limb buds appear
6		Upper lip and nose formed Pinna and EAM begin forming Feet begin forming
7		Eyelids begin to form Fingers begin to form Outer ear continues to form Genitalia begin to form
8		Fingers distinct Toes begin forming Wrist, elbows, and knees discernible Pinna formed but low on head Genitalia continue to form

segmental patterning in the hindbrain and spinal cord is dominated by the transcripts of homeotic genes.

Homeosis conveys the idea that certain body parts appear to be duplicates of other body parts or have features typical of other body parts within the same organism or across species (Favier & Dolle, 1997). For example, the features of one somite region (one segment) may be duplicated in an adjacent second somite region. Indeed, many segments of the body axis appear to be new iterations of an existing segment or only slight modifications of one another.

REGIONALIZATION OF THE NERVOUS SYSTEM

Segmentation also regionalizes the nervous system; however, in this case, segmentation subdivides the newly formed neural tube rather than forming new tissue segments in a manner comparable to somite formation. Spatially ordered molecular gradients and timed molecular signals are responsible for inducing regional divisions of the brain. Examples of regionalization include the subdivision of cranial and caudal ends of the nervous system (i.e., brain regions versus spinal cord regions), gross subdivision of the brain (forebrain or prosenencephalon, midbrain or mesencephalon, hindbrain or rhombencephalon), slightly more refined subdivision of brain regions (telencephalon, diencephalon, metencephalon, myelencephalon), and the elaboration of segments in the hindbrain (rhombomeres, Figure 4–9). These regions are schematically illustrated for the embryo at various stages of development in Figure 4–10 and for the adult in Figure 4–11.

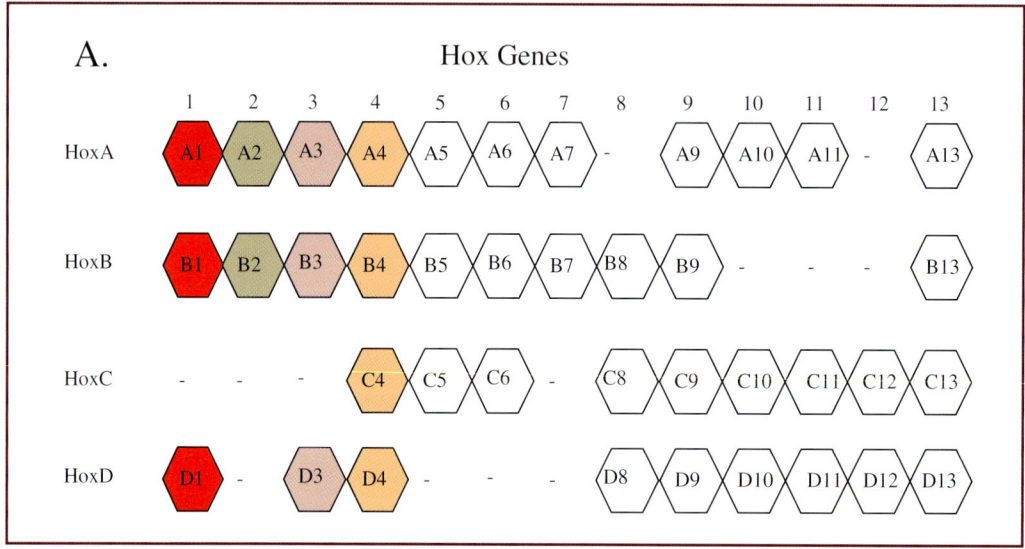

Figure 4–9. Hox genes and segmentation in cranial regions of the embryo. **A.** Vertebrate Hox gene complexes. Represented are four chromosomes each containing a unique Hox gene complex with designations of HoxA, HoxB, HoxC, and HoxD. Thirteen vertical columns represent the 13 paralagous Hox genes found on each chromosome. The genes are shown from left to right in sequence from the most 3' locus to the most 5' end of each chromosomal complex. continues

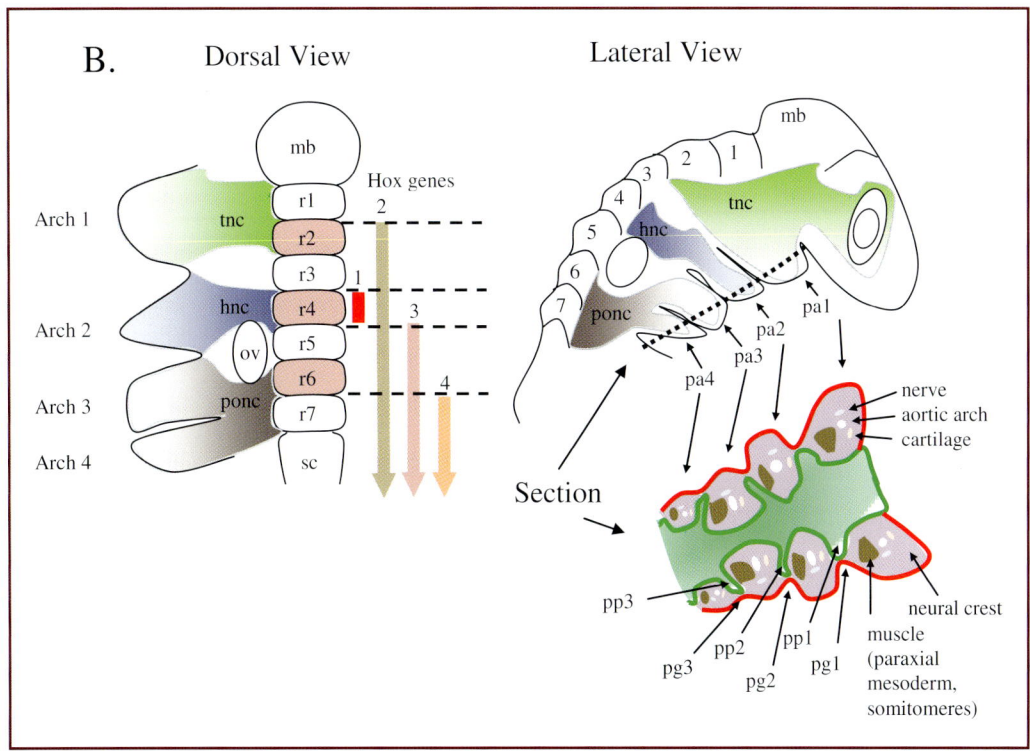

Figure 4–9. continued *The first four* Hox *gene paralogs are colored red, brown, purple, and orange, respectively. The most anterior borders of expression for these four genes are illustrated in the dorsal view, which highlights the unique combinations of* Hox *genes for each axial segment of the embryonic hindbrain.* Hox *genes 5 through 13 are expressed successively in more caudal regions establishing axial segments of the spinal cord. Note that some genes are not present in all chromosomes. There are 39 known* Hox *genes in vertebrate animals and all are listed in the diagram.* **B.** *Dorsal and lateral views of vertebrate embryo. Three streams of neural crest cells are shown including: trigeminal* (tnc), *hyoid* (hnc) *and post-otic* (ponc). *Rhombomeres are labeled 1 through 7 (r1 to r7 dorsal view). In the dorsal view,* Hox *genes expressed in the hindbrain are shown using colors matching those used in the top diagram of paralagous* Hox *genes (Hox1 [red], Hox 2 [brown], Hox3 [purple], Hox4 [orange]. The sharp anterior boundaries of each* Hox *gene are marked by a dashed line. These demark the unique combinations of overlapping fields of expression, which impart a distinct developmental identity to each axial segment. The alternately colored (white and pink) odd and even rhombomeres represent the separate rhombomere compartments, which restrict cellular migration across rhombomere boundaries. Four pharyngeal arches are also represented (Arch 1 through Arch 4) [mb = midbrain, ov = otic vesicle, sc = spinal cord].* **Section,** *view of section through the pharyngeal arches (dotted line). Ectoderm (red), endoderm (green), incipient muscle associated with paraxial mesoderm (somitomeres, brown), and neural crest tissue (purple) identify the emerging tissues of pharyngeal arches (pa1, pa2, pa3, and pa4). The tips of the pharyngeal pouches (pp1, pp2, and pp3) are formed between the arches where the endoderm and ectoderm nearly come into contact. The pharyngeal grooves (pg 1, pg 2, pg 3) are formed by ectoderm.*

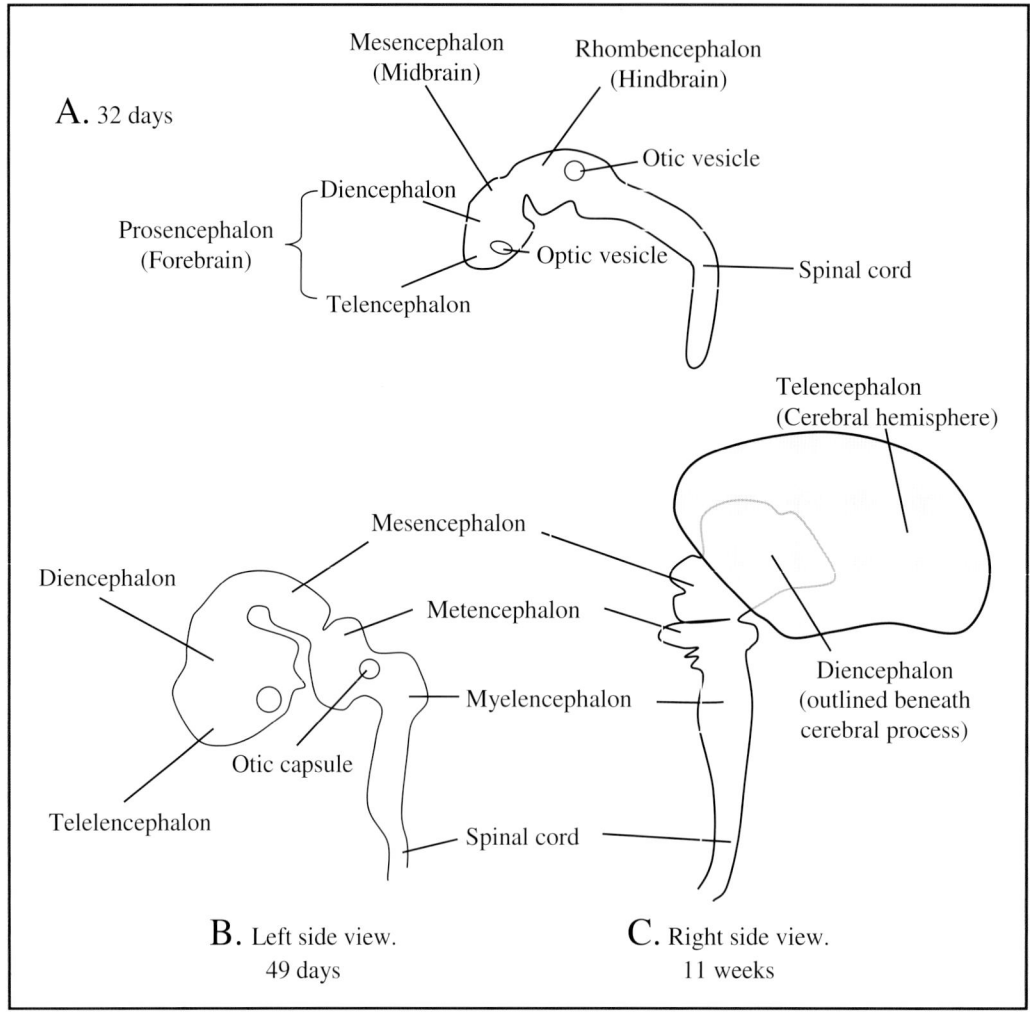

Figure 4–10. Regionalization of the brain. Not to scale. Highly schematic representation of developing brain regions as annotated. Human embryo is represented. **A.** Embryo at 32 days gestation. **B.** 49-day embryo shown from the left side view. **C.** 11-week fetus shown from the right side. Note the diencephalon is outlined below the outermost cerebral process (gray area), which has been rendered partially transparent here.

There also is evidence supporting the neuromere model of segmentation of the nervous system (Aboitiz & Montiel, 2007). Grossly, neuromeres are segments of thickened neuroepithelial tissue separated by slightly narrower constrictions that appear transiently during the course of neurulation. This includes rhombomeres (seven segments of the hindbrain labeled r1 to r7, e.g., Figure 4–9B), the mesomere (a single segment the midbrain), and prosomeres (six segments in the prosencephalon labeled p1 to p6). We will emphasize the rhombomeres of the hindbrain and the broad regional divisions of the prosencephalon, mesencephalon, and metencephalon described above.

Basic Concepts in Embryology 111

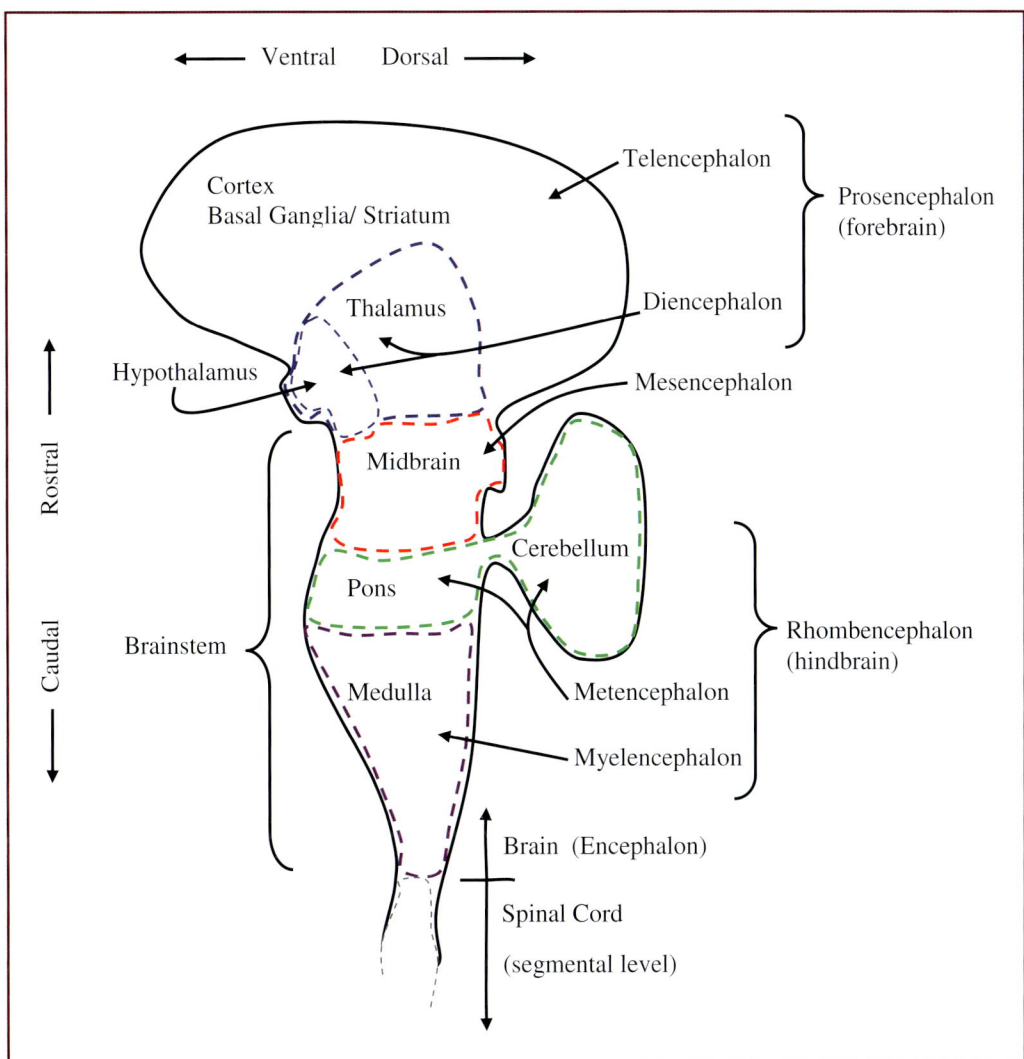

Figure 4–11. Regionalization of the mature, adult brain.

Recall that early patterning signals arising from the AVE and prechordal mesoderm lead to the adoption of anterior (cranial) neural characteristics in the neural plate. On the other hand, signals from the nodal organizer and chordamesoderm lead to the formation of more caudal (posterior) domains during gastrulation and neurulation (e.g., Beddington & Robertson, 1998; Wilson & Houart, 2004). The AVE and the prechordal mesoderm (often referred to as head organizing regions) in concert with the notochord appear to be involved early in the process of subdividing regions of the forebrain and midbrain from regions of the hindbrain and spinal cord (Figures 4–6 and 4–12). Regions elaborating signals that lead to the adoption of more cranial structures such as the prosencephalon and mesencephalon are said to

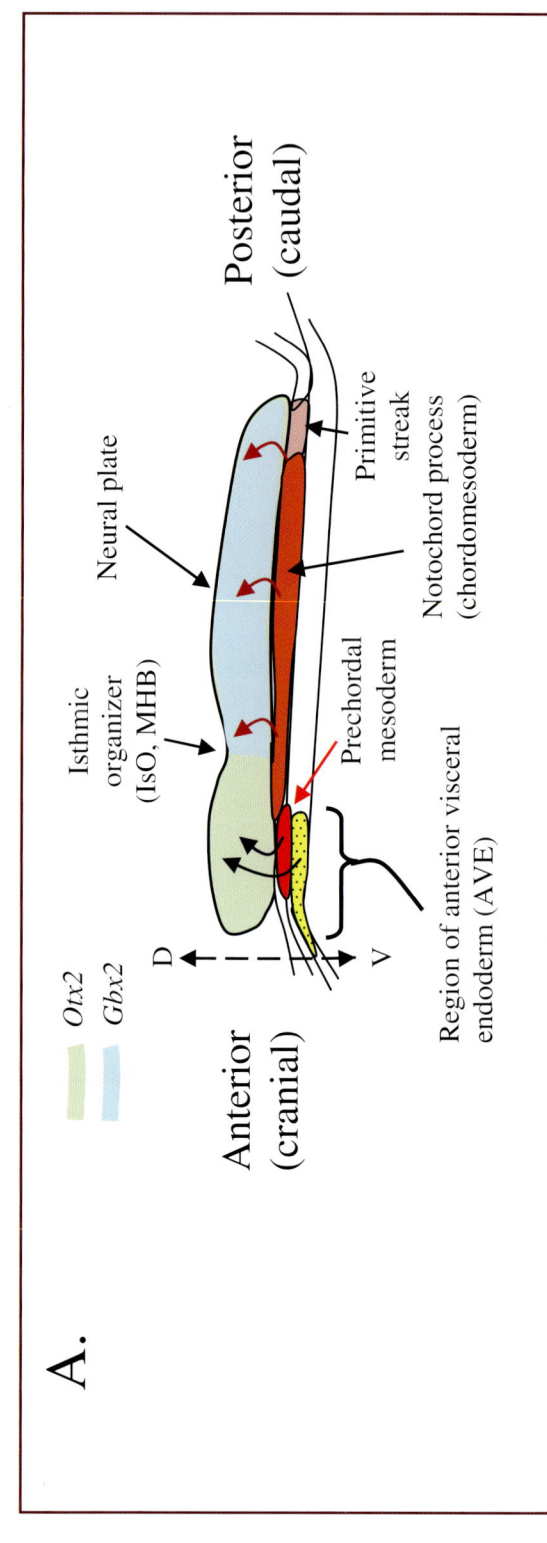

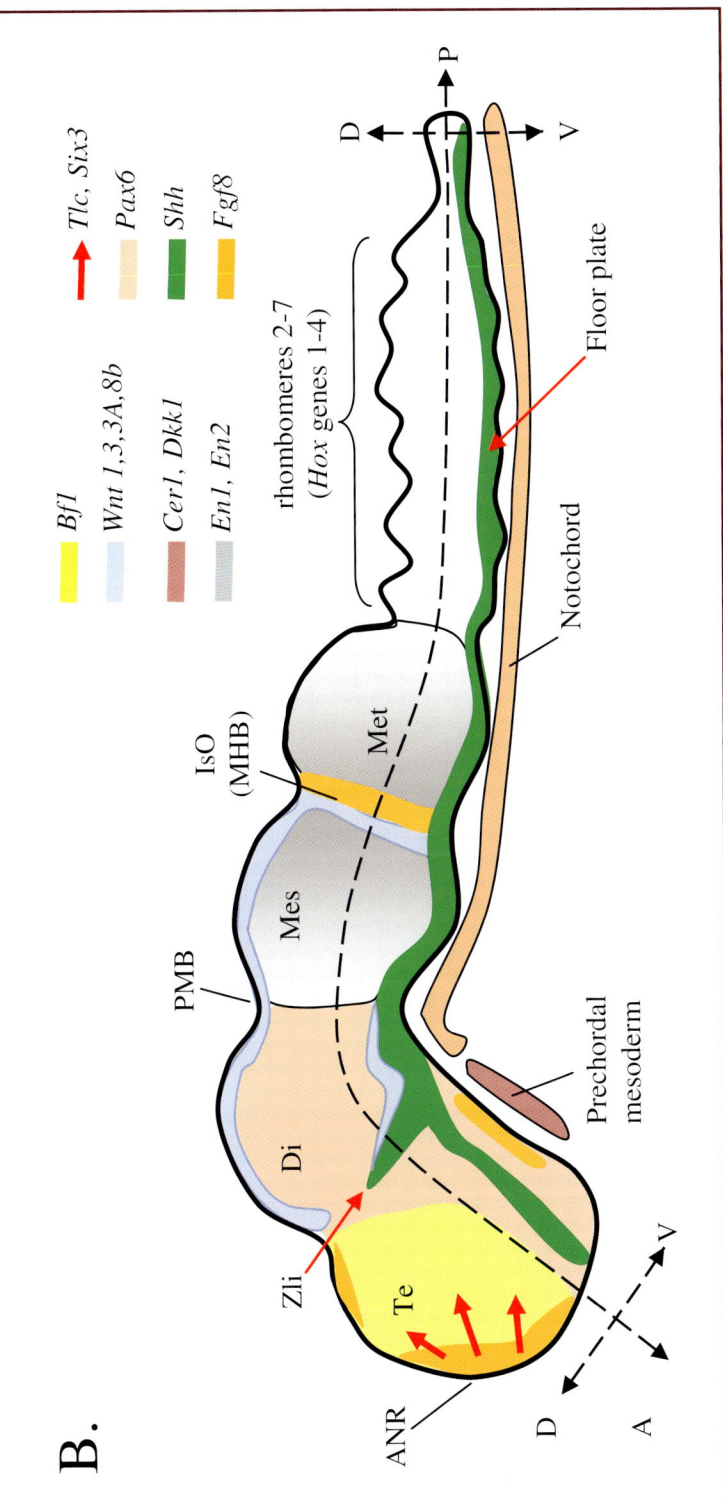

*Figure 4–12. Signaling and transcription factors involved in the regionalization and patterning of the nervous system. **A**. Signals patterning the anterior-posterior axis in the neural plate. Key signals arising early during neural plate formation from the node and notochord include noggin, chordin and follistatin. Wnt signals from the region of the node have powerful posteriorizing effects on rostral tissues (not shown, see Figure 4–13). Cer1 and Dkk1 are key signals from the AVE and prechordal mesoderm that preserve the anterior fate for the rostral neural plate by inhibiting Wnt signaling. Low levels of Wnt in anterior regions lead to the expression of Otx2 and Gbx2 and the formation of the midbrain hindbrain border (MHB) as well as the formation of the isthmic organizer (IsO). **B**. Signaling in the neural tube at later stages. Longitudinal axis (dashed line: anterior [A], posterior [P]) and dorsal-ventral axis (dashed line perpendicular to longitudinal axis: dorsal [D], ventral [V]). Four boundaries are formed based on signaling patterns: Metencephalon (Met)-rhomboencephalon (rhombomeres 2–7), Mesencephalon (Mes)-metencephalon (IsO, MHB), prosencephalon-mesencephalon (PMB), Dorsal diencephalon-telencephalon/ventral diencephalon (zona limitans intrathalamicus, Zli). ANR: anterior neural ridge. Te: telencephalon.*

have an "anteriorizing" or "cranializing" influence on the neural plate, whereas regions leading to the formation of more caudal structures like the rhombencephalon and spinal cord are often referred to as "posteriorizing" or "caudalizing" domains. The interplay between the signals of these regions as well as between other patterning signals is believed to result in the partitioning of structures along the anterior-posterior axis in the nascent nervous system. Other dorsalizing and ventralizing signals serve to pattern neural structures in the dorsal-ventral axis. Many of these signals interact and serve to induce changes in the shape and form of the brain and thus are commonly referred to as morphogenic factors and the regions under such control are termed morphogenic fields.

The exact nature of all the signals responsible for the ontogeny of the nervous system is simply not known. How signals are orchestrated to produce the elaborate complexity of the brain remains an exciting mystery to solve. However, there has been an explosion of new information about developmental mechanisms due to the successes of research in molecular biology. We will examine, in simplified form, a few general models that describe major signaling factors responsible for establishing regional identities in neural structures along the major anterior-posterior and dorsal-ventral axes of the embryo.

Regional Anterior-Posterior Patterning

Hindbrain

Segmentation in the hindbrain involves the formation of rhombomeres. The hindbrain becomes segmented into seven pairs of rhombomeres, one on either side of midline. The spatiotemporal expression of HOX genes is a major contributor to rhombomere formation (see Figure 4–9). Extensive studies in humans and mice have revealed 39 HOX genes located on four different chromosomes. The HOX genes form a single complex (cluster) on each of four chromosomes (labeled A, B, C and D in Figure 4–9). The 39 HOX genes occupying loci across the four different chromosomes have been divided into 13 groups based on the similarity of coding sequences. Those genes most closely related to each other across chromosomes form each distinct group. The groups are numbered in order from 1 to 13 according to their position in the DNA complex where 1 is at the most 3' region and 13 is at the most 5' region of each gene cluster. The HOX gene groups (1 to 13) on the corresponding four chromosomes are termed paralogs. Paralogous genes are thought to arise initially as duplicates of the original ancestral gene. Over time (e.g., millions of years), the particular base pair sequences of individual paralogous genes may change and they are said to "diverge." The appearance of four paralogous complexes HoxA, HoxB, HoxC, and HoxD suggest that the original gene clusters were duplicated four times during evolution. This may have been accomplished by two successive genome-wide duplications around 500 million years ago (Tümpel, Wiedemann, & Krumlauf, 2009).

The gene clusters HOXA, HOXB, HOXC, and HOXD are found on the corresponding chromosomes 7, 17, 12, and 2 in the human (Carlson, 2004). The genes are expressed in sequential order from 1 to 13. As an example HOXA1, HOXB1, and HOXD1 are paralogs of the first gene in the clusters and the earliest genes expressed during development. HOX gene products are transcription factors that control genes critical for proper morphological development of the body plan. The HOX gene family serves to specify a particular regional identity to cells occupying each of a series of segmental compartments systematically arranged along the anterior-posterior axis of the embryo. The basic function of genes is to partition undifferentiated cells into compartments with specific structural fates. Each region gives rise to specific structures based on the combination of HOX genes expressed in each segment. Each segment leads to the formation of a metemere appropriate for its position along the anterior-posterior axis.

HOX genes are essential for the normal formation of hindbrain cranial segments. Figure 4–9 illustrates the spatial expression within the hindbrain and spinal cord from a rodent model. *Hox* genes appearing in the hindbrain include paralogs of *Hox1, Hox2, Hox3,* and *Hox4.* The colinear expression of these *Hox* genes is directly linked to segmentation of the hindbrain.

During early periods of gastrulation, there is little restriction of cellular movements along the anterior-posterior axis in the prospective hindbrain. Cells readily intermingle across the future boundaries of rhombomeres. However, this free mixing of cells disappears later with the formation of hindbrain rhombomeres. Indeed, within each rhombomeric compartment, cells move and mix readily whereas movements across adjacent rhombomere boundaries are restricted. The boundaries to movement form because the cells within each rhombomere begin to express distinct cellular adhesion molecules, thus identifying the cells of each rhombomere as a particular group with particular adhesion properties. Each compartment thus contains cells intermingled from various populations that sort according to the nature of their attraction and adhesion to one another. Some aspects of compartmentalization may be due in part to the expression of ephrin ligands in even numbered rhombomeres 2, 4, and 6 and their corresponding receptors only in odd numbered rhombomeres 3 and 5 (see Figure 4–9).

Thus, during hindbrain segmentation, cells acquire specific identities such that the cells of each rhombomere have unique characteristics and collectively they are able to respond to external signals in a manner that differs from neighboring segments. These unique characteristics provide for the development of diverse neuronal components in the hindbrain. Rhombomeres give rise to brain structures but they also provide structural order to the cranial ganglia, branchiomotor nerves, and cranial neural crest stream paths (see Pharyngeal Arches, Figure 4–9).

Although the patterning of *Hox* gene expression is essential for rhombomeric sequences, several transcription factors play a role in regional control of *Hox* expression in the hindbrain. *Krox20* (zinc finger transcription factor) directly regulates *Hoxb2* and *Hoxa2* in rhombomeres 3 and 4. The Kreisler gene (*Kr*) directly activates *Hoxa3* and *Hoxb3* in rhombomere 5, but in rhomobomere 6 *Kr* upregulates only *Hoxa3*. Retinoic acid (RA) and its receptors also are involved in regulating hindbrain *Hox* genes especially *Hoxd4, Hoxa1,* and *Hoxb1* (Deschamps et al., 1999). Other signals such as Notch, BMP and Wnt serve to guide cell differentiation and growth at rhombomere boundaries. Rhombomeres also can give rise to signals such as FGF, which may influence the development of adjacent rhombomeres and provide a local organizing function (Tümpel et al., 2009).

Metencephalon, Mesencephalon, and Diencephalon

The AVE and prechordal axial mesoderm secrete the gene products Cer1, Dkk1, and frizzled-related protein (Frzb1), which act to pattern anterior neural plate and structures of the forebrain. One crucial action of these signals is to inhibit Wnt signaling in anterior regions of the neural plate. Wnt signals originate in the node located at the posterior edge of the neural plate. Levels of Cer1 and Dkk1 are highest in anterior regions of the brain and decrease progressively at positions approaching the posterior end of the neural plate. This produces a gradient of Wnt signaling, which is minimal at the most anterior regions and maximal at posterior locations (Figures 4–12 and 4–13). The Wnt gradient leads to the activation of the transcription factor *Xiro1* (Aboitiz & Montiel, 2007; Glavic, Gomez-Skarmeta, & Mayor, 2002) and in turn the expression and partitioning of the markers orthodenticle homologue 2 (*Otx2*, a homeodomain transcription factor) and gastrulation brain homeobox 2 (*Gbx2*, a homeotic gene). The expression domains of *Otx2* and *Gbx2* define the prospective midbrain-hindbrain border,

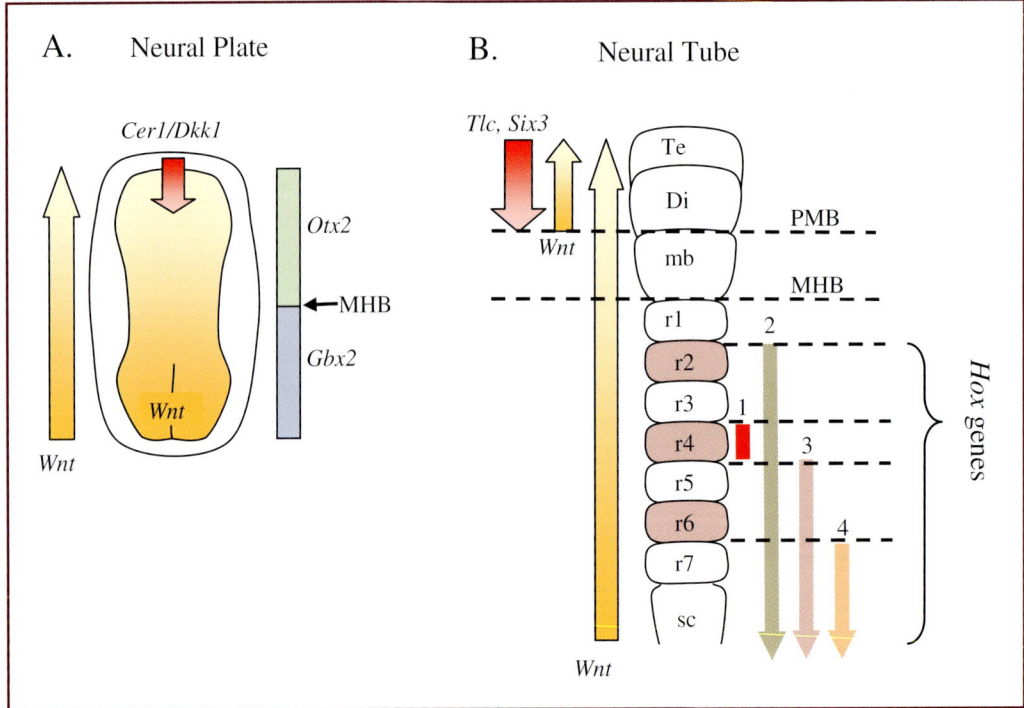

Figure 4–13. Major anterior-posterior expression domains and gradients in the neural plate (**A**) and neural tube (**B**). Combinations of expressed genes and levels serve to create regional distinctions in tissues and form distinct cellular compartments. Key signaling factors include posteriorizing signals such as Wnt, Gbx2, and anteriorizing signals such as Cer1, Dkk1, Tlc, Six3, Otx2. Segmentation signals include Hox genes 1 through 4. Di = diencephalon, mb = midbrain, MHB = midbrain-hindbrain border, PMB = prosencephalon-mesencephalon border, r1–r7 = rhombomeres 1 through 7, sc = spinal cord, Te = telencephalon.

which is located in the vicinity of the isthmus. The tissue of the midbrain-hindbrain border becomes a major organizing influence on surrounding regions (see Figures 4–12B and 4–13), and for this reason, it is also referred to as the isthmic organizer. Formation of the midbrain-hindbrain border begins during gastrulation and is well resolved in the neural plate (Aboitiz & Montiel 2007; Ciani & Salinas, 2005).

Initially, regions of the prospective prosencephalon and mesencephalon express *Otx2* whereas the rhombencephalon and spinal cord express *Gbx2* (see Figures 4–12 and 4–13). It is likely that these two expression domains specify the respective identities of the mesencephalon and metencephalon as well as their anterior-posterior location in the brainstem. The sharp border between *Otx2* and *Gbx2* domains is a result of the mutual repression each exerts on the other. The relative expression levels of the two factors determine the position of the boundary. The interaction of *Otx2* and *Gbx2* at the boundary also leads to the expression of *Fgf8* in a thin strip of tissue on the caudal side of the border. *Fgf8* in turn induces gradients of engrailed (*En1* and *En2*) on either side of the border. These signals, as well as *Wnt1*, *Pax2*, and *Pax5*, play a role in patterning and maintaining the mesencephalon and metencephalon and their rostrocaudal orientation. The metencephalon arises largely from rhombomere 1 and further caudal extension of the metencephalon is prevented by the expression of *Hoxa2* in rhombomere 2 (see Figures 4–12B and 4–13). On the other hand, the rostral extent of the mesencephalon is restricted in part by the interaction of *Pax6* with engrailed genes and *Pax2* (Araki & Nakamura, 1999; Carlson, 2004; Glavic et al., 2002; Kobayashi et al., 2002; Matsunaga, Araki, & Nakamura, 2000; Wurst & Bally-Cuif, 2001). These interactions establish the location of the border between the mesencephalon and the diencephalon (prosencephalon-mesencephalon border).

Forebrain

The forebrain arises from the anterior neural plate during gastrulation. As the neural tube closes, head regions of the embryo begin to bend ventrally (see Figures 4–10 and 4–12). The central neuraxis thus curves, creating a radial component to anatomic axes. This curvature can lead to confusion regarding the description of locations for forebrain structures in relation to standard anatomical axes. To reduce such problems, we will adopt the common convention of aligning the anterior-posterior axis with the central neural tube and the rostro-caudal (longitudinal) axis as shown in Figure 4–12. The anterior-posterior axis in the forebrain runs from the telencephalon (the most rostral structure) to the end of the caudal border of the diencephalon. The dorsal-ventral axis runs perpendicular to the anterior-posterior axis and at the most rostral level encompasses the telencephalon and hypothalamus.

Initially, the neural plate is regionalized by strong caudalizing (e.g., *Wnt*) and rostralizing (e.g., *Cer1*, *Dkk1*, and *Frzb1*) signals that form interactive concentration gradients along the anterior-posterior axis (see Figure 4–13). There is evidence of additional sources of Wnt signals located in prospective diencephalic and mesencephalic regions of the anterior neural plate (see Figures 4–12 and 4–13). These regions secrete Wnt1, Wnt3, Wnt3a, and Wnt8b (Braun, Etheridge, Bernard, Robertson, & Roelink, 2003; Buckles et al., 2004; Ciani & Salinas 2005; Houart et al., 2002; Roelink & Nusse, 1991). In the prospective forebrain, overexpression of *Wnt* signaling has a posteriorizing effect, tending to enhance caudal diencephalic structures. Signals countering the *Wnt* effects are *Tlc* (homologue of Frizzled-related protein) and the homeodomain transcription factor *Six3*. Blocking or downregulating *Wnt* signals from the midbrain and diencephalon with *Tlc* or *Six3*, for example, promotes the telencephalon (Houart et al., 2002). The formation of anterior forebrain structures is critically dependent on both *Tlc* and *Six3* expression. The source of *Tlc* and *Six3* is in the most rostral region of the forebrain called the anterior neural ridge (ANR, also referred to as the anterior neural plate boundary (ANB, see Figure 4–12). The ANR is an important organizer region for forebrain structures. Graded levels of *Tlc* are thought to provide a gradient of *Wnt* inhibition, which leads to the regionalization of the telencephalon and diencephalon (see Figure 4–13).

Another anterior-posterior organizing region in the forebrain is the zona limitans intrathalamica (ZLi). At the most rostral limit of the floor plate (see Figure 4–13) and underlying notochord (level of diencephalon), *Shh* expression turns dorsally and follows the ZLi to the edge of the dorsal telencephalon (see Figure 4–12). The *Shh* domain in the ZLi divides the caudal thalamus (which lies more dorsal) from the more rostral thalamus (which lies more ventral), the eye fields and the telencephalon. The location of this anterior-posterior division of the prosencephalon is likely determined by two genes, *Irx3* and *Six3*. *Irx3* is a vertebrate homologue of the Drosophila iroquois complex (iro-C) and its expression domain is sharply bounded anteriorly by the ZLi and extends caudally through the midbrain in the developing neural tube. At the same time, the expression domain of *Six3* extends posteriorly from the most anterior regions of the forebrain and ends abruptly at the ZLi adjacent to the *Irx3* domain. Products from these genes repress each other and, therefore, may be responsible for forming a sharp diencephalic boundary (Braun et al., 2003; Ciani & Salinas, 2005; Kobayashi et al 2002). One model proposes that *Wnt* expression induces *Irx3* posteriorly and thus specifies a posterior fate for the dorsal thalamus. Wnt antagonists secreted from the prechordal plate, such as Dkk1, act to reduce Wnt levels anteriorly, thus leading to the expression of *Six3* as well as the formation of the *Irx3-Six3* border and ensuring an anterior fate for the rostral forebrain. As noted above, *Tlc* expression also promotes the anteriorization of the rostal forebrain (see Figures 4–12 and 4–13). *Tlc* in turn is thought to be controlled in part by BMP signaling that arises from the ectoderm at the margins of the neural plate (Houart et al., 2002). The influence of BMP represents an example where anterior-posterior signals interact with dorso-ventral signaling to produce complex patterning effects as described below for more caudal regions of the brain and spinal cord. Shh also is secreted by the ventral telencephalon, hypothalamus, and floor plate of the neural tube and provides a ventralizing signal throughout most of the neuraxis. These signals are considered in more detail below.

Dorsal-Ventral Patterning

Generally, BMPs expressed from the dorsal ectoderm have a dorsalizing influence on neural tube structures, whereas Shh has a ventralizing influence. BMP forms a dorsoventral concentration gradient in the neural tube (Figure 4–14). The BMP gradient acts to maintain or increase levels of Pax3, Pax7, Msx1, and Msx2 (Carlson, 2004). The gradients result in the forma-

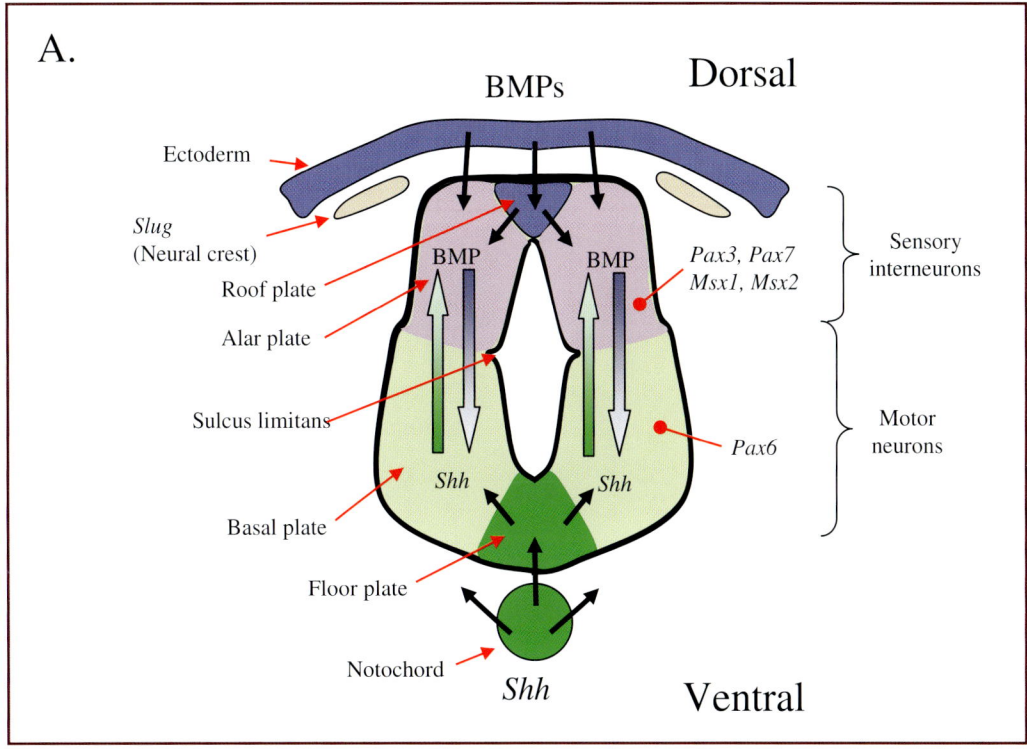

Figure 4–14. **A.** *Schematic illustration of a cross section through the neural plate illustrating the dorsoventral signaling counter gradients of BMPs (blue arrow) and Shh (green arrow). The resulting combination of signal concentrations in concert with anterior-posterior signal gradients (e.g., Wnt) lead to regional patterning and dorsal neural (e.g., sensory interneurons) versus ventral neural (e.g., motor neurons) fates.* continues

tion of the roof plate, the alar plate, and sensory interneurons in the spinal cord. The roof plate itself becomes a signal source for BMPs and Wnts (Aboitiz & Montiel, 2007). Shh, on the other, hand is initially secreted from the notochord and suppresses the dorsal *Pax* genes (*Pax3* and *Pax7*), which induces the floor plate. The floor plate in turn secretes Shh, which forms a ventrodorsal concentration gradient in the neural tube (see Figure 4–14). This Shh gradient has the property of inducing ventral fates such as motor neurons in nearby regions. The combined gradients of BMP and Shh lead to the repression and induction of many other transcription factors, thus forming unique combinations of control signal levels throughout the dorsoventral plane. It is thought that the combination of these more focal dorsoventral signals with rostrocaudal patterning signals (e.g., Wnt) produces a large variety of neural cell phenotypes in the central nervous system (Carlson, 2004).

Dorsal-Ventral Patterning of the Forebrain

The dorsal-ventral axis of the forebrain runs from the telencephalon to the hypothalamus (see Figure 4–14). Signaling regions such as the ANR, ZLi and the ventral floor of the telencephalon are believed

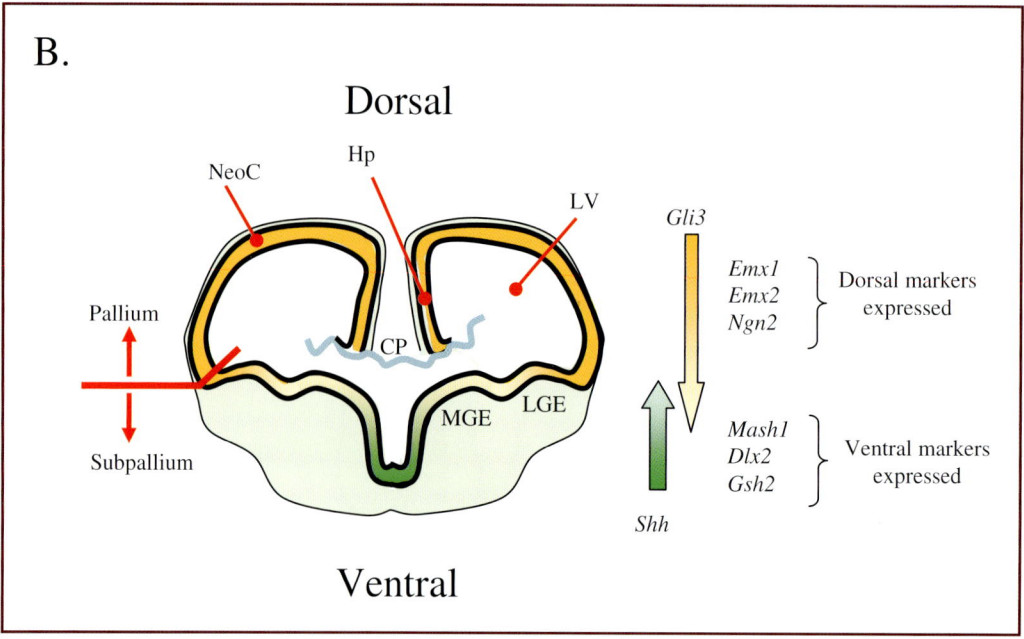

Figure 4–14. continued **B.** *Schematic cross section through the rostral forebain of the neural tube. Ventral sources of* Shh *(green) produce decreasing concentrations at more dorsal regions whereas dorsal sources of Gli3 (orange) produce a gradient of decreasing concentration at more ventral locations. Orange and green arrows to the right reflect the dorsoventral gradient scheme. The combination of gradients results in the appearance of specific dorsal (Emx1, Emx2, Ngn2) and ventral (Mash1, Dlx2, Gsh2) markers each playing a particular role in local dorso-ventral patterning of the telencephalon. CP = choroid plexus. Hp = hypothalamus (medial pallium). LV = lateral ventricle. LGE = lateral ganglionic eminence. MGE = medial ganglionic eminence. The LGE and MGE give rise to the globus pallidus and corpus striatum in the adult. NeoC = neocortex (dorsal pallium). Pallium and subpallium are the major dorsal and ventral subdivisions in the embryonic telencephalon, respectively.*

to play important roles in the regionalization and morphogenesis of the forebrain. *Fgf8* is a contributing factor in the patterning and elaboration of the frontal cortex and basal ganglia. The ANR secretes Fgf8, which induces brain factor-1 (*Bf1*, also known as *Foxg1*). *Bf1* is a regulator of telencephalic development and is essential for ventral forebrain and eye formation (Aboitiz & Montiel, 2007; Carlson, 2004). The ANR expresses three BMP antagonists including *Fgf8*, *Chrd,* and *Nog*. Losses in *Chrd* and *Nog* expression lead to various deficits in forebrain structure.

Shh also is believed to be important in dorsoventral patterning of the telencephalon and hypothalamus in concert with *Gli3* (see Figure 4–14B), a role reminiscent of its role in more caudal regions of the brain (see Figure 4–14A). Shh is secreted by the ZLi and by anterioroventral regions of the forebrain. The ventralizing effects of Shh on telencephalon and hypothalamus depend on the competence of cells rather than Shh concentration. Tissue response competence to Shh varies with age. *Gli3* is a repressor of *Shh* and its distribution complements that of *Shh* extending from the most dorsal to the most ventral third of the telencephalon and hypothalamus (Rallu et al., 2002). *Gli3* upregulates BMP and *Wnt* expression, and it is likely that its dorsalizing effects are mediated through these signals. Losses in Gli3 signaling result in absent choroid plexus and hippocampus (among other things) and a general ventralization of the cerebral hemisphere (see Figure 4–14).

CRANIAL PLACODES

Olfactory, visual, and auditory sensory organs as well as cranial sensory ganglia develop from small regions of thickened ectoderm (rather than from the neural plate) at the anterior end of the embryo (see review by Baker & Bronner-Fraser, 2001). These regions are called cranial placodes (Figure 4–15). Olfactory neurons and epithelium are formed from the olfactory placode, the lens of the eye develops from the lens placode, and the inner ear develops from the otic placode (discussed in Chapter 6). The hypophyseal placode forms the anterior portion of the pituitary gland. The trigeminal placode forms the sensory neurons of cranial nerve V (sensation for the face, jaw, and teeth). The epibranchial placodes form the sensory components of cranial nerves VII, IX, and X (taste and visceral sensation).

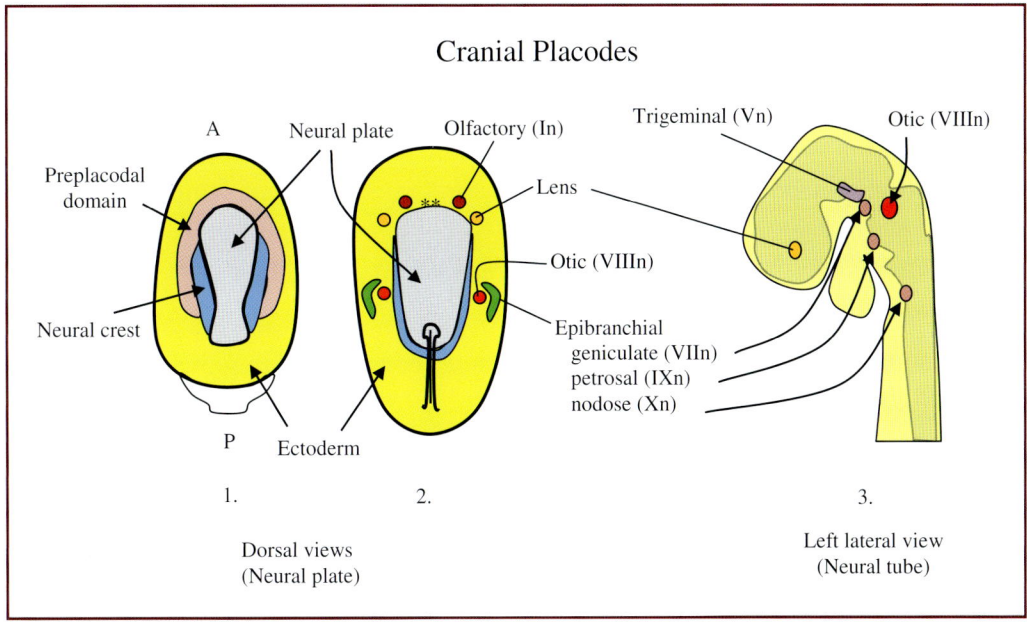

Figure 4–15. Schematic representation of cranial placode locations in the ectoderm. Dorsal view 1 (left) illustrates the horseshoe-shaped preplacodal domain appearing early in the formation of the neural plate. Dorsal view 2 depicts locations at the neural plate stage. Cranial placodes represented include olfactory (I_n), lens, otic (VIII_n) and epibranchial placodes (VII_n, IX_n, X_n). The double asterisk indicates the future position of the hypophyseal placode (pituitary) although it will lie ventrally and will not be visible in this view. Panel 3 (right) shows the neural tube stage from the lateral view identifying the trigeminal (V_n), geniculate (VII_n), petrosal (IX_n), and nodose (X_n) placodes and corresponding cranial nerves.

Current hypotheses suggest that cranial placodes develop from a two-step process (Baker & Bonner-Fraser, 2001; Ohyama, Groves, & Martin, 2007). The first step in placode formation is the appearance of a horseshoe-shaped preplacodal domain at the border of the cranial neural plate (see Figure 4–15). The preplacodal domain identifies the cranial location where all of the sensory placodes will develop. The expression of several genes from the *Dlx, Six, Eya, Bmp, Foxi,* and *Msx* families is thought to identify the preplacodal region since mutations in some of these genes have been shown to alter the size or shape of the preplacodal region (see review by Schlosser, 2006).

The second step in placode development is the formation of the specific sensory placodes at discrete locations (see Figure 4–15). This second step is thought to be due to local inducing signals where specific genes will be expressed for each sensory placode. For example, *Pax6* is critical for development of the lens placode and *Pax8* may be the earliest marker for the otic placode. Chapter 6 discusses the otic placode in detail.

PHARYNGEAL APPARATUS

The pharyngeal apparatus will develop during week 4 and will eventually become the cranio-facial elements (e.g., nasal cavities, jaw, mouth), the neck, laryngeal structures (e.g., hyoid, pharynx, larynx), thyroid and parathyroid glands, as well as portions of the ear. The pharyngeal apparatus, also known as the branchial apparatus, appears on the lateral surface on each side of the head region during week 4. The arches are numbered sequentially in the cranial to caudal direction (see Figure 4–9). Six arches will develop; however, the fifth arch is often absent and components of arches 4 and 6 often fuse. Therefore, arches one through four are the most prominent and are discussed here. Each arch is separated from the adjacent arch(es) by a pharyngeal groove on the outer surface and a pharyngeal pouch on the inner surface of the embryo. Indeed, the development of the pharyngeal pouches initiates the visible appearance of the arches as the endodermal cells that line the inner surface of the embryo push laterally to meet the surface ectoderm. The grooves and pouches also are numbered sequentially such that arches one and two are separated by pharyngeal groove one (outer surface) and pharyngeal pouch one (inner surface). The outer surface of the pharyngeal apparatus is lined with ectoderm, innermost with endoderm, and sandwiched between is mesenchyme (derived from mesoderm and neural crest cells). How do neural crest cells become part of the pharyngeal arches when they develop at the midline during neurulation? The neural crest cells that comprise a portion of the pharyngeal arch mesenchyme migrate from the rhombomeres via neural crest streams (see Figures 4–7 and 4–9). Neural crest cell streams from rhombomeres one and two migrate to pharyngeal arch 1, rhombomere 4 to pharyngeal arch 2, and rhombomeres 6 and 7 to pharyngeal arches 3 and 4.

Pharyngeal Arches

Figure 4–9 shows the composition of the pharyngeal arches. Each arch contains cartilage (neural crest mesenchyme) and muscle (mesoderm mesenchyme), and is supplied by a cranial nerve and an aortic arch (blood vessel). The first pharyngeal arch (also called the mandibular arch) contains Meckel's cartilage, and is supplied by the maxillary and mandibular portions of the trigeminal nerve (cranial nerve V) and the first aortic arch. Mandibular arch components will develop into ossicles and ligaments of the middle ear (see Chapter 5), muscles of mastication, the tensor tympani in the middle ear, and the maxillary arteries. Pharyngeal arch 2 (also called the hyoid arch) contains Reichert's cartilage and is supplied by the facial nerve (cranial nerve VII) and the second aortic arch. Hyoid arch components will develop into ossicles of the middle ear, portions of the temporal and hyoid bones, muscles for facial expression, the stapedius muscle of the middle ear, and the stapedial artery. Pharyngeal arch 3 is supplied by the glossopharyngeal nerve (cranial nerve IX) and the third aortic arch. Its muscle and cartilage components will become portions of the hyoid bone and muscles of the pharynx. The third aortic arch will become the common carotid and internal carotid arteries. Pharyngeal arch 4 is supplied by branches of the vagus nerve (cranial nerve X) and the fourth aortic arch, which will develop into muscles and cartilage of the pharynx and larynx, and portions of the aorta and subclavian arteries.

Pharyngeal Grooves

The pharyngeal grooves separate the pharyngeal arches on the lateral surface of the embryo (see Figure 4–9). The first pharyngeal groove ectoderm develops into the external acoustic meatus and outer layer of the tympanic membrane (discussed in Chapter 5). Grooves 2 through 4 merge to form the cervical sinus and then typically disappear during development of the neck.

Pharyngeal Pouches

As previously stated, the pharyngeal pouches are forming on the endodermal surface of the embryo before the pharyngeal arches are visible on the ectodermal surface. Indeed, the pharyngeal endoderm provides the inductive genetic signal to establish and pattern the pharyngeal apparatus and directs the neural crest cell streams (see reviews by Graham, 2003, 2008; Graham, Gegbie, & McGonnell, 2004; Trainor & Krumlauf, 2001). Pharyngeal pouch one eventually will form the middle ear cavity and the inner layer of the tympanic membrane. The second pharyngeal pouch will become the palatine tonsil. Cells that comprise the third pouch proliferate to eventually form the thymus, a primary organ for lymphocyte production during gestation. The third and fourth pouches will become the inferior and superior parathyroid glands, respectively.

Spatiotemporal expression of many genes is critical for the development and patterning of the pharyngeal apparatus. HOX genes and growth factors (TGFs and FGFs) play major roles (Cohen, 2002; Graham, 2008; Greene & Pisano, 2004;). Table 4–3 lists several genes that are critical for development of the first and second pharyngeal arches. Mutations in these genes can produce deficits in craniofacial, vascular, and auditory structures resulting in syndromic forms of hearing loss. Indeed, craniofacial anomalies are 1 of 11 important risk factors associated with permanent hearing loss in childhood (Joint Committee on Infant Hearing [JCIH], 2007).

FETAL PERIOD

During embryogenesis, one might be hard pressed to select the human embryo from an array of embryos representing various vertebrate species. By the end of week 8, the human embryo has a distinctive human baby appearance and the primordia for all major organ systems have formed.

Table 4–3. Some of the Genes, and Known Expression Patterns That Are Important for Normal Development of the Pharyngeal Arches, Particularly the First and Second Arches

Gene	Locus	Pharyngeal Apparatus Expression	Human Disorder
Eya1	8q13.3		Branchio-Oto-Renal (BOR) syndrome[1]
Tcof1	5q32-q33.1	Primarily arch 1, portion of arch 2	Treacher Collins syndrome[2]
Pax1	20p11.2	Dorsal tip of each pouch	
Bmp7	20q13.1-13.3	Posterior margin—pouches 2, 3, and 4	
Fgf8	10q24	Anterior margin—pouches 2, 3, and 4	Kallman syndrome 6[3]
Shh	7q36	Posterior margin—pouches 2 and 3	Holoprosencephaly 3[4]
Hoxa2	7p15-p14	Neural crest of arch 2	Microtia, hearing impairment, and cleft palate[5]
Edn1	6p24-p23	All germ layers of pharyngeal apparatus	
Ednra	4q31.2		
Dlx2	2q32	Proximal regions arches 1 and 2	
Dlx3	17q21.3-q22	Mesenchyme of pharyngeal arches	Trichodenotoosseous syndrome[6]
Tbx1	22q11.2	Mesenchyme of arches 3 and 4	Velocardiofacial/DiGeorge syndrome[7]
Gsc	14q32.1	Mesenchyme of groove 1	
Tfap2a	6p24	Neural crest cells of the arches	Branchio-oculofacial syndrome[8]

[1]Abdelhak et al. (1997); [2]Treacher Collins Syndrome Collaborative Group (1996); [3]Falardeau et al. (2008); [4]Roessler et al. (1996); [5]Alasti et al. (2008); [6]Price, Bowden, Wright, Pettenati, and Hart (1998); [7]Yagi et al. (2003), Merscher et al. (2001); [8]Milunsky et al. (2008).

At the start of week 9, the embryo is called a fetus and the fetal period of development will extend from 9 weeks postfertilization until birth (normally 38 weeks postfertilization or 40 weeks after the last normal menstrual period). During the fetal period, tissues and organs will continue to differentiate and will undergo considerable growth. At week 9, fetal length from crown to rump averages 50 mm and weight averages 8 grams (Moore & Persaud, 2003, p. 103). Over the next 30 to 32 weeks, the fetus will grow over 300 mm and 3300 grams. By week 38, the fetus averages 360 mm long and 3400 grams (Moore & Persaud, 2003, p. 103). The rate of growth for length remains relatively steady throughout the fetal period; however, weight gain favors the second half of the fetal period (6 grams/day from weeks 9 to 20 compared to 24 grams/day from weeks 22 to 38). Normal full-term birth weight averages 3400 grams. Low birth weight averages 2500 grams and very low birth weight is 1500 grams or less.

Early during the fetal period, the eyelids close (around week 9) and remain closed until weeks 26 to 28. From weeks 9 to 12, the intestines and urogenital system are developing such that by week 12, the gender often can be determined via ultrasound. The cranium and bones of the limbs begin to ossify during weeks 9 to 12, followed by ossification of the skeleton from weeks 13 to 16. Even though the eyes are closed, they are now positioned anterior on the face and slow eye movements can be observed by week 16. The pinnae are also at or near their final head position by week 16. During weeks 17 to 20, hair and eyebrows are visible; however, the skin is covered with vernix and fine, downy hair called lanugo. From weeks 21 to 25, the fetus appears pink or reddish in color because thin skin overlying the capillary beds reveals the reddish hue of the blood. Rapid eye movements can be seen, and the eyelids soon will open. The central nervous system and respiratory system are still developing, but by weeks 26 to 28, many fetuses survive if premature birth occurs and appropriate medical intervention occurs. The final weeks of development (weeks 30 to 38 postfertilization) reveal significant increase in body fat, weight gain, and sufficient maturation of organ systems for neonatal viability.

Although the focus of this textbook is on development during the prenatal period, it is important to recognize that development does not stop at birth. Considerable maturation of auditory and vestibular systems takes place during infancy (birth to 12 months) and childhood (13 months to puberty). For example, timing of synaptic transmission and neural conduction along the auditory pathways improves substantially during infancy and early childhood along with maturation of motor control systems and neural synchrony for upright walking and balance.

ENVIRONMENTAL INFLUENCES

Genes control the course of normal development, and genetic disorders can disrupt development. This is not to say that environment plays no role. Indeed, environmental factors alone or environment interacting with genetic factors may also disrupt development. Genetic disorders and multifactorial inheritance were discussed in Chapter 3 using examples of both syndromic and nonsyndromic hearing loss. Other common examples where genes and/or environment may lead to birth defects include cleft lip, cleft palate, and neural tube defects such as spina bifida.

Environmental factors alone may account for up to 10% of human birth defects (Brent, 2004). Environmental agents that produce a congenital anomaly or raise the incidence of an anomaly in the population are called teratogens. Teratogens can be classified into five broad categories: (1) maternal factors; (2) infectious agents; (3) drugs and chemicals; (4) physical agents; and (5) mechanical factors. Table 4–4 lists several examples, including those that may affect development of the ear (Dyer, Strasnick, & Jacobson, 1998; Strasnick & Jacobson, 1995). In addition,

Table 4–4. Some Examples of Teratogenic Agents and Internet Resources

Maternal Health	Diabetes
	Endocrine Deficits (e.g., hypothyroidism)
	Phenylketonuria
	Nutritional Deficits (e.g., folic acid, starvation)
Infectious Agents	
Viral	Cytomegalovirus
	Rubella
	Herpes Simplex
	Varicella Zoster
Bacterial	Syphilis
Parasitic	Toxoplasmosis
Drugs	Alcohol
	Nicotine
	Antibiotics (e.g., aminoglycosides)
	Anticoagulants (e.g., warfarin)
	Anticonvulsants (e.g., Dilantin)
	Chemotherapeutics (e.g., cisplatin)
	Renin-Angiotensin Inhibitors
Chemicals	Mercury
	Lead
	Polychlorinated Biphenyls (PCBs)
Physical Agents	Radiation
	Hyperthermia
Mechanical Factors	Oligohydramnios (reduced amniotic fluid)
	Amniotic Bands
	Umbilical Cord Constrictions
Web sites	Organization of Teratology Information Specialists (OTIS) (http://www.otispregnancy.org)
	TOXNET: Toxicology Data Network (http://toxnet.nlm.nih.gov/)

Sources: Information from Brent (2004), Dyer, Strasnick, and Jacobson (1998), Frias and Gilbert-Barness (2008), Roizen (2003), and Strasnick and Jacobson (1995).

some Internet resources listed in Table 4–4 allow one to look up information on known or suspected teratogens.

For an agent to be classified as a teratogen, it must meet several criteria. First, the exposure must be during a critical period of development in order to affect that organ system. Critical periods vary across organ systems. Exposure to teratogens during highly sensitive periods may cause major structural abnormalities, indeed arrest development altogether. During less sensitive periods structures may be relatively unaffected but functional abnormalities may still occur. For the ear, the highly sensitive critical period for teratogenic exposure is during weeks 3 to 10 as this is the time frame during which morphogenesis and significant sensory cell differentiation is occurring (discussed further in Chapters 5 and 6). The brain, on the other hand, has an extended critical period from weeks 3 to 20. Therefore, mental retardation is a birth defect often listed for a number of environmental teratogens. In addition to a critical period for exposure, drugs or chemicals must evidence a dose-response relationship such that increasing exposure levels results in more severe birth defects. Third, the embryo's genetic makeup can predispose it to be more sensitive to environmental agents. Finally, interactions can occur among multiple environmental factors (e.g., mother's health coupled with antibiotic exposure) producing a birth defect that is more severe than any single factor alone might produce.

CONCLUSIONS

Embryology involves a complex, orderly sequence of gene expression, much of which is shared by many organisms. All major organ systems undergo significant development during the embryonic period with significant refinements and maturation occurring during the fetal period. Although genes control development, maternal and environmental influences also can play a significant role in development, particularly during critical periods. As we will learn in the coming chapters, development of the ear follows this general schema with significant development occurring during embryogenesis with maturational refinements during the fetal period.

GENERAL REFERENCES

Carlson, B. M. (2004) *Human embryology and developmental biology* (3rd ed.). Philadelphia, PA: Mosby.

Ebbert, J. D., & Sussex, I.M. (1965). *Interacting systems in development*. Evanston, IL: Holt-McDougal.

Moore, K. L., & Persaud, T. V. N. (2003). *The developing human: Clinically oriented embryology* (7th ed.). Philadelphia, PA: Saunders.

Raff, R. A. (1996). *The shape of life: Genes, development, and the evolution of animal form.* Chicago, IL: University of Chicago Press.

Sanes, D. H., Reh, T. A., & Harris, W. A. (2000). *Development of the nervous system.* New York, NY: Academic Press.

LITERATURE CITED

Abdelhak, S., Kalatzis, V., Heilig, R., Compain, S., Samson, D., Vincent, C., Weil, D., . . . Petit, C. (1997). A human homologue of the Drosophila eyes absent gene underlies branchio-oto-renal (BOR) syndrome and identifies a novel gene family. *Nature Genetics, 15,* 157–164.

Aboitiz, F., & Montiel, J. (2007). Co-option of signaling mechanisms from neural induction to telencephalic patterning. *Reviews in the Neurosciences, 18,* 311–342.

Alasti, F., Sadeghi, A., Sanati, M. H., Farhadi, M., Stollar, E., Somers, T., & Van Camp, G. (2008). A mutation in HOXA2 is responsible for autosomal-recessive microtia in an Iranian family. *American Journal of Human Genetics, 82,* 982–991.

Araki, I., & Nakamura, H. (1999). *Engrailed* defines the position of dorsal di-mesencephalic boundary by repressing diencephalic fate. *Development, 126,* 5127–5135.

Baker, C. V. H., & Bronner-Fraser, M. (2001). Vertebrate cranial placodes. I. Embryonic induction. *Developmental Biology, 232,* 1–61.

Basch, M., & Bronner-Fraser, M. (2006). Neural crest inducing signals. In J. P. Saint-Jeannett (Ed.), *Neural crest induction and differentation* (pp. 24–31). New York, NY: Landes Bioscience and Springer + Business Media.

Beddington, R. S. P., & Robertson, E. J. (1998). Anterior patterning in mouse. *Trends in Genetics, 14*(7), 277–284.

Braun, M. M., Etheridge, A., Bernard, A., Robertson, C. P., & Roelink, H. (2003). Wnt signaling is required at distinct stages of development for the induction of the posterior forebrain. *Development, 130,* 5579–5587.

Brent, R. L. (2004). Environmental causes of human congenital malformations: The pediatrician's role in dealing with these complex clinical problems caused by a multiplicity of environmental and genetic factors. *Pediatrics, 113,* 957–968.

Buckles, G. R., Thorpe, C. J., Ramel, M. C., & Lekven, A. C. (2004). Combinatorial Wnt control of zebrafish midbrain-hindbrain boundary formation. *Mechanisms of Development, 121,* 437–447.

Carlson, B. M. (2004) *Human embryology and developmental biology* (3rd ed.) Philadelphia, PA: Mosby.

Ciani, L., & Salinas, P.C. (2005). Wnts in the vertebrate nervous system: From patterning to neuronal connectivity. *Nature Reviews Neuroscience, 6,* 351–362.

Cohen, M. M. (2002). Malformations of the craniofacial region: Evolutionary, embryonic, genetic, and clinical perspectives. *American Journal of Medical Genetics (Seminars in Medical Genetics), 115,* 245–268.

Copp, A. J., Greene, N. D. E., & Murdoch, J. N. (2003). Genetic basis of mammalian neurulation. *Nature Reviews Genetics, 3,* 784–793.

Deschamps, J., van den Akker, E., Forlani, S., de Graaff, W., Oosterveen, T., . . . Roelfsema, J. (1999). Initiation, establishment and maintenance of Hox gene expression patterns in the mouse. *International Journal of Developmental Biology, 43,* 635–650.

di-Gregorio, A., Sancho, M., Stuckey, D.W., Crompton, L. A., Godwin, J., Mishina, Y., . . . Rodriquez, T. A. (2007). *Development, 134,* 3359–3369.

Dyer, J. J., Strasnick, B., & Jacobson, J. T. (1998). Teratogenic hearing loss: A clinical perspective. *American Journal of Otology, 19,* 671–678.

Falardeau, J., Chung, W. C. J., Beenken, A., Raivio, T., Plummer, L., Sidis, Y., . . . Pitteloud, N. (2008). Decreased FGF8 signaling causes deficiency of gonadotropin-releasing hormone in humans and mice. *Journal of Clinical Investigation, 118,* 2822–2831.

Favier, B., & Dolle, P. (1997). Developmental function of mammalian HOX genes. *Molecular Human Reproduction, 3*(2), 115–131.

Frias, J. L., & Gilbert-Barness, E. (2008). Human teratogens: Current controversies. *Advances in Pediatrics, 55,* 171–211.

Glavic, A., Gomez-Skarmeta, J. L., & Mayor, R. (2002). The homeoprotein *Xiro1* is required for midbrain-hindbrain boundary formation. *Development, 129,* 1609–1621.

Graham, A. (2003). Development of the pharyngeal arches. *American Journal of Medical Genetics, 199A,* 251–256.

Graham, A. (2008). Deconstructing the pharyngeal metamere. *Journal of Experimental Zoology (Molecular Development and Evolution), 310B,* 336–344.

Graham, A., Gegbie, J., & McGonnell, I. (2004). Significance of the cranial neural crest. *Developmental Dynamics, 229,* 5–13.

Greene, R. M., & Pisano, M. M. (2004). Perspectives on growth factors and orofacial development. *Current Pharmaceutical Design, 10,* 2701–2717.

Houart, C., Caneparo, L., Heisenberg, C. P., Barth, K. A., Take-Uchi, M., et al. (2002). Establishment of the telencephalon during gastrulation by local antagonism of Wnt signaling. *Neuron, 35,* 255–265.

Joint Committee on Infant Hearing. (2007). Year 2007 position statement: Principles and guidelines for early hearing detection and intervention programs. Available from http://www.asha.org/policy.

Kobayashi, D., Kobayashi, M., Matsumoto, K., Ogura, T., Nakafuku, M., & Shimamura, K. (2002). Early subdivisions in the neural plate define distinct competence for inductive signals. *Development, 129,* 83–93.

Levine, A. J., & Brivenlou, A. H. (2007). Proposal of a model of mammalian neurulation. *Developmental Biology, 308,* 247–256.

Matsunaga, E., Araki, I., & Nakamura, H. (2000). *Pax6* defines the di-mesencephalic boundary by repressing *En1* and *Pax2. Development, 127,* 2357–2365.

Merscher, S., Funke, B., Epstein, J. A., Heyer, J., Puech, A., Lu, M. M., . . . Kuchertapati, R. (2001). TBX1 is responsible for cardiovascular defects in velo-cardio-facial/DiGeorge syndrome. *Cell, 104,* 619–629.

Milunsky, J. M., Maher, T. A., Zhao, G., Roberts, A. E., Stalker, H. J., . . . Lin, A. E. (2008). TFAP2A mutations result in branchio-oculo-facial syndrome. *American Journal of Human Genetics, 82*(5), 1171–1177.

Moore, K. L., & Persaud, T. V. N. (2003). *The developing human: Clinically oriented embryology* (7th ed.). Philadelphia, PA: Saunders.

Nakatsu, T., Uwabe, C., & Shiota, K. (2000). Neural tube closure in humans initiates at multiple sites: Evidence from human embryos and implications for the pathogenesis of neural tube defects. *Anatomy and Embryology, 201,* 455, 466.

Ohyama, T., Groves, A. K., & Martin, K. (2007). The first steps towards hearing: Mechanisms of otic placode induction. *International Journal of Developmental Biology, 51,* 463–472.

Price, J. A., Bowden, D. W., Wright, J. T., Pettenati, M. J., & Hart, T. C. (1998). Identification of a mutation in DLX3 associated with tricho-dento-osseous (TDO) syndrome. *Human Molecular Genetics, 7,* 563–569.

Rallu, M., Machold, R., Gaiano, N., Corbin, J. G., McMahon, A. P., & Fisheli, G. (2002). Dorsoventral patterning is established In the telencephalon of mutants lacking both Gli3 and Hedgehog signaling. *Development, 120,* 4963–4974.

Roelink, H., & Nusse, R. (1990). Expression of two members of the Wnt family during mouse development—restricted temporal and spatial patterns in the developing neural tube. *Genes and Development, 5,* 381–388.

Roessler, E., Belloni, E., Gaudenz, K., Jay, P., Berta, P., Scherer, S. W., et al. (1996). Mutations in the human Sonic hedgehog gene cause holoprosencephaly. *Nature Genetics, 14,* 357–360.

Roizen, N. J. (2003). Nongenetic causes of hearing loss. *Mental Retardation and Developmental Disabilities Research Reviews, 9,* 120–127

Schlosser, G. (2006). Induction and specification of cranial placodes. *Developmental Biology, 294,* 303–351.

Smukler, S. R., Runciman, S. B., Xu, S. & van der Kooy, D. (2006). Embryonic stem cells assume a primitive neural stem cell fate in the absence of extrinsic influences. *Journal of Cell Biology, 172*(1), 79–90.

Strasnick, B., & Jacobson, J. T. (1995). Teratogenic hearing loss. *Journal of the American Academy of Audiology, 6,* 28–38.

Trainor, P. A., & Krumlauf, R. (2001). Hox genes, neural crest cells and branchial arch patterning. *Current Opinion in Cell Biology, 13,* 698–705.

Treacher Collins Syndrome Collaborative Group. (1996). Positional cloning of a gene involved in the pathogenesis of Treacher Collins syndrome. *Nature Genetics, 12,* 130–136.

Tropepe, V., Hitoshi, S., Sirard, C., Mak, T. W., Rossant, J., & van der Kooy, D. (2001). Direct neural fate specification from embryonic stem cells: A primitive mammalian neural stem cell stage acquired through a default mechanism. *Neuron, 30,* 65–78.

Tümpel, S., Wiedemann, L. M., & Krumlauf, R. (2009). *Hox* genes and segmentation of the vertebrate hindbrain. *Current Topics in Developmental Biology, 88,* 103–137.

Wilson, S. W., & Houart, C. (2004). Early steps in the development of the forebrain. *Developmental Cell, 6,* 167–181.

Wurst, W., & Bally-Cuif, L. (2001). Neural plate patterning: Upstream and downstream of the isthmic organizer. *Nature Reviews Neuroscience, 2,* 99–108.

Yagi, H., Furutani, Y., Hamada, H., Sasaki, T., Asakawa, S., Minoshima, S., Ichida, F., et al. (2003). Role of TBX1 in human del22q11.2 syndrome. *Lancet, 362,* 1366–1373.

Embryogenesis of the Outer and Middle Ear

KEY POINTS

1. The temporal bone develops from mesodermal and neural crest mesenchyme via intramembranous and endochondral ossification. Once developed, it will continue to ossify throughout childhood up to puberty.
2. The pinna develops from six auricular hillocks that form around the first pharyngeal groove during the fifth week after fertilization.
3. The first pharyngeal groove invaginates to form the external auditory meatus.
4. The invaginating first pharyngeal groove meets the protruding first pharyngeal pouch ultimately forming the tympanic membrane. The three layers of the tympanic membrane are derived from the three germ layers: ectoderm on the outer surface, endoderm on the inner surface, and mesoderm in the middle.
5. Expansion of the first pharyngeal pouch along with resorption of mesenchyme forms the middle ear cavity with the innermost portion of the pouch extending to the pharynx via the Eustachian tube. All are lined with endodermal mucous membrane.
6. The middle ear ossicles are formed from the first and second pharyngeal arches. The malleus and incus are derived from the first arch mesenchyme. The stapes is derived from the second arch mesenchyme.
7. Muscles, ligaments, and neural supply to the middle ear also are derived from the first and second pharyngeal arches.

INTRODUCTION

The outer ear and middle ear develop from the first and second pharyngeal arches. Recall from Chapter 4 that the pharyngeal arches first appear during week 4 after fertilization under the influence of genetic cues from the endoderm and migrating neural crest cells. Patterning of the pharyngeal apparatus involves spatiotemporal expression of many genes, several of which are also involved in patterning of the body axis and brain such as members of the *Fgf*, *Bmp*, and *Hox* gene families as well as a number of other homeobox transcription factors (see Table 4–3). *Shh* and *Wnt* play important roles as well.

Table 5–1 lists several genes that are important for outer and middle ear development. You will recognize a number of them from Chapter 4. The molecular cues and interactions necessary for normal outer and middle ear development have only recently been studied in depth in animal

Table 5–1. Some of the Genes Important for Outer and Middle Ear Development

Outer Ear	Middle Ear	Gene	Locus	Human Disorder
Pinna		Bmp5	6p12.1	
Pinna		Fgf3	11q13	Congenital deafness, inner ear agenesis, microtia, and microdontia[1]
Pinna		Fgfr2	10q26	Apert syndrome, Crouzon syndrome[2]
Pinna		Fgfr3	4p16.3	Achondroplasia[3]
Pinna		Fgf10	5p13-p12	LADD syndrome[4]
Pinna, EAM		Nkx5.3 (Hmx1)	4p16.1	Oculo-auricular syndrome[5]
Pinna	Stapes	Pbx1	1q23	
Pinna?	Stapes	Bcl2	18q21.3	
Pinna	Stapes	Hoxa1	7p15.3	Bosley-Salih-Alorainy/Athabaskan brainstem dysgenesis syndromes[6]
Pinna	Malleus, Stapes	Hoxa1/Hoxb1	17q21-q22	
Pinna	Incus, Stapes	RARs α, β, γ	17q21.1, 3p24, 12q13	
Pinna	Middle Ear	Hoxa2	7p15-p14	Microtia, Hearing impairment, and Cleft palate[7]

Table 5–1. *continued*

Outer Ear	Middle Ear	Gene	Locus	Human Disorder
Pinna	Middle ear	*Rax*	18q21.3	
Pinna, EAM	All ossicles	*Dlx5/Dlx6*	7q22/7q22	
Pinna, EAM	All ossicles	*Edn1*	6p24-p23	
Pinna, EAM	All ossicles	*Ednra*	4q31.2	
Pinna, EAM	All ossicles	*Tbx1*	22q11.2	DiGeorge/Velocardiofacial syndrome[8]
Pinna	All ossicles	*Ece1*	1p36.1	Hirschsprung disease[9]
Pinna, EAM	All ossicles	*Eya1*	8q13.3	Branchial Oto-Renal (BOR) syndrome[10]
EAM	Malleus-manubrium	*Gsc*	14q32.1	Isolated(?) Microtia[11]
EAM	All ossicles	*Prrx1*	1q24	
	Malleus-gonial bone	*Pitx1*	5q31	
	Malleus-processus brevis	*Msx1*	4p16.1	Tooth agenesis; Orofacial cleft; Witkop syndrome[12]
	Stapes	*Dlx5*	7q22	
	Stapes	*Dlx1*	2q32	
	Stapes	*Wnt1/Wnt3a*	12q12-q13/1q42	
	Malleus, incus	*Fgf8*	10q24	Kallmann syndrome 6[13]
	Incus, stapes	*Dlx2*	2q32	
	All ossicles	*Ap2*	6p24	Branchiooculofacial syndrome[14]
	Middle ear	*Nkx3.2 (Bapx1)*	4p16.1	Spondylo-Megaepiphyseal-Metaphyseal-dysplasia (SMMD)[15]

RAR: Retinoic acid receptors.

[1]Tekin et al. (2007, 2008); [2]Jabs et al. (1994), Wilkie et al. (1995); [3]Shiang et al. (1994); [4]Milunsky, Zhao, Maher, Colby, & Everman (2006); Rohmann et al. (2006); [5]Schorderet et al. (2008); [6]Tischfield et al. (2005); [7]Alasti et al. (2008), [8]Yagi et al. (2003), Arnold et al. (2006); [9]Hofstra et al. (1999); [10]Abdelhak et al. (1997); [11]Zhang, Zhang, & Yin (2010); [12]Vastardis, Karimbux, Guthua, Seidman, & Seidman (1996), Jumlongras et al. (2001), Jezewski et al. (2003); [13]Falardeau et al. (2008); [14]Milunsky et al. (2008); [15]Hellemens et al. (2009).

models, particularly mutant mouse strains. Therefore, we use the animal nomenclature for gene names, but emphasize human developmental ages and the fact that mutations in many of these genes result in human disorders (see Table 5–1). For some structures, widely varying ages have been published and some controversies exist with regard to tissue origins. Several of these discrepancies and controversies are noted throughout the discussion.

TEMPORAL BONE

The temporal bone, one of several bones that make up the cranium, consists of the squamous, tympanic, petrous, and mastoid portions, as well as the zygomatic and styloid processes (reviewed in Chapter 1). The tympanic, mastoid, and petrous portions surround the middle ear whereas the petrous portion encases the inner ear. Formation of the skull requires several steps that are well described in embryology texts (see General Resources) and the scientific literature (e.g., Castriota-Scanderbeg, 2005; Coburne, 2000; Cohen, 2006; Hall & Miyake, 2000; Jiang, Iseki, Maxson, Sucov, & Morriss-Kay, 2002; McBratney-Owen, Iseki, Bamforth, Olsen, & Morriss-Kay, 2008; Morriss-Kay, 2001; Morriss-Kay & Wilkie, 2005; Szabo-Rogers, Smithers, Yakob, & Liu, 2010; Wilkie & Morriss-Kay, 2001). First, the cells that will become cartilage or bone must migrate to the appropriate site. For example, neural crest cells migrate into the pharyngeal arches (reviewed in Chapter 4) and will contribute to the facial skeleton and skull bones. Second, epithelium and mesenchyme interact to induce condensation of the mesenchymal cells, which is the third step in skeletogenesis. Nearly 30 genes or gene products are known to be involved in the formation of condensations controlling their initiation, growth, boundaries, and size (reviewed by Hall & Miyake, 2000). Condensation malformations can have a dramatic impact on bone development. Genes and proteins known to impact condensations include growth factors, cell adhesion and extracellular matrix molecules, *Hox* genes, and a variety of transcription factors (Table 5–2). During the fourth step in skeletogenesis, the condensations differentiate into cartilage or bone. The ossification process that first forms a cartilaginous matrix and then lays bone onto the matrix is called endochondral ossification. Bone that forms directly from mesenchyme without a cartilaginous stage is termed intramembranous ossification. The skull develops via both endochondral and intramembranous ossification.

Like the skull, the temporal bone develops from intramembranous and endochondral ossification and has both mesodermal and neural crest origins. The squamous portion, zygomatic process, and tympanic ring (tympanic portion) of the temporal bone are of neural crest origin and undergo intramembranous ossification. The remainder of the temporal bone (petrous and mastoid portions, styloid process) originates from mesoderm and undergoes endochondral ossification.

A range of fetal ages for skull development is presented in the literature. This is likely due to the fact that the cranium consists of several bones that develop and ossify over a broad time range. In addition, the specific processes of skeletogenesis (e.g., appearance of condensations, appearance of ossification centers) can vary in their developmental time course for different regions of the skull. Table 5–3 displays the time course for development of the various portions of the temporal bone as well as the other components of the outer and middle ear. Development

Table 5–2. Some of the Genes That Contribute to Condensation Formation During Bone Development

Gene	Locus	Human Disorder
Pax1	20p11.2	
Pax2	10q24.3-q25.1	Renal Coloboma Syndrome[1]
Pax9	14q12-q13	Selective Tooth Agenesis[2]
Hoxa2	7p15-p14	Microtia, Hearing Impairment, and Cleft Palate[3]
Hoxd11	2q31-q32	
Hoxd12	2q31-q32	
Hoxd13	2q31-q32	Synpolydactyly[4]
Msx1	4p16.1	Tooth Agenesis; Orofacial Cleft; Witkop Syndrome[5]
Msx2	5q34-q35	Craniosynostosis, Parietal Foramina[6]
Sox9	17q24.3-q25.1	Campomelic Dysplasia[7]
Runx2	6p21	Cleidocranial Dysplasia[8]
Bmp2	20p12	
Bmp4	14q22-q23	Microphthamia Syndrome 6, Orofacial Cleft 11[9]
Bmp5	6p12.1	
Bmp7	20q13.1-q13.3	
Fgf2	4q25-q27	
Tgfb1	19q13.1	Camurati-Engelmann Disease[10]
Foxc2	16q24.3	Lymphedema-Distichiasis Syndrome[11]
Ncam	11q23.1	
Nog	17q22	Proximal Symphalangism, Multiple Synostoses Syndrome, Stapes Ankylosis Syndrome without Symphalangism[12]
Prrx1	1q24	

[1]Sanyanusin et al. (1995); [2]Stockton, Das, Goldenberg, D'Souza, & Patel (2000); [3]Alasti et al. (2008); [4]Akarsu, Stoilov, Yilmaz, Sayli, & Sarfarazi (1996); [5]Vastardis, Karimbux, Guthua, Seidman, & Seidman (1996), Jumlongras et al. (2001), Jezewski et al. (2003); [6]Ma, Golden, Wu, & Maxson (1996), Wilkie et al. (2000); [7]Foster et al. (1994); [8]Mundlos et al. (1997); [9] Bakrania et al. (2008), Suzuki et al. (2009); [10]Kinoshita et al. (2000), Janssens et al. (2000); [11]Fang et al. (2000); [12]Gong et al. (1999); Brown et al. (2002).

of the temporal bone begins during the embryonic period and progresses throughout the fetal period. Indeed, for some elements, complete growth, ossification, and fusion of sutures will continue after birth extending to puberty and adulthood.

Table 5–3. Timeline of Major Events for Human Middle Ear Development

Event	\multicolumn{36}{c}{Age in Weeks}

Event	1	2	3	4	5	6	7	8	9	10	11	12	13	14	15	16	17	18	19	20	21	22	23	24	25	26	27	28	29	30	31	32	33	34	35	36
TB																																				
TR							▓	▓	▓	▓	▓	▓	▓	▓	▓	▓	▓	▓	▓	▓	▓	▓	▓	▓	▓	▓	▓	▓	▓	▓	▓	▓	▓	▓	▓	▓
Squamous							▓	▓	▓	▓	▓	▓	▓	▓	▓	▓	▓	▓	▓	▓	▓	▓	▓	▓	▓	▓	▓	▓	▓	▓	▓	+	+	+	+	+
Petrous							▓	▓	▓	▓	▓	▓	▓	▓	▓	▓	▓	▓	▓	▓	▓	▓	▓	▓	▓	▓	▓	▓	▓	▓	▓	+	+	+	+	+
Mastoid							▓	▓	▓	▓	▓	▓	▓	▓	▓	▓	▓	▓	▓	▓	▓	▓	▓	▓	▓	▓	▓	▓	▓	▓	▓	+	+	+	+	+
Pinna					▓	▓	▓	▓	▓	▓	▓	▓	▓	▓	▓	▓	▓	▓	▓	▓																
EAM								▓	▓	▓	▓	▓	▓	▓	▓	▓	▓	▓	▓	▓	▓	*	▓													
TM								▓	▓	▓	▓	▓	▓	▓	▓	▓	▓	▓	▓	▓	▓	▓	▓	▓												
Ossicles																																				
Malleus							▓	▓	▓	▓	▓	▓	▓	▓	▓	▓	▓	▓	▓	▓	▓	▓	▓	▓	▓	▓										
Incus							▓	▓	▓	▓	▓	▓	▓	▓	▓	▓	▓	▓	▓	▓	▓	▓	▓	▓	▓	▓										
Stapes							▓	▓	▓	▓	▓	▓	▓	▓	▓	▓	▓	▓	▓	▓	▓	▓	▓	▓	▓	▓	▓	▓								
ME Cavity																																				
Primary								▓	▓	▓	▓	▓	▓	▓	▓	▓	▓	▓	▓	▓	▓	▓	▓	▓	▓	▓	▓	▓	▓	▓						
Epitymp.																						▓	▓	▓	▓	▓	▓	▓	▓	▓						
Mastoid																													▓	▓	▓	+	+			
ET																						▓	▓	▓	▓	▓	▓	▓	▓	▓	▓	▓	▓	▓	▓	▓

Note: Age approximations are from published data on humans (cited in the text) or estimated from mouse models. Youngest ages represent onset of morphological appearance or expression of distinct genetic markers.

TB: temporal bone; TR: tympanic ring; EAM: external acoustic meatus; TM: tympanic membrane; ME: middle ear; ET: Eustachian tube; Epitymp.: Epitympanum.

*Some discrepancy in the literature regarding opening of the EAM (see text for details). The bony portion of the EAM continues to ossify until age 7.

+Development extends beyond birth. See text for details.

According to the Atlas and Database on Human Developmental Anatomy (http://www.ana.ed.ac.uk/anatomy/database/humat/), *Gray's Anatomy* (Standring, 2005), general embryology texts (see General Resources), Nemzek et al. (1996), Streeter (1918), and Bardeen (1910), the otic capsule (or periotic capsule), which ultimately develops into the petrous and mastoid portions of the temporal bone, is apparent at 5 weeks of age and will become cartilaginous during weeks 6 to 7. The region of the otic capsule surrounding the developing vestibular system will chondrify before the cochlear part. By the end of week 7, the otic capsule joins with other cartilages to form the cranial base (Figure 5–1). As the otic capsule becomes cartilaginous, the oval and round windows are apparent as they remain membranous while the surrounding tissue chondrifies. Because the otic capsule contributes to both the petrous and mastoid portions of the temporal bone, it is often described as the petromastoid bone during development and it is difficult to find developmental ages specified for the petrous and mastoid portions separately. Landmarks on the mastoid portion (mastoid foramen and groove for the transverse sinus) have been described for embryos at 8 weeks (Lewis, 1920). By 12 weeks, the cartilaginous cranial base is nearly complete but no bone has yet appeared (Figure 5–1B). Several ossification centers, numbering anywhere from 6 to 14, will appear at various locations on the petrous and mastoid regions from 16 to 20 weeks. Ossification will occur quite rapidly until the labyrinth is surrounded by bone sometime between 22 and 24 weeks. Despite the fact that the otic capsule is considered to be adult size by 24 weeks, the internal acoustic meatus and vestibular aqueduct will continue to grow until about 3 years of age (Nemzek et al., 1996). Furthermore, the petromastoid will not join with the squamous portion or tympanic ring until after birth. The mastoid is also not fully formed at birth, rather it is generally accepted that the mastoid antrum is present at birth, but the mastoid process develops during the first two to three years after birth and becomes completely pneumatized during late childhood to puberty (Cinamon, 2009).

At 6 weeks, the condensation that will become the styloid process is present and becomes cartilaginous by week 7. During week 7, the squamous portion and zygomatic process of the temporal bone first appear as an ossification center at the base of the zygoma. The squamous portion will ossify but will not join with the tympanic ring until shortly before birth, and will join with the petromastoid portion during the first months after birth. The squamous portion also will articulate eventually with the occipital, sphenoid, and parietal bones of the skull (see Figure 5–1C). At birth, the cranial sutures are not closed but the plates are widely coupled with connective tissue to accommodate the growing brain during childhood. Although some of the sutures close in the first year or two of life, others will finally ossify in early adulthood (Castriota-Scanderbeg, 2005).

PINNA

The pinna develops from the auricular hillocks, which appear on the first and second pharyngeal arches (on either side of the first pharyngeal groove) during week 5 after fertilization. The auricular hillocks typically are numbered 1 through 6 or labeled "a" through "g." The first number or letter begins with the ventralmost hillock on the first pharyngeal arch,

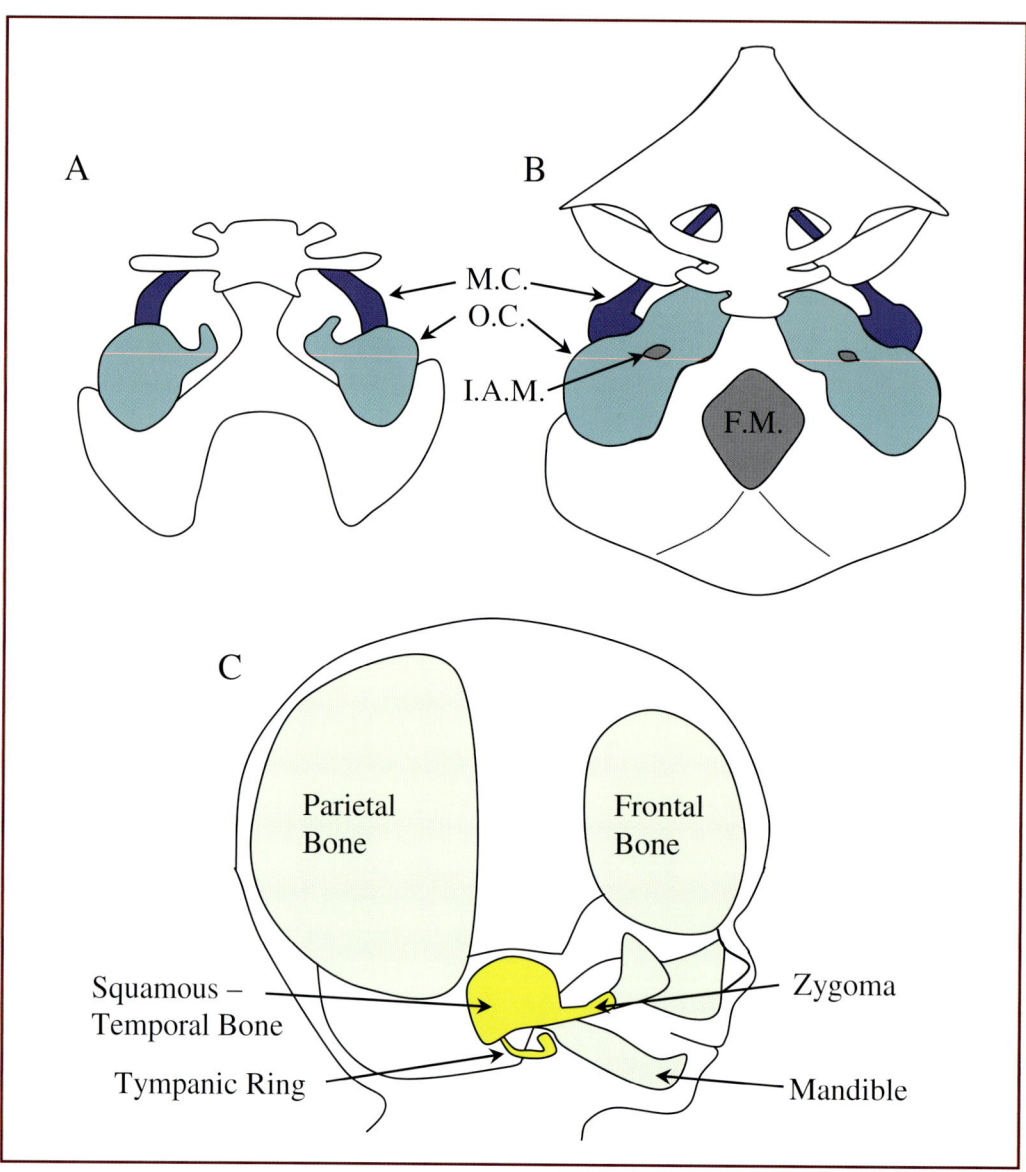

Figure 5–1. Development of the skull base (**A** and **B**) and the skull vault (**C**) highlighting those portions relevant to the temporal bone and the ear. **A.** Schematic of the cartilaginous embryonic skull base at 8 weeks. Meckel's cartilage (M.C.) is highlighted in dark blue. The otic capsule (O.C.) is shown in light blue. Notable for the otic capsule is the cartilage surrounding the developing cochlea (hook shaped portion of the otic capsule), which lags behind the vestibular portion. By 12 weeks (**B**), the cochlea has reached its full number of turns and the otic capsule envelops the entire inner ear. The otic capsule is also now joined with the other cartilages that comprise the skull base. The internal acoustic meatus (I.A.M.) is visible and the foramen magnum (F.M.) is formed. **C.** The skull vault is formed from membranous bone (undergoes intramembranous ossification), which includes the squamous portion of the temporal bone (highlighted in yellow), parietal, and frontal bones (light green). Development at 12 weeks of age is shown. The sutures between the bones of the skull vault are not fused at birth to accommodate brain growth during childhood. In addition to the parietal and frontal bones, the other bones of the skull that develop by intramembranous ossification are shaded including the tympanic ring (highlighted in yellow). Based on Bardeen (1910).

increasing sequentially in the dorsal direction, then to the second arch moving in the ventral direction (Figure 5–2).

The anterior three hillocks ultimately will form anterior portions of the pinna (tragus, helix) whereas the posterior hillocks will form the posterior portions that comprise the majority of the pinna overall (antihelix, crua of the anithelix, scapha, antitragus, earlobe). During week 6, the underlying mesenchyme will begin forming cartilage as the hillocks continue to

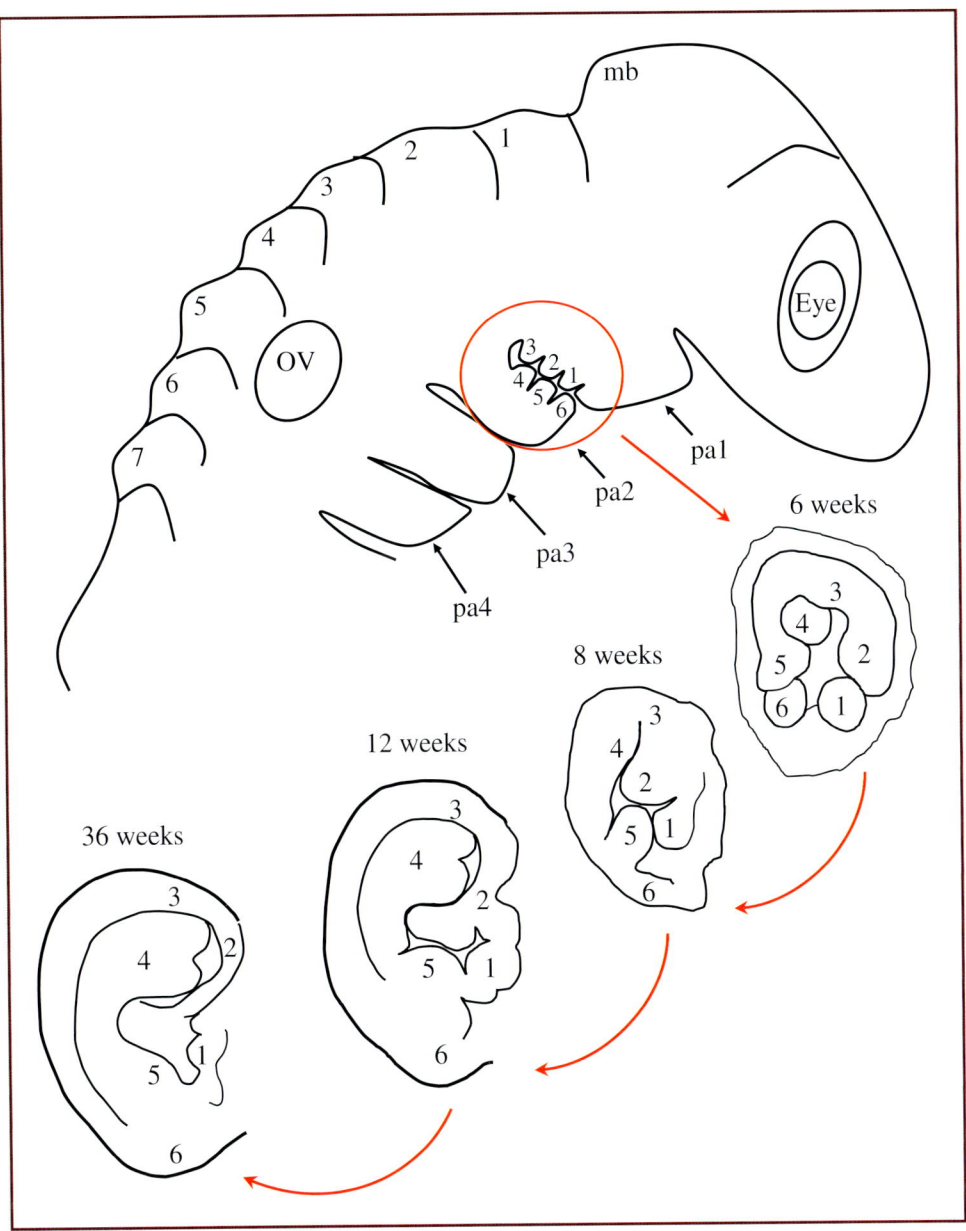

Figure 5–2. Morphogenesis of the pinna. Based on McMurrich (1923, p. 455).

undergo rearrangement and form the pinna shape (see Figure 5–2). By week 11 to 12, the pinna has developed its anatomical features, but will move from a ventromedial position at week 7 to a lateral position on the head by week 20.

Pinna Malformations

Pinna malformations represent a continuum from anotia (absent pinna) to various stages of microtia (small or malformed pinna). Atresia (absent external auditory meatus) or stenosis (narrowing of the ear canal) often accompanies microtia. Other anomalies that can be associated with pinna malformations include craniofacial, cardiac, eye, limb, and renal defects. The prevalence of anotia/microtia varies from 0.8 to 4.2 per 10,000 live births (reviewed by Alasti & Van Camp, 2009), but, depending on geographic area and ethnic group, higher estimates have been reported (Harris, Källén, & Robert, 1996; Klockars, Suutarla, Kenala, Ala-Mello, & Rautio, 2007; National Birth Defects Prevention Network [NBDPN], 2009; Suutarla et al., 2007). For example, Castilla and Orioli (1986) reported a prevalence of microtia as high as 17.4 per 10,000 in Quito, Equador. Microtia typically is unilateral, males are more affected than females, and the right ear is affected more often than the left.

Microtia often is due to environmental factors, but genetic mutations or multifactorial inheritance also is considered causal. Genetic causes of microtia include chromosomal abnormalities or single gene mutations (see Table 5–1). Many of the single gene mutations result in syndromes where microtia is part of the clinical phenotype. For example, Alasti et al. (2008) described a missense mutation in *Hoxa2* in a large Iranian family with an autosomal recessive microtia that also included stenotic ear canal, malformed ossicular chain, severe to profound mixed hearing loss, and cleft palate. Isolated microtia (i.e., microtia without other clinical abnormalities) that appeared to follow an autosomal dominant mode of inheritance was described by Zhang, Zhang, and Yin (2010) in several members of a Chinese family with bilateral microtia. The microtia in this family appeared to be isolated (or nonsyndromic) as the affected individuals had normal medical examinations with no other syndromic stigmata (e.g., craniofacial, cardiovascular, or renal abnormalities), and they reportedly had normal hearing. In several affected individuals, the investigators identified a deletion in exon 3 of the goosecoid gene (*Gsc*) resulting in a frameshift mutation. The authors stated, however, that environmental factors or multifactorial inheritance may have played a causal role for some family members and that further study was necessary. As we will see, *Gsc* is a critical gene for outer and potentially middle ear development.

EXTERNAL AUDITORY MEATUS/TYMPANIC MEMBRANE

The external auditory meatus (EAM) develops from the first pharyngeal groove. A required structure for EAM development is the tympanic ring, which first appears as a condensation in the mesenchyme of the first pharyngeal arch during week 7 to 8 (Figure 5–3). The tympanic ring will elongate and extend around the first pharyngeal groove to the second pharyngeal arch forming a C-shape or incomplete circle that eventually will be integrated into the temporal bone and surround the circumference of the tympanic membrane

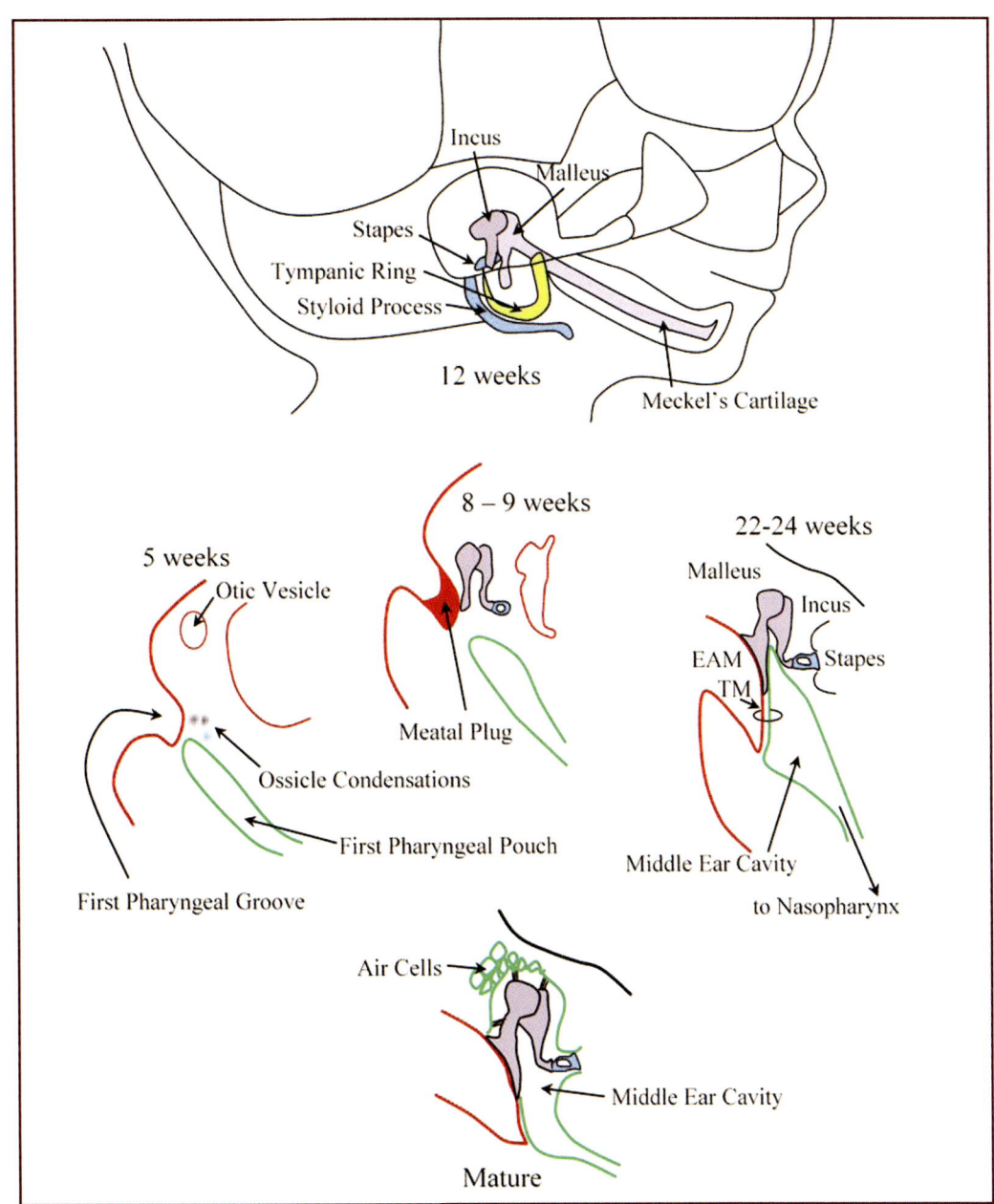

Figure 5–3. *Development of the external acoustic meatus (EAM) and the middle ear from 5 weeks to maturity. Illustrations for 5 weeks, 8–9 weeks, 22–24 weeks, and maturity represent coronal sections through the middle ear. The EAM is lined with ectoderm (shown in red). The middle ear cavity and eventual mastoid air cells are lined with endoderm (shown in green). Ectoderm, endoderm, and mesoderm contribute to the tympanic membrane. The lateral view depicted for 12 weeks shows the orientation of the middle ear ossicles relative to the tympanic ring (yellow), Meckel's cartilage (purple), and the styloid process (blue). The squamous portion of the temporal bone is not colored. At 12 weeks, Meckel's cartilage is continuous with the head of the malleus. Meckel's cartilage extends into and is enveloped by the developing bone of the mandible. The ossicles derived from the first pharyngeal arch (malleus and incus) are shown in purple and the stapes, derived from the second pharyngeal arch, is shown in blue. An alternative to this classical theory for ossicular derivatives is depicted in Figure 5–4.*

(Anson, Bast, & Richany, 1955; Mallo & Gridley, 1996). The tympanic ring will begin ossifying around 10 weeks, but will not join with the other portions of the temporal bone (squamous and petromastoid) until shortly before birth (squamous portion) or in the first months after birth (petromastoid). In addition, the tympanic ring will continue to grow and ossify until about age 7 to 10 years forming the bony portion of the EAM.

During development, the tympanic ring exerts significant influence on the development of the EAM, tympanic membrane, and manubrium of the malleus. If the tympanic ring is absent, the EAM will not form (Mallo, 1997; Mallo & Gridley, 1996; Mallo, Schrewe, Martin, Olson, & Ohnemus, 2000; Martin, Bradley; & Olson, 1995; Rivera-Perez, Wakamiya, & Behringer, 1995; Yamada et al., 1995). Malformed tympanic rings also have been associated with atresia in humans (Lambert & Dodson, 1996).

Two genes have been identified that are critical for full development of the tympanic ring. First, *Gsc* encodes a paired homeobox protein that acts as a transcription factor in the cells where it is expressed. *Gsc* expression contributes to the elongated, curved portion of the tympanic ring, and thereby is critical for formation of the EAM. Second, NK3 homeobox 2 (*Nkx3.2*) encodes an NK family homeobox protein that is critical for skeletal development. *Nkx3.2* ultimately contributes to middle ear patterning, but its expression overlaps with *Gsc* in the anterior region of the developing tympanic ring (Tucker, Watson, Lettice, Yamada, & Hill, 2004). This anterior portion will form the gonial bone, which ultimately becomes incorporated with the malleus as the anterior process (Rodriguez Vázquez, Mérida Velasco, & Jiménez Collado, 1991). Interestingly, the gonial bone forms by intramembranous ossification whereas the remainder of the malleus is formed by endochondral ossification. Mutations in other genes also can affect the tympanic ring (see Table 5–1). For example, mutations in paired related homeobox gene 1 (*Prrx1*) and Endothelin-1 (*Edn1*) also result in an absent tympanic ring, but cause a number of other craniofacial skeletal abnormalities (Mallo & Gridley, 1996). Therefore, the roles for *Prrx1* and *Edn1* in tympanic ring formation may be secondary to their broader roles in skeletogenesis. *Edn1* is thought to contribute to epithelial mesenchymal interactions, and along with *Fgf8*, may induce *Gsc* and *Prrx1* (Mallo, 2001). As described above, *Gsc* is critical for tympanic ring formation. *Prrx1* appears to be a critical factor in formation of mesenchymal condensations (Martin et al., 1995), and the size and growth of preskeletal condensations can have a major impact on bone development.

Another important event during EAM development is the appearance of the meatal plug (Anson et al., 1955; Nishimura & Kumoi, 1992). The first pharyngeal groove invaginates between weeks 5 to 6, and as the tympanic ring forms, the ectoderm of the invaginating EAM becomes "associated" with the tympanic ring. The medial surface of the ectoderm will flatten at the level of the tympanic ring and briefly appose the endoderm of the protruding first pharyngeal pouch (see Figure 5–3). The mesoderm, however, soon thickens between the ectoderm-endoderm boundaries spreading the two germ layers apart. Indeed, the three germ layers at the medial end of the developing EAM eventually will form the tympanic membrane.

The medial extremity produced by the inward elongation of the EAM becomes completely filled with proliferating ectodermal cells around 8 to 9 weeks forming the meatal plug (see Figure 5–3; Nishimura & Kumoi, 1992). At week 13, the meatal plug shows some evidence of cell death at the medial most margin leaving an epithelial layer on the outer surface of the developing tympanic membrane (Anson et al., 1955; Nishimura & Kumoi, 1992). The meatal plug will continue to split from a medial to lateral direction opening the lumen of the EAM by week 18. By weeks 21 to 22, the tympanic membrane is formed with an ectodermal outer layer, an endodermal inner layer, and a mesodermal fibrous layer sandwiched between; however, the medial layer is described as still "relatively thick" by Anson et al. (1955, p. 810). Further, Anson et al. reported that the tympanic membrane has not yet become seated in the tympanic ring at 22 weeks, but presumably does so by birth. Nishimura and Kumoi (1992) stated that the tympanic membrane is fully formed at 21 weeks. There also is discrepancy in the literature regarding the age at which the EAM becomes fully patent as some textbooks suggest that patency is achieved at 28 weeks (e.g., Keith, 1921; Gasser, 1975), whereas the published data cited above suggest the lumen of the EAM is open by 22 weeks. The EAM does continue to widen and lengthen during late fetal stages and into childhood. The medial portion of the EAM remains cartilaginous until the tympanic ring fully ossifies around 7 to 10 years of age.

The development of the lateral wall of the tympanic membrane is influenced by the developing tympanic ring, whereas the medial wall appears to develop passively via expansion of the first pharyngeal pouch (Mallo, 1997, 1998, 2001). The exact mechanism(s) by which the tympanic ring or EAM controls or influences the developing tympanic membrane remain to be discovered. Mallo (1998) suggested that the tympanic ring pulls the epithelium of the EAM towards it shaping the medial surface of the invaginating meatus while pulling it to its appropriate position in the developing ear. The tympanic ring and EAM also influence the development of the manubrium of the malleus.

OSSICLES

Two theories exist with regard to the developmental origins of the ossicles. One theory proposes that the malleus and incus develop from neural crest mesenchyme in the first pharyngeal arch and the stapes develops from neural crest in the second pharyngeal arch. The second theory suggests that the head of the malleus and the body of the incus are derived from the first arch whereas the remaining portions of the two ossicles are derived from the second arch. Figure 5–4 illustrates the two theories. General embryology texts and much of the scientific literature appear to support the first theory. Rodriquez, Shah, and Kenna (2007) support the latter theory. Whyte et al. (2009) provide a succinct review of the controversy and presented some evidence for dual origins for the malleus and incus. Whyte et al. reported that prior to embryonic week 8, mesenchyme separates the manubrium from the head of the malleus, and separates the long process of the incus from its body. By week 8, the two portions of each ossicle have fused. Although no data were shown, the investigators stated that the separate portions of the malleus and incus arose from the first and second pharyngeal arches.

Regardless of the origins, the condensations that will form the ossicles first appear during week 5 at the proximal end of Meckel's cartilage and near Reichert's

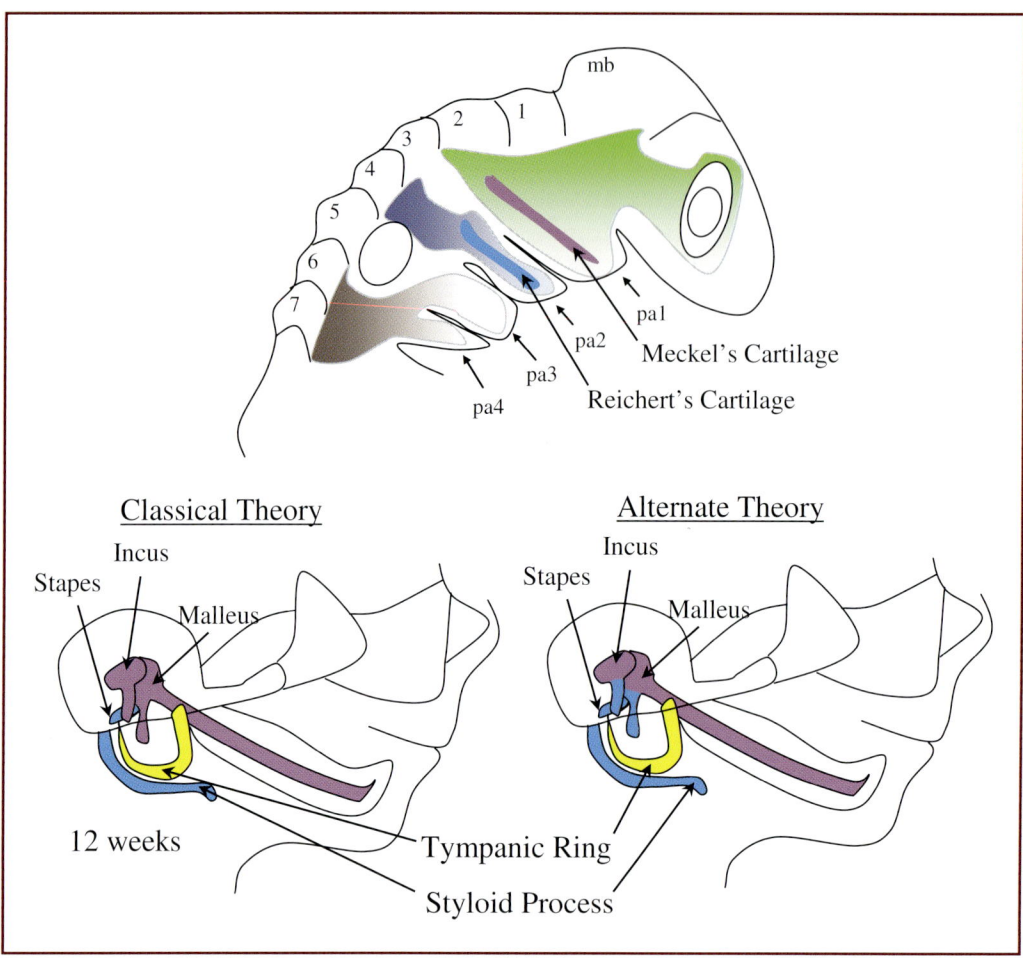

Figure 5–4. *Schematic representation of two theories for the origins of the middle ear ossicles. The classical theory describes the malleus and incus as derivatives of Meckel's cartilage (shown in purple) from pharyngeal arch 1 (pa1), while the stapes is derived from pharyngeal arch 2 (pa2). The alternative theory contends that there are dual origins for the malleus and incus with portions being derived from both the first and second pharyngeal arches.*

cartilage (see Figure 5–3). The condensation associated with Meckel's cartilage will separate soon after it is formed, and the two condensations ultimately will develop into the malleus and incus. The stapes develops from the second arch condensation; however, there is some thought that only the crua and head of the stapes develop from the second arch whereas the footplate develops from the mesenchyme of the otic capsule (e.g., Rodriquez et al., 2007). Rodriguez Vazquez (2005) followed stapes development in the human embryo from 32 to 56 days and concluded that the stapes was completely derived from the second pharyngeal arch. Whyte et al. (2009) concurred with this conclusion.

By weeks 7 to 8, the ossicles become cartilaginous. Around 15 weeks, the malleus and incus are adult size and begin to ossify. By 20 to 26 weeks, the ossification process is complete although portions of the ossicles are still surrounded by mesenchyme. The manubrium of the malleus will be embedded in the layers of the tympanic membrane by 17 weeks (Anson et al., 1955). Several studies suggest that development of the EAM and manubrium may be coordinated. In experimental animals where the EAM was absent, the manubrium was missing or underdeveloped (Mallo et al., 2000; Martin et al., 1995; Yamada et al., 1995). The same consequence may occur in humans. Ishimoto, Ito, Kondo, Yamasoba, and Kaga (2004) evaluated the status of manubrium development in 47 clinical cases of atresia and 8 cases of stenosis. They reported that the manubrium was missing in all cases of atresia, but was present in those with stenosis where an abnormal, but patent EAM is present. In animals, the manubrium also will develop normally, even when large portions of the malleus are malformed or absent, if the EAM is present (Mallo, 1997; Mallo et al., 2000). Mallo (2001) suggested that the EAM may induce manubrium development.

Initially, the head of the malleus and body of the incus are fused, but separate as the incudomallear joint forms. Morphological features of the incudomallear joint first appear during weeks 11 to 12 (Amin & Tucker, 2005). Early gene markers for joint development include *Gdf5* and *Bapx1 (Nkx3.2)* (Amin & Tucker, 2005; Tucker et al., 2004). The joint then develops along with the ossicles and becomes fully formed by 20 to 26 weeks (Whyte et al. 2002).

At the initial stages of ossicular development, the stapes is ring shaped, but will become more stirrup shaped around week 10 and begin ossifying around week 18. The footplate is ossified by week 26 except for a portion facing the oval window. The head and crura of the stapes are nearly ossified by week 29. The incudostapedial joint develops over this same time course to 29 weeks. The articulation of the stapes footplate with the oval window appears to develop by week 12, at which time the annular ligament is differentiated (Whyte et al., 2002).

Many genes can affect development of the middle ear and ossicles, several of which are listed in Table 5–1. Genes required for skeletogenesis (listed in Tables 5–1 and 5–2), neural crest formation, and migration can impact middle ear development along with other skeletal elements of the face and cranium. Mallo (1998, 2001) and Fekete (1999) have reviewed a number of these "general" genes as well as those more specific to middle ear components. In general, genes expressed in the first two pharyngeal arches and the neural crest streams contributing to these arches play important roles in patterning the middle ear. Outer and middle ear phenotypes resulting from genetic mutations may vary widely, which may be due to redundancy in the system (one or more genes assume the role of nonfunctional genes) or secondary consequences (altered expression of one gene affects expression of other genes). The complex molecular requirements including spatiotemporal sequence of gene expression and gene interactions for normal outer and middle ear development are just beginning to be elucidated.

MUSCLES, LIGAMENTS, AND NEURAL ELEMENTS

The muscles of the middle ear are thought to be derived from the first and second arches. The tensor tympani develops from the first pharyngeal arch, whereas the stapedius muscle develops from the second arch. However, Rodriguez Vaszquez (2009), and Rodriguez Vazquez, Mérida

Velasco, and Verdugo Lopez (2010) contend that the stapedius muscle develops from two independent origins, one for the tendon (interhyale) and another for the belly of the muscle (second arch). The stapedius muscle is visible by weeks 10 to 11 and is continuous with the tendon, which is attached to the head of the cartilaginous stapes. At weeks 12 to 14, the mesenchyme condenses to form the pyramidal eminence (bony channel on the posterior wall of the middle ear that surrounds the stapedius muscle), and between weeks 15 to 17 the pyramidal eminence differentiates.

The ligaments that ultimately suspend the ossicles in the middle ear space first appear at the end of the third month but are embedded in the mesenchyme that surrounds the ossicles. Fibrils will appear during the fourth month and by the seventh month, collagenous fibers are visible. At birth the ligaments appear adult-like (Hartwein & Rauchfuss, 1987). The anterior ligament attached to the malleus is a remnant of Meckel's cartilage that was connected to the malleus during development (Anson et al., 1955; Cesarani, Tombolini, Fagnani, & Domenach Matau, 1991).

The nerves that innervate the tensor tympani and stapedius muscles develop from the first and second pharyngeal arches, respectively. Ultimately, the tensor tympani muscle is innervated by the trigeminal nerve and the stapedius by the facial nerve.

TYMPANIC CAVITY/EUSTACHIAN TUBE

The tympanic cavity and the Eustachian tube develop from the first pharyngeal pouch (see Figure 5–3). The pouch will begin protruding toward the ectoderm of the first pharyngeal groove around 3 to 4 weeks. As the pouch expands, it will form a portion of the tympanic cavity and by week 22 the terminal end of the pouch will form the inner layer of the tympanic membrane. The remainder of the tympanic cavity will be expanded via resorption of the mesenchyme. Therefore, two processes aerate the middle ear cavity: (1) expansion of the first pharyngeal pouch; and (2) mesenchyme resorption. The epitympanum does not clear until late fetal ages (34+ weeks) along with the mastoid antrum right before birth. As the tympanic cavity becomes clear of mesenchyme, endoderm will cover all surfaces in the middle ear including the mastoid air cells once they become pneumatized during childhood.

The Eustachian tube forms over the same time course as the tympanic cavity (reviewed by Proctor, 1967). From weeks 7 to 9, the second pharyngeal arch grows constricting the first pharyngeal pouch. The region lateral to the constriction will form the tympanic cavity while the medial portion will become the Eustachian tube. During weeks 16 to 20, the mesoderm surrounding the tube will form cartilage. As the petromastoid and squamous portions of the temporal bone ossify during the seventh month, the bony portion of the Eustachian tube also will ossify.

CONCLUSIONS

The outer ear and middle ear develop from the first and second pharyngeal arches. The time course for development extends over several weeks with some components continuing to develop postnatally. The major elements of the outer and middle ear (EAM, tympanic membrane, and ossicles) appear to be sufficiently developed to transmit sound during the third trimester of development. Development is coordinated among various structures (e.g., EAM and manubrium development) in a spatiotemporal manner. The number of studies regarding the genetics of outer and middle ear development appears to lag behind genetic studies of inner ear development (Chapter 6); however, molecular studies are advancing in this area.

GENERAL RESOURCES

Carlson, B. M. (2004). *Human embryology and developmental biology* (3rd ed.). Philadelphia, PA: Mosby.

Gasser, R. F. (1975). *Atlas of human embryos*. New York, NY: Harper and Row.

Ham, A. W. (1974). *Histology* (7th ed.) Philadelphia, PA: J. B. Lippincott.

Keibel, F., & Mall, F. P. (Eds.). (1910). *Manual of human embryology* (Vol. 1). Philadelphia, PA: J. B. Lippincott.

Keith, A. (1921). *Human embryology and morphology*. London, UK: Edward Arnold.

Moore, K. L., & Persaud, T. V. N. (2003). *The developing human: Clinically oriented embryology* (7th ed.) Philadelphia, PA: Saunders.

Standring, S. (Ed.). (2005). *Gray's anatomy: The anatomical basis of clinical practice* (39th ed.) London, UK: Churchill Livingstone.

LITERATURE CITED

Abdelhak, S., Kalatzis, V., Heilig, R., Compain, S., Samson, D., Vincent, C., . . . Petit, C. (1997). A human homologue of the Drosophila eyes absent gene underlies branchio-oto-renal (BOR) syndrome and identifies a novel gene family. *Nature Genetics, 15,* 157–164.

Akarsu, A. N., Stoilov, I., Yilmaz, E., Sayli, B. S., & Sarfarazi, M. (1996). Genomic structure of HOXD13 gene: A nine polyalanine duplication causes synpolydactyly in two unrelated families. *Human Molecular Genetics, 5,* 945–952.

Alasti, F., Sadeghi, A., Sanati, M. H., Farhadi, M., Stollar, E., Somers, T., & Van Camp, G. (2008). A mutation in HOXA2 is responsible for autosomal-recessive microtia in an Iranian family. *American Journal of Human Genetics, 82,* 982–991.

Alasti, F., & Van Camp, G. (2009). Genetics of microtia and associated syndromes. *Journal of Medical Genetics, 46,* 361–369.

Amin, S., & Tucker, A. S. (2006). Joint formation in the middle ear: Lessons from the mouse and guinea pig. *Developmental Dynamics, 235,* 1326–1333.

Anson, B. J., Bast, T. H., & Richany, S. R. (1955). The fetal and early postnatal development of the tympanic ring and related structures in man. *Annals of Otology, Rhinology and Laryngology, 64*(3), 802–823.

Arnold, J. S., Braunstein, E. M., Ohyama, T., Groves, A. K., Adams, J. C., Brown, M. C., & Morrow, B. El. (2006). Tissue specific roles of Tbx1 in the development of the outer, middle, and inner ear, defective in 22q11DS patients. *Human Molecular Genetics, 15*(10), 1629–1639.

Bakrania, P., Efthymiou, M., Klein, J. C., Salt, A., Bunyan, D. J., Wyatt, A., . . . Ragge, N. K. (2008). Mutations in BMP4 cause eye, brain, and digit developmental anomalies: Overlap between the BMP4 and hedgehog signaling pathways. *American Journal of Human Genetics, 82,* 304–319.

Bardeen, C. R. (1910). The development of the skeleton and of the connective tissues. In F. Keibel & F. P. Mall (Eds.), *Manual of human embryology* (Vol. 1, pp. 318–453). Philadelphia, PA: J. B. Lippincott.

Brown, D. J., Kim, T. B., Petty, E. M., Downs, C. A., Martin, D. M., Strouse, P. J., . . . Lesperance, M. M. (2002). Autosomal dominant stapes ankylosis with broad thumbs and toes, hyperopia, and skeletal anomalies is caused by heterozygous nonsense and frameshift mutations in NOG, the gene encoding noggin. *American Journal of Human Genetics, 71,* 618–624.

Castilla, E. E., & Orioli, I. M. (1986). Prevalence rates of microtia in South America. *International Journal of Epidemiology, 15,* 364–368.

Castriota-Scanderbeg, A. (2005). Skull. In A. Castriota-Scanderbeg & D. Dallipicolla (Eds.), *Abnormal skeletal phenotypes: From simple signs to complex diagnoses* (pp. 3–109). Berlin, Germany: Springer-Verlag.

Cesarani, A., Tombolini, A., Fagnani, E., & Domenach Matau, J. M. (1991). The anterior ligament of the human malleus. *Acta Anatomica, 142,* 313–316.

Cinamon, U. (2009). The growth rate and size of the mastoid air cell system and mastoid bone: A review and a reference. *European Archives of Otorhinolaryngology, 266,* 781–786.

Coburne, M. T. (2000). Construction for the modern head: Current concepts in cranio-facial development. *Journal of Orthodontics, 27,* 307–314.

Cohen, M. M. (2006). The new bone biology: Pathologic, molecular, and clinical correlates. *American Journal of Medical Genetics Part A, 140A,* 2646–2706.

Falardeau, J., Chung, W. C. J., Beenken, A., Raivio, T., Plummer, L., Sidis, Y., . . . Pitteloud, N. (2008). Decreased FGF8 signaling causes deficiency of gonadotropin-releasing hormone in humans and mice. *Journal of Clinical Investigation, 118,* 2822–2831.

Fang, J., Dagenais, S. L., Erickson, R. P., Arlt, M. F., Glynn, M. W., Gorski, J. L., . . . Glover, T. W. (2000). Mutations in FOXC2 (MFH-1), a forkhead family transcription factor, are responsible for the hereditary lymphedema-distichiasis syndrome. *American Journal of Human Genetics, 67,* 1382–1388.

Fekete, D. M. (1999). Development of the vertebrate ear: Insights from knockouts and mutants. *Trends in Neuroscience, 22,* 263–269.

Foster, J. W., Dominguez-Steglich, M. A., Guioli, S., Kwok, C., Weller, P. A., . . . Goodfellow, P. N. (1994). Campomelic dysplasia and autosomal sex reversal caused by mutations in an SRY-related gene. *Nature, 372,* 525–530.

Gasser, R. F. (1975). *Atlas of human embryos.* New York, NY: Harper and Row.

Gong, Y., Krakow, D., Marcelino, J., Wilkin, D., Chitayat, D., Babul-Hirji, R., . . . Warrman, M. L.. (1999). Heterozygous mutations in the gene encoding noggin affect human joint morphogenesis. *Nature Genetics, 21,* 302–304.

Hall, B. K., & Miyake, T. (2000). All for one and one for all: Condensations and the initiation of skeletal development. *BioEssays, 22,* 138–147.

Harris, J., Källén, B., & Robert, E. (1996). The epidemiology of anotia and microtia. *Journal of Medical Genetics, 33,* 809–813.

Hartwein, J. H. J., & Rauchfuss, A. (1987). The development of the ossicular ligaments in the human middle ear. *Archives of Otorhinolaryngology, 244,* 23–25.

Hellemans, J., Simon, M., Dheedene, A., Alanay, Y., Mihci, E., Rifai, L., . . . Mortier, G. (2009). Homozygous inactivating mutations in the NKX3-2 gene result in spondylo-megaepiphyseal-metaphyseal dysplasia. *American Journal of Human Genetics, 85,* 916–922.

Hofstra, R. M. W., Valdenaire, O., Arch, E., Osinga, J., Kroes, H., Loffler, B.-M., . . . Buys, C. Hl. (1999). A loss-of-function mutation in the endothelin-converting enzyme 1 (ECE-1) associated with Hirschsprung disease, cardiac defects, and autonomic dysfunction. *American Journal of Human Genetics, 64,* 304–308.

Ishimoto, S., Ken, I., Kondo, K., Yamasoba, T., & Kaga, K. (2004). The role of the external auditory canal in the development of the malleal manubrium in humans. *Archives of Otolaryngology-Head and Neck Surgery, 130,* 913–916.

Jabs, E. W., Li, X., Scott, A. F., Meyers, G., Chen, W., Eccles, M., . . . Jaye, M. (1994). Jackson-Weiss and Crouzon syndromes are allelic with mutations in fibroblast growth factor receptor 2. *Nature Genetics, 8,* 275–279.

Janssens, K., Gershoni-Baruch, R., Guanabens, N., Migone, N., Ralston, S., Bonduelle, M., . . . Van Hul, W. (2000). Mutations in the gene encoding the latency-associated peptide of TGF-beta-1 cause Camurati-Engelmann disease. *Nature Genetics, 26,* 273–275.

Jezewski, P. A., Vieira, A. R., Nishimura, C., Ludwig, B., Johnson, M., O'Brien, S. E., . . . Murray, J. C. (2003). Complete sequencing shows a role for MSX1 in non-syndromic cleft lip and palate. *Journal of Medical Genetics, 40,* 399–407.

Jiang, X., Iseki, S., Maxson, R. E., Sucov, H. M., & Morriss-Kay, G. M. (2002). Tissue origins and interactions in the mammalian skull vault. *Developmental Biology, 241,* 106–116.

Jumlongras, D., Bei, M., Stimson, J. M., Wang, W.-F., DePalma, S. R., Seidman, C. E., . . . Olsen, B. R. (2001). A nonsense mutation in MSX1 causes Witkop syndrome. *American Journal of Human Genetics, 69,* 67–74.

Keith, A. (1921). *Human embryology and morphology.* London, UK: Edward Arnold.

Kinoshita, A., Saito, T., Tomita, H., Makita, Y., Yoshida, K., Ghadami, M., . . . Yoshiura, K. (2000). Domain-specific mutations in TGFB1 result in Camurati-Engelmann disease. *Nature Genetics, 26,* 19–20.

Klockars, T., Suutarla, S., Kenala, E., Ala-Mello, S., & Rautio, J. (2007). Inheritance of microtia in the Finnish population. *International Journal of Pediatric Otorhinolaryngology, 71,* 1783–1788.

Lambert, P. R., & Dodson, E. E. (1996). Congenital malformations of the external auditory canal. *Otolaryngology Clinics of North America, 29,* 741–760.

Lewis W. H. (1920). The cartilaginous skull of a human embryo twenty one millimeters in length. *Contributions to embryology* (Vol. 9, pp. 299–3240. Washington, DC: Carnegie Institution of Washington.

Ma, L., Golden, S., Wu, L., & Maxson, R. (1996). The molecular basis of Boston-type craniosynostosis: The pro148-to-his mutation in the N-terminal arm of the MSX2 homeodomain stabilizes DNA binding without altering nucleotide sequence preferences. *Human Molecular Genetics, 5,* 1915–1920.

Mallo, M. (1997). Retinoic acid disturbs mouse middle ear development in a stage specific fashion. *Developmental Biology, 184,* 175–186.

Mallo, M. (1998). Embryological and genetic aspects of middle ear development. *International Journal of Developmental Biology, 42,* 11–22.

Mallo, M. (2001). Formation of the middle ear: Recent progress on the developmental and molecular mechanisms. *Developmental Biology, 241,* 410–419.

Mallo, M., & Brändlin, I. (1997). Segmental identity can change independently in the hindbrain and rhombencephalic neural crest. *Developmental Dynamics, 210,* 146–156.

Mallo, M., & Gridley, T. (1996). Development of the mammalian ear: Coordinate regulation of formation of the tympanic ring and the external acoustic meatus. *Development 122,* 173–179.

Mallo, M., Schrewe, H., Martin, J. F., Olson, E. N., & Ohnemus, S. (2000). Assembling a functional tympanic membrane: Signals from the external acoustic meatus coordinate development of the malleal manubrium. *Development, 127,* 4127–4136.

Martin, J. F., Bradley, A., & Olson, E. N. (1995). The paired-like homeobox gene Mhox is required for early events of skeletogenesis in multiple lineages. *Genes and Development, 9,* 1237–1249.

McBratney-Owen, B., Iseki, S., Bamforth, S. D., Olsen, B. R., & Morriss-Kay, G. M. (2008). Development and tissue origins of the mammalian cranial base. *Developmental Biology, 322,* 121–132.

McMurrich, J. P. (1923). *The development of the human body: A manual of human embryology.* Philadelphia, PA: P. Blakiston's Son & Co.

Milunsky, J. M., Maher, T. A., Zhao, G., Roberts, A. E., Stalker, H. J., Zori, R. T., . . . Lin, A. E. (2008). TFAP2A mutations result in branchio-oculo-facial syndrome. *American Journal of Human Genetics, 82,* 1171–1177.

Milunsky, J. M., Zhao, G., Maher, T. A., Colby, R., & Everman, D. B. (2006). LADD syndrome is caused by FGF10 mutations. *Clinical Genetics, 69,* 349–354.

Morriss-Kay, G. M. (2001). Derivation of the mammalian skull vault. *Journal of Anatomy, 199,* 143–151.

Morriss-Kay, G. M., & Wilkie, A. O. M. (2005). Growth of the normal skull vault and its alteration in craniosynostosis: Insights from human genetics and experimental studies. *Journal of Anatomy, 207,* 637–653.

Mundlos, S., Otto, F., Mundlos, C., Mulliken, J. B., Aylsworth, A. S., Albright, S., . . . Olsen,

B. R. (1997). Mutations involving the transcription factor CBFA1 cause cleidocranial dysplasia. *Cell, 89,* 773–779.

National Birth Defects Prevention Network. (2009). Population-based birth defects surveillance data from selected states, 2002–2006. *Birth Defects Research (Part A), 85,* 939–1004.

Nemzek, W. R., Brodie, H. A., Chong, B. W., Babcock, C. J., Hecht, S. T., Salamat, S., . . . Seibert, J. A. (1996). Imaging findings of the developing temporal bone in fetal specimens. *American Journal of Neuroradiology, 17,* 1467–1477.

Nishimura, Y., & Kumoi, T. (1992). The embryologic development of the human external auditory meatus. *Acta Otolaryngologica, 112,* 496–503.

Proctor, B. (1967). Embryology and anatomy of the Eustachian tube. *Archives of Otolaryngology, 86,* 503–514.

Rivera-Perez, J. A., Wakamiya, M., & Behringer, R. R. (1999). Goosecoid acts cell autonomously in mesenchyme-derived tissues during craniofacial development. *Development, 126,* 3811–3821.

Rodriquez, K., Shah, R. K., & Kenna, M. (2007). Anomalies of the middle and inner ear. *Otolaryngologic Clinics of North America, 40,* 81–96.

Rodriguez Vázquez, J. F. (2005). Development of the stapes and associated structures in human embryos. *Journal of Anatomy, 207,* 165–173.

Rodriguez Vázquez, J. F. (2009). Development of the stapedius muscle and pyramidal eminence in humans. *Journal of Anatomy, 215,* 292–299.

Rodriguez Vázquez, J. F., Mérida Velasco, J. R., & Jiménez Collado, J. (1991). A study of the os goniale in man. *Acta Anatomica (Basel), 142,* 186–192.

Rodriguez Vázquez, J. F., Mérida Velasco, J. R., & Verdugo Lopez, S. (2010). Development of the stapedius muscled and unilateral agenesia of the tendon of the stapedius muscle in a human fetus. *Anatomical Record, 293,* 25–31.

Rohmann, E., Brunner, H. G., Kayserili, H., Uyguner, O., Nurnberg, G., Lew, E. D., . . . Wollnik, B. (2006). Mutations in different components of FGF signaling in LADD syndrome. *Nature Genetics, 38,* 414–417.

Sanyanusin, P., Schimmenti, L. A., McNoe, L. A., Ward, T. A., Pierpont, M. E. M., Sullivan, M., . . . Eccles, M. R. (1995). Mutation of the PAX2 gene in a family with optic nerve colobomas, renal anomalies and vesicoureteral reflux. *Nature Genetics, 9,* 358–364.

Schorderet, D. F., Nichini, O., Boisset, G., Polok, B., Tiab, L., Mayeur, H., . . . Munier, F. L. (2008). Mutation in the human homeobox gene NKX5-3 causes an oculo-auricular syndrome. *American Journal of Human Genetics, 82,* 1178–1184.

Shiang, R., Thompson, L. M., Zhu, Y.-Z., Church, D. M., Fielder, T. J., Bocian, M., . . . Wasmuth, J. J. (1994). Mutations in the transmembrane domain of FGFR3 cause the most common genetic form of dwarfism, achondroplasia. *Cell, 78,* 335–342.

Standring, S. (Ed.). (2005). *Gray's anatomy: The anatomical basis of clinical practice* (39th ed.) London, UK: Churchill Livingstone.

Stockton, D. W., Das, P., Goldenberg, M., D'Souza, R. N., & Patel, P. I. (2000). Mutation of PAX9 is associated with oligodontia. *Nature Genetics, 24,* 18–19.

Streeter, G. L. (1918). The histogenesis and growth of the otic capsule and its contained periodic tissue spaces in the human embryo. *Contributions to embryology* (Vol. 7, pp. 5–54). Washington, DC: Carnegie Institution of Washington.

Suutarla, S., Rautio, J., Ritvanen, A., Ala-Mello, S., Jero, J., & Klockars, T. (2007). Microtia in Finland: Comparison of characteristics in different populations. *International Journal of Pediatric Otorhinolaryngology, 71,* 1211–1217.

Suzuki, S., Marazita, M. L., Cooper, M. E., Miwa, N., Hing, A., Jugessur, A., . . . Murry, J. C. (2009). Mutations in BMP4 are associated with subepithelial, microform, and overt cleft lip. *American Journal of Human Genetics, 84,* 406–411.

Szabo-Rogers, H. L., Smither, L. E., Yakob, W., & Liu, K. J. (2010). New directions in craniofacial morphogenesis. *Developmental Biology, 341,* 84–94.

Tekin, M., Hismi, B. O., Fitoz, S., Ozdag, H., Cengiz, F. B., Sirmci, A., . . . Akar, N. (2007). Homozygous mutations in fibroblast growth factor 3 are associated with a new form of syndromic deafness characterized by inner ear agenesis, microtia, and microdontia. *American Journal of Human Genetics, 80,* 338–344.

Tekin, M., Ozturkmen, A. H., Fitoz, S., Birnbaum, S., Cengiz, F. B., Sennaroglu, L., . . . Duman, D. (2008). Homozygous FGF3 mutations result in congenital deafness with inner ear agenesis, microtia, and microdontia. *Clinical Genetics, 73,* 554–565.

Tischfield, M. A., Bosley, T. M., Salih, M. A. M., Alorainy, I. A., Sener, E. C., Nester, M. J., . . . Engle, E. C. (2005). Homozygous HOXA1 mutations disrupt human brainstem, inner ear, cardiovascular and cognitive development. *Nature Genetics, 37,* 1035–1037.

Tucker, A. S., Watson, R. P., Lettice, L. A., Yamada, G., Hill, R. E. (2004). Bapx1 regulates patterning in the middle ear: Altered regulatory role in the transition from the proximal jaw during vertebrate evolution. *Development, 131,* 1235–1245.

Vastardis, H., Karimbux, N., Guthua, S. W., Seidman, J. G., & Seidman, C. E. (1996). A human MSX1 homeodomain missense mutation causes evidence tooth agenesis. *Nature Genetics, 13,* 417–421.

Whyte, J., Cisneros, A., Jus, C., Fraile, J., Obon, J., & Vera, A. (2009). Tympanic ossicles and pharyngeal arches. *Anatomia Histologia Embryologia, 38,* 31–33.

Whyte, J. R., Gonzalez, L., Cisneros, A. I., Jus, C., Torres, A., & Sarrat, R. (2002). Fetal development of the human tympanic ossicular chain articulations. *Cells Tissues Organs, 171,* 241–249.

Wilkie, A. O. M., & Morriss-Kay, G. M. (2001). Genetics of craniofacial development and malformation. *Nature Reviews Genetics, 2,* 458–468.

Wilkie, A. O. M., Slaney, S. F., Oldridge, M., Poole, M. D., Ashworth, G. J., Hockley, A. D., et al. (1995). Apert syndrome results from localized mutations of FGFR2 and is allelic with Crouzon syndrome. *Nature Genetics, 9,* 165–172.

Wilkie, A. O. M., Tang, Z., Elanko, N., Walsh, S., Twigg, S. R. F., Hurst, J. A., . . . Maxson, Re. E. Jr. (2000). Functional haploinsufficiency of the human homeobox gene MSX2 causes defects in skull ossification. *Nature Genetics, 24,* 387–390.

Yagi, H., Furutani, Y., Hamada, H., Sasaki, T., Asakawa, S., Minoshima, S., . . . Matsuoka, R. (2003). Role of TBX1 in human del22q11.2 syndrome. *Lancet, 362,* 1366–1373.

Yamada, G., Mansouri, A., Torres, M., Stuart, E. T., Blum, M., Schultz, M., . . . Gruss, P. (1995). Targeted mutation of the murine goosecoid gene results in craniofacial defects and neonatal death. *Development, 121*(9), 2917–2922.

Zhang, Q., Zhang, J., & Yin, W. (2010). Pedigree and genetic study of a bilateral, congenital, microtia family. *Plastic and Reconstructive Surgery, 125*(3), 979–987.

Embryogenesis of the Inner Ear

> ### KEY POINTS
>
> 1. The inner ear will develop from the otic placode, which appears during week three postfertilization.
> 2. The otic placode and eventual otocyst are influenced by several molecular pathways that pattern the otocyst into various regions of genetic expression (e.g., distinguishing dorsal from ventral, anterior from posterior, medial from lateral).
> 3. Semicircular canals and cristae will develop from dorsal regions of the otocyst, macular structures at or near the dorsal-ventral boundary, and cochlear structures from ventral regions.
> 4. In general, development of the inner ear proceeds from dorsal to ventral, and as such, vestibular development precedes cochlear development. The semicircular canals begin to form during week 5, the maculae during week 6, and the cochlea reaches one half turn during week 6.
> 5. Cochlear development generally proceeds from midbasal regions to the base and apex and from inner hair cell regions to outer hair cell regions.
> 6. An orchestrated sequence of gene expression is required during inner ear development to specify neural ganglion and sensory patches, differentiate hair cells and supporting structures, and initiate and maintain inner ear homeostasis.

INTRODUCTION

Much of the knowledge about the genes and molecular mechanisms that contribute to inner ear development is derived from studies using animal models. The number of identified genes has increased substantially over the past several years and will likely continue to increase. In addition, the complex interactions among the molecules and signaling pathways are also being elucidated but are not yet fully understood. The following discussion strives to summarize major events and the molecular processes underlying those events; however, some of the mechanisms described here may be revised as research progresses.

Most of the information presented in this chapter is based on studies using chick or mouse models, and the developmental ages reflect mouse embryonic age in days. Human developmental ages in weeks postfertilization are presented in parentheses based on Bredberg (1968), Deschesne (1992), Pujol, Lavigne-Rebillard, and Lenoir (1998), Jeffery and Spoor (2004), Sans and Deschesne (1985), and Sulik and Cotanche (2004). Tables 6–1 through 6–9 display a time line for human developmental events in weeks postfertilization, list some of the important genes that likely contribute to each developmental event, and identify human disorders associated with genetic mutations. Many of the disorders listed in the tables include hearing loss as part of the phenotype. Although the discussion and tables are not exhaustive for every possible gene that may be expressed or signaling pathway that may be involved during inner ear development, they do highlight a large number of genes as does the discussion below. Guidelines for gene nomenclature in mice have been followed as carefully as possible. Therefore, genes are written in italicized text with the first letter capitalized. When not italicized, the term is referring to the protein product that would be used by the cell for development. Finally, Table 6–10 highlights some of the germ layer derivatives for the inner ear structures.

FORMATION OF THE OTIC PLACODE AND OTIC VESICLE

The inner ear is formed from primordial ectodermal cells near rhombomeres five and six in the embryonic hindbrain. These cells form a flat, thickened patch of ectoderm called the otic placode (Figure 6–1) on embryonic day 8.5 (E8.5) for the mouse and during week 3 for humans. The otic placode appears morphologically as the first step in the formation of the inner ear; however, its appearance is preceded by molecular inducing signals arising from the mesoderm, hindbrain, or both. Inducing signals include fibroblast growth factors (FGF), bone morphogenetic proteins (BMP), and wingless (Wnt) signaling pathways (antagonist for BMP signals) to form a preplacodal field. In other words, FGFs, BMPs, and Wnt signaling induce cells of the preplacodal region to express specific genes. Recall from Chapter 4 that the preplacodal field expresses several genes including members of *Dlx, Six, Eya, Bmp,* and *Foxi* families. Subsequent expression of *Pax2* and perhaps *Pax8* defines a region of tissue committed to form the otic placode, and eventual Wnt signaling separates the otic placode from epidermal tissue (reviewed by Ohyama, Groves, & Martin, 2007). According to Groves (2005), the cells ultimately forming the placode are not simply gathered from adjacent cells, but rather it appears likely that the cells migrate from a wide area and somehow collect to form the placode.

Table 6–1. Timeline of Major Events for Human Inner Ear Development. Age approximations are from published data on human specimens (Bredberg, 1968; Deschesne, 1992; Jeffery & Spoor, 2004; Pujol, Lavigne-Rebillard, & Lenoir, 1998; Sans & Deschesne. 1985; Sulik & Cotanche, 2004). Youngest ages represent onset of morphological appearance or expression of distinct genetic markers. Ages for development of stria vascularis, basilar and Reissner's membranes are unknown but likely parallel cochlear development. Blue represents undifferentiated regions or cristae development, red represents the macular regions, and gray represents the cochlea.

Event	Age in Weeks
	1 2 3 4 5 6 7 8 9 10 11 12 13 14 15 16 17 18 19 20 21 22 23 24 25 26 27 28 29 30 31 32 33 34 35 36
Placode	
Otic Vesicle	
Delamination	
Labyrinth	
Canals	
Vestibule	
Cochlea	
Prosensory Patches	
Cristae	
Maculae	
Cochlea	
Membranes	
Cupula	
Otoconial (+)	
Tectorial	

continues

Table 6–1. *continued*

Event	Age in Weeks

Event	1	2	3	4	5	6	7	8	9	10	11	12	13	14	15	16	17	18	19	20	21	22	23	24	25	26	27	28	29	30	31	32	33	34	35	36
Hair Cells Differentiate																																				
Cristae							▓	▓	▓	▓	▓	▓	▓	▓																						
Maculae							▓	▓	▓	▓	▓	▓	▓	▓																						
Cochlea																																				
IHC											▓	▓	▓																							
OHC								▓	▓	▓	▓	▓	▓	▓																						
Stereocilia																																				
Cristae								▓	▓	▓	▓	▓	▓	▓	▓																					
Maculae										▓	▓	▓	▓	▓	▓																					
Cochlea																																				
IHC																				▓	▓	▓	▓													
OHC																						▓	▓													
Neurons in Epithelia																																				
Cristae						▓	▓																													
Maculae						▓	▓																													
Cochlea							▓																													
Afferent Synapses																																				
Cristae							▓	▓	▓	▓	▓	▓	▓	▓	▓	▓	▓	▓	▓	▓	▓	▓	+													
Maculae							▓	▓	▓	▓	▓	▓	▓	▓	▓	▓	▓	▓	▓	▓	▓	▓	+													
Cochlea																																				
IHC									▓	▓	▓	▓	▓	▓	▓	▓	▓	▓	▓	▓	▓	▓	▓	▓	+											
OHC									▓	▓	▓	▓	▓	▓	▓	▓	▓	▓	▓	▓	▓	▓	▓	▓	+											

Table 6–1. *continued*

Event	1	2	3	4	5	6	7	8	9	10	11	12	13	14	15	16	17	18	19	20	21	22	23	24	25	26	27	28	29	30	31	32	33	34	35	36
Efferent Synapses																																				
Cristae									▓	▓	▓	▓	▓	▓	▓	▓	▓	▓	▓	▓	▓	▓	▓	▓	+											
Maculae									▓	▓	▓	▓	▓	▓	▓	▓	▓	▓	▓	▓	▓	▓	▓	▓	+											
Cochlea																																				
IHC																							+													
OHC																														+						
Nonsensory Structures																																				
Pillar Cells																								▓	▓	▓	▓	▓	▓	▓	▓	▓	▓	▓	▓	▓
T. of Corti									▓	▓	▓	▓	▓	▓	▓	▓	▓	▓	▓	▓	▓	▓	▓	▓	▓	▓	▓	▓	▓	▓	▓	▓	▓	▓	▓	▓
Sp. of Nuel									▓	▓	▓	▓	▓	▓	▓	▓	▓	▓	▓	▓	▓	▓	▓	▓	▓	▓	▓	▓	▓	▓	▓					
ISC																				▓	▓	▓	▓	▓	▓	▓	▓	▓	▓	▓	▓					
SV																																				
RM																																				
BM																																				

IHC = inner hair cell; OHC = outer hair cell; T. of Corti = tunnel of Corti; Sp. of Nuel = space of Nuel; ISC = inner sulcus cells; SV = stria vascularis; RM = Reissner's membrane; BM = basilar membrane

"+" indicates that refinements likely occur for some time after the period indicated by the bar

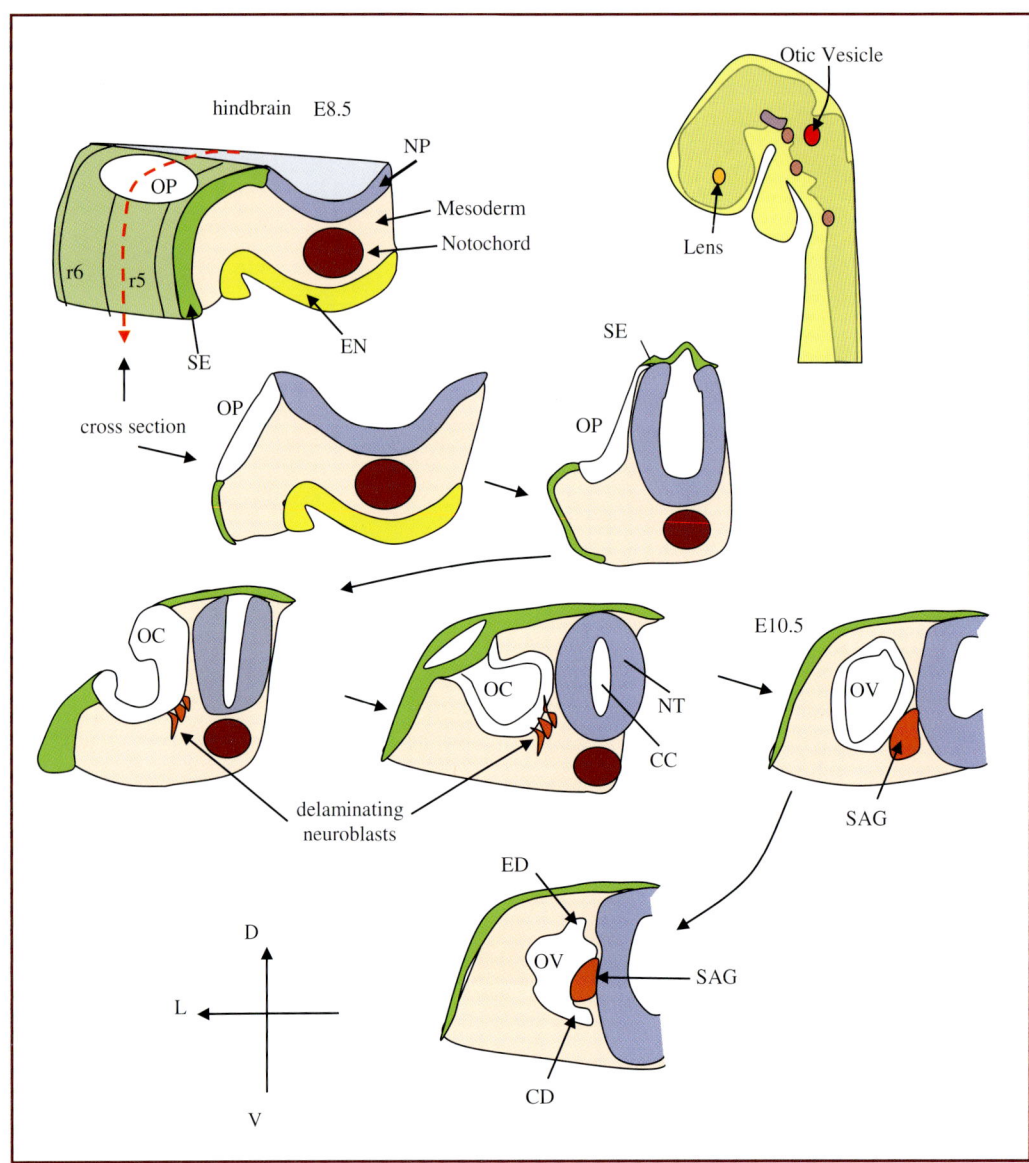

Figure 6–1. Formation of the otic vesicle from the otic placode over a period from approximately E8.5 to just beyond E10.5 in mouse. ED = endolymphatic duct; EN = endoderm; CC = central canal; CD = cochlear duct; D = dorsal; L = lateral; Mesoderm = mesenchyme presumptive mesoderm; NP = neural plate; NT = neural tube; OC = otic cup; OP = otic placode; OV = otic vesicle; r5-r6 = rhombomeres 5 and 6; SAG = statoacoustic ganglion; SE = surface ectoderm (epidermis); V = ventral.

Cells of the otic placode give rise to the otic vesicle (also called otocyst) at E10.5 for the mouse (week 4 for human) from which sensory, nonsensory, and most neural cells of the inner ear will be derived (see Figure 6–1). The otic vesicle, however, does not give rise to olivocochlear efferents or autonomic innervation of blood vessels. Shortly after forming, the otic placode begins to bend and fold inward producing a shallow dimple. This invagination of ectoderm continues until a deep pit is formed, which is called the otic pit or otic cup (see Figure 6–1). The dorsal edge of the otic tissue remains in close approximation to the hindbrain/neural plate during this process. This association favors molecular signaling and formation of asymmetric molecular signal fields. Ultimately, the otic cup closes to form the otic vesicle. Closure of the otic cup and the proximal neural groove (thus forming the neural tube locally) occurs at about the same time during week 4.

During and after its formation, the otic vesicle is influenced by signals from the hindbrain as well as the mesenchyme surrounding the otic vesicle (known as the periotic mesenchyme or mesoderm) that becomes the bony labyrinth of the inner ear. Several important signaling molecules play a dominant role in establishing the dorsal-ventral morphological axis within the otocyst: (1) sonic hedgehog (*Shh*), required for formation of the ventral structures such as the cochlea and statoacoustic ganglion as well as restriction of the prosensory domain; (2) *Wnt*, important for both dorsal and ventral structures; and (3) *Fgf*, also important for both dorsal and ventral structures (Figure 6–2).

The dorsal-ventral boundary of the otic vesicle distinguishes the dorsal vestibular sensors from the ventral cochlear sensors. *Shh, Wnt,* and *Fgf* molecular signals are thought to form gradients of influence along the dorsal-ventral extent of otic tissue (Schneider-Maunoury & Pujades, 2007). Tissue elements at any given position experience a unique combination of signal levels from each source. Each combination of signals has the potential to favor one program of development or another. *Wnt* signals arise from

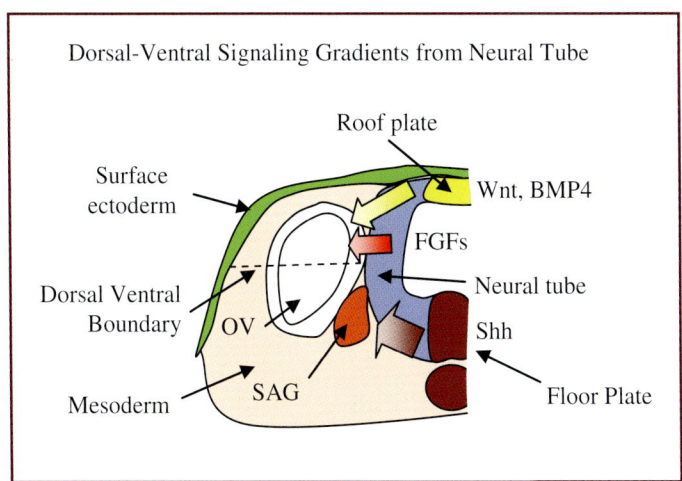

Figure 6–2. *Dorsoventral signaling during the formation of the otic vesicle. OV = otic vesicle; SAG = statoacoustic ganglion; Signaling molecules include FGFs, Shh (ventral), Wnt, and BMPs (dorsal).*

the dorsal neural tube whereas cells expressing *Shh* are located in the floor plate of the neural tube and notochord (see Figure 6–2). *Fgf* signals arise from the neural tube (hindbrain) as well as surrounding mesenchyme and endoderm. Thus, *Wnt* signals maximally affect cells of the dorsal otic cup and vesicle. Some of the dorsal otic cells responding to *Wnt* signals are thought to migrate to ventral regions after *Wnt* exposure. Once they arrive in ventral regions of the otic cup and vesicle, these cells are exposed to *Shh* signaling that gives rise to ventral structures including the statoacoustic ganglion and the cochlea. Another candidate signal includes the *Bmp* family. Bmps are secreted from the roof of the hindbrain and dorsal ectoderm. *Bmp* signals may influence *Shh* and thus play a role in adjusting signal levels in the otic placode and vesicle. These gradients and influences are reminiscent of dorsal-ventral signaling gradients operating generally in the brain and spinal cord as described in Chapter 4. Sine oculus homeobox homolog (*Six1*), a gene expressed in the ventral portion of the otic vesicle, has also been shown to play a major role in patterning the otic vesicle and influencing expression of other genes within the otic vesicle (Ozaki et al., 2004).

Molecular signals responsible for the order and layout of structures along the anteroposterior and mediolateral dimensions of the otic vesicle are less clear. The *Tbx1* gene is expressed within the otocyst and is thought to play some role in anterior-posterior axis determination. However, it is plausible that these axes are also identified initially by hindbrain signals such as *Fgf* and *Wnt*, owing to their early three-dimensional signal gradients in the otic placode and cup (Schneider-Maunoury & Pujades, 2007).

Other genes reported to be expressed in the otic vesicle include orthodenticle 1 (*Otx1*), orthodenticle 2 (*Otx2*), Jagged 1 (*Jag1*), lunatic fringe (*Lfng*), fibroblast growth factor 3 (*Fgf3*), SRY (sex determining region Y)-box 2 (*Sox2*), eyes absent homolog 1 (*Eya1*), sine oculus homeobox homolog 4 (*Six4*), paired box gene 2 (*Pax2*), distal-less homeobox 5 (*Dlx5*), H6 family homeobox 3 (*Hmx3*), dachshund homologs 1 and 2 (*Dach1* and *Dach2*), T-box 1 (*Tbx1*), and retinoic acid enzymes (e.g., Baker & Bronner-Fraser, 2001; Bok, Chang, & Wu, 2007; Cantos, Cole, Acampora, Simeone, & Wu, 2000; Fekete & Wu, 2002; Ozaki et al., 2004; Romand, 2003; Romand, Dollé, & Hashino, 2006;). Some of these molecules play a role in the order and layout of structures along the antero-posterior and mediolateral dimensions of the otic vesicle whereas others regulate epithelial/mesenchymal interactions or contribute to future developmental events (e.g., delamination or formation of the prosensory domain).

DELAMINATION AND FORMATION OF THE STATOACOUSTIC GANGLION

Under the influence of Shh proteins, proneural cells (neuroblasts) detach from and leave the epithelium of the otic cup and vesicle and then migrate medially and ventrally into the mesenchyme ultimately to form the statoacoustic ganglion (Figures 6–1 and 6–3). This process is called delamination; it begins in epithelial regions of the presumptive utricle and cristae as early as E10.5 (week 4 in humans), and represents the beginning of neurogenesis. The epithelial neuroblasts divide and ultimately differentiate into auditory and vestibular neurons. The proneural gene neurogenin 1 (*Ngn1*) is essential for this process (Ma, Anderson, & Fritzsch, 2000). The gene neurogenic differentiation 1

Embryogenesis of the Inner Ear   161

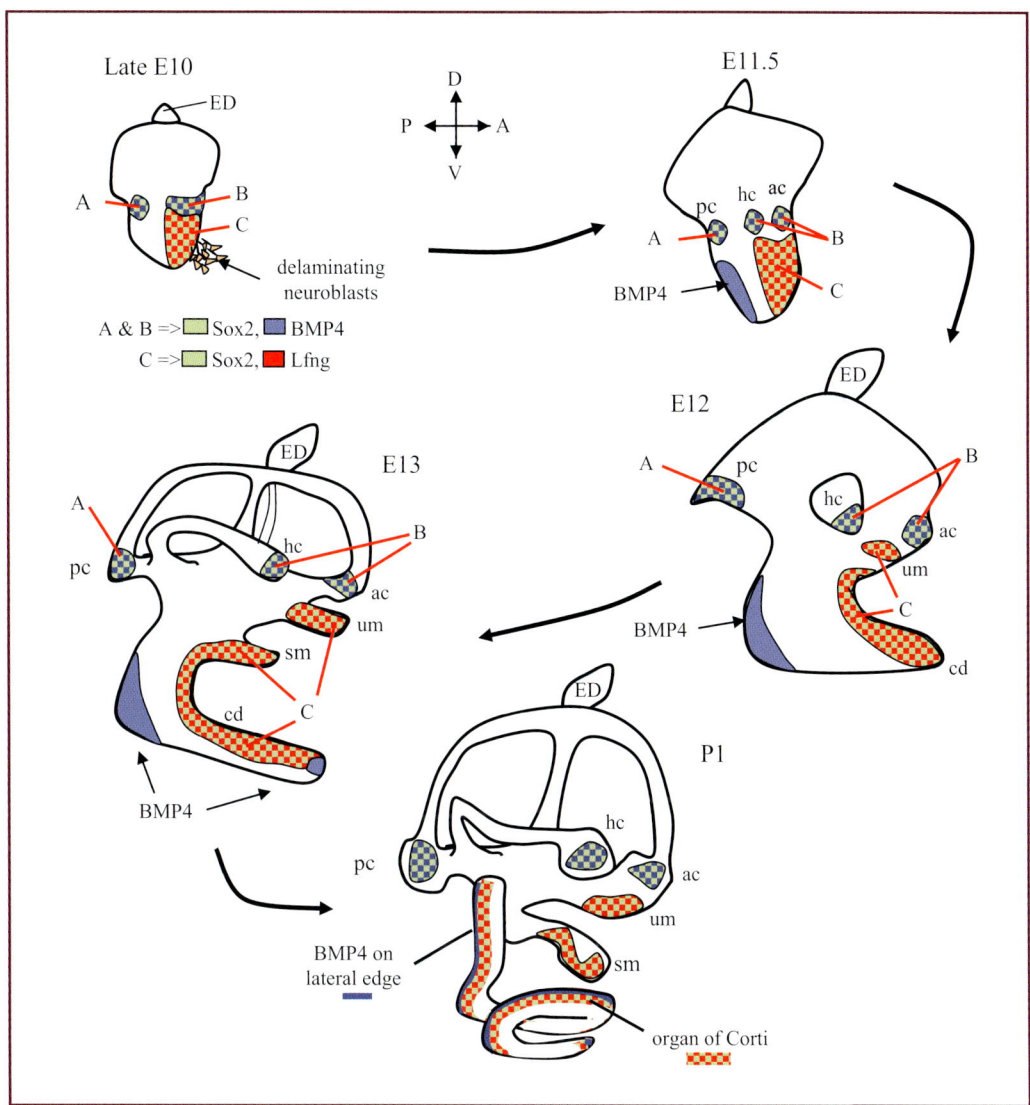

Figure 6–3. *Lateral views of locations for important markers of prosensory and sensory domains during development. Regions forming the most dorsal sensory epithelia (cristae of semicircular canals) are distinguished early by the expression of* Sox2 *and* Bmp4 *(labels A and B), whereas the more ventral regions are marked with* Sox2 *and lunatic fringe (*Lfng*, labeled C). The ventral prosensory domains elaborate the sensory epithelia of the utricular macula (um), saccular macula (sm) and cochlear duct (cd) in that order.* Bmp4 *alone marks nonsensory regions of the cochlear duct (lateral or strial edge). ac = anterior crista; hc = horizontal (lateral) crista; pc = posterior crista; E days are equivalent to days post conception (dpc); P1 = postnatal day 1. Modified from Bok, Chang, and Woo (2007). Reproduced with permission from* The International Journal of Developmental Biology, 51, *526.*

(*NeuroD*) also plays a role and contributes to the survival of neurons formed. Delamination expands to include prosensory regions within or adjacent to the sacculus and eventually the cochlea. Delamination continues through periods as late as E17 (week 9 in humans). Further development of the statoacoustic ganglion is covered later in the chapter.

FORMATION OF PROSENSORY PATCHES

For warm-blooded vertebrates, there are six or seven sensory patches in the mature inner ear (utricle, saccule, cochlea, three cristae, and the macula lagena in birds). This list does not include the very small sensory patch called the crista neglecta, about which little is known (Montandon, Gacek, & Kimura, 1970). The sensory epithelia form in close association with the nonsensory components (i.e., cristae with canals, maculae with vestibule, and cochlea with cochlear duct). There is evidence that molecular signals from the sensory patches guide surrounding nonsensory structures in morphogenesis. For example, *Fgf10* is expressed in the cristae, and its receptor, *Fgfr*, is expressed in surrounding nonsensory canal tissue. Knock out of the *Fgf10* gene that is expressed in the cristae results in malformed cristae and canal structures suggesting cooperation between the developing cristae and the surrounding nonsensory tissue (reviewed by Bok et al., 2007).

Before such cooperative signaling can occur, the proneural and prosensory cells must be specified and distinguished from the surrounding epithelium. These precursor "neurosensory" cells (cells that yet may take on a neural or sensory cell fate) are specified early in the development of the otocyst. A neurosensory domain can be identified at late E10 (week 4) with the expression of *Sox2*, *Bmp4,* and *Lfng* (see Figure 6–3). These genes may be viewed as markers for cells destined to become neurons or sensory cells (Bok et al., 2007; Morsli, Choo, Ryan, Johnson, & Wu, 1998). Inactivation of these genes or of corresponding pathways interferes with the formation of the statoacoustic ganglion and sensory organs. The patterns of expression for these genes at the earliest stages mark the neurosensory domain as illustrated in Figure 6–3. Those cells destined to become neural cells will soon delaminate under the influence of *Ngn1* as described above while the remaining cells will become the sensory epithelia. Changes in expression over time are also shown in Figure 6–3.

Dorsal Sensory Patches

At late E10 (week 4), *Sox2*-positive regions of the otic vesicle outline the neurosensory domain including presumptive cristae, maculae, and cochlear prosensory fields as well as the proneural fields where neuroblasts delaminate (see Figures 6–3A, 6–3B, and 6–3C). This includes a single patch of *Sox2* label at the posterior otic border approximately midway between the dorsal and ventral boundaries of the otocyst (mid-dorsoventral) (see Figure 6–3A) and a wide anteroventral (rostroventral) band running from a mid-dorsoventral line to the ventral otic border (see Figure 6–3C). Superposed on the dorsal most *Sox2* fields is a *Bmp*-positive strip that runs along the anteroposterior axis and includes the single posterior *Sox2* patch (see Figures 6–3A and 6–3B). This *Bmp*-positive domain corresponds to all three presumptive cristae, where the focal patch (A) marks the presumptive posterior cristae. *Lfng*-positive cells are located throughout the anteroventral *Sox2* band ventral to the *Bmp4*-positive domain (see Figure 6–3C). Significantly, *Bmp4* and *Lfng* fields overlap only at the most distal tip of the cochlea.

The patterns mark and distinguish which domains become cristae (*Bmp4* expression) and which become macular and cochlear sensors (*Lfng* expression, *Notch* signaling). At the earliest stages, *Bmp4*, for example, is a marker for presumptive cristae and these labeled epithelial regions appear in the dorsolateral segment of the otocyst (see Figures 6–3A and 6–3B). Two genes (*Dlx5* and *Hmx3*) have been implicated in the formation of canals and are expressed in the same regions. *Wnt* signaling is important in producing a vestibular fate for dorsal regions and it regulates *Dlx* but not *Hmx* in the formation of cristae and canals. As noted above, other genes also play a role in the development of cristae and canals (i.e., *Sox2*, *Jag1*, and *Fgf10*). Interestingly, the gene *Otx1* is expressed only in the presumptive lateral (horizontal) crista and canal, and its expression is required for the formation of the lateral canal.

Ventral Sensory Patches

Anteroventral regions of the otocyst are marked by *Lfng* expression and form in sequence the utricle, then the saccule, and finally the cochlea. As noted above, neuroblasts delaminate from this region and migrate anteriorly to form the statoacoustic ganglion over the period from E10.5 to E17 (weeks 4 to 9). Initially, there is no distinction between presumptive macular and cochlear regions within the *Lfng* field (see Figure 6–3C). However, development proceeds from dorsal to ventral revealing first the more dorsal maculae and then the cochlea. By E12 (week 5), the utricle is clearly segregated and by E13 (late week 5) the saccule is easily recognized, but a clear segregation from cochlear fields is evident only after E13.5 (week 6) (Morsli et al., 1998). Segregation of these macular and cochlear sensory patches requires *Lmx1a*, which is a homeodomain transcription factor of the Islet-Lim family and is expressed in select nonsensory regions of the otic vesicle after E10.5 (week 4) (Nichols et al., 2008). In the absence of *Lmx1a*, the utricle, saccule, and cochlea form a combined malformed single mosaic sensory epithelium. The progressive segregation of sensory epithelia also appears to be linked to the simultaneous delamination process as cells are removed from the prosensory regions to become neurons of the statoacoustic ganglion (Fritzsch et al., 2002).

FORMATION OF THE COCHLEA AND ZONE OF NONPROLIFERATION (ZNP)

The emerging cochlear duct can be seen initially as a lengthening of the ventral pole of the otocyst beginning at about E11 in the mouse (week 5 in humans) (see Figure 6–3). The duct continues to extend first ventromedially, then it turns abruptly, anteriorly forming a curved hook by late E12 (late week 5) (Morsli et al., 1998). This represents initiation of the cochlear coil, and by E13 (week 6) a full half turn is achieved (see Figure 6–3). The cochlea will continue to elongate and increase the number of coils, but let's first look at changes happening inside the cochlear duct. The reader is referred to Figure 6–4A regarding cochlear axes, orientations, and terminology.

At the early stages of cochlea formation (week 5), an asymmetric thickening of the duct wall is present where the floor occupies a large portion of the duct volume (Figure 6–4B). The duct in cross section tends to be ovoid or triangular in shape, and the thick floor region extends side to side and occupies nearly the bottom half of the duct. The floor is composed of undifferentiated epithelial cells, which form a ridge or sheet (four to five cells thick) that extends from the cochlear base to apex. As noted below, the shape of the

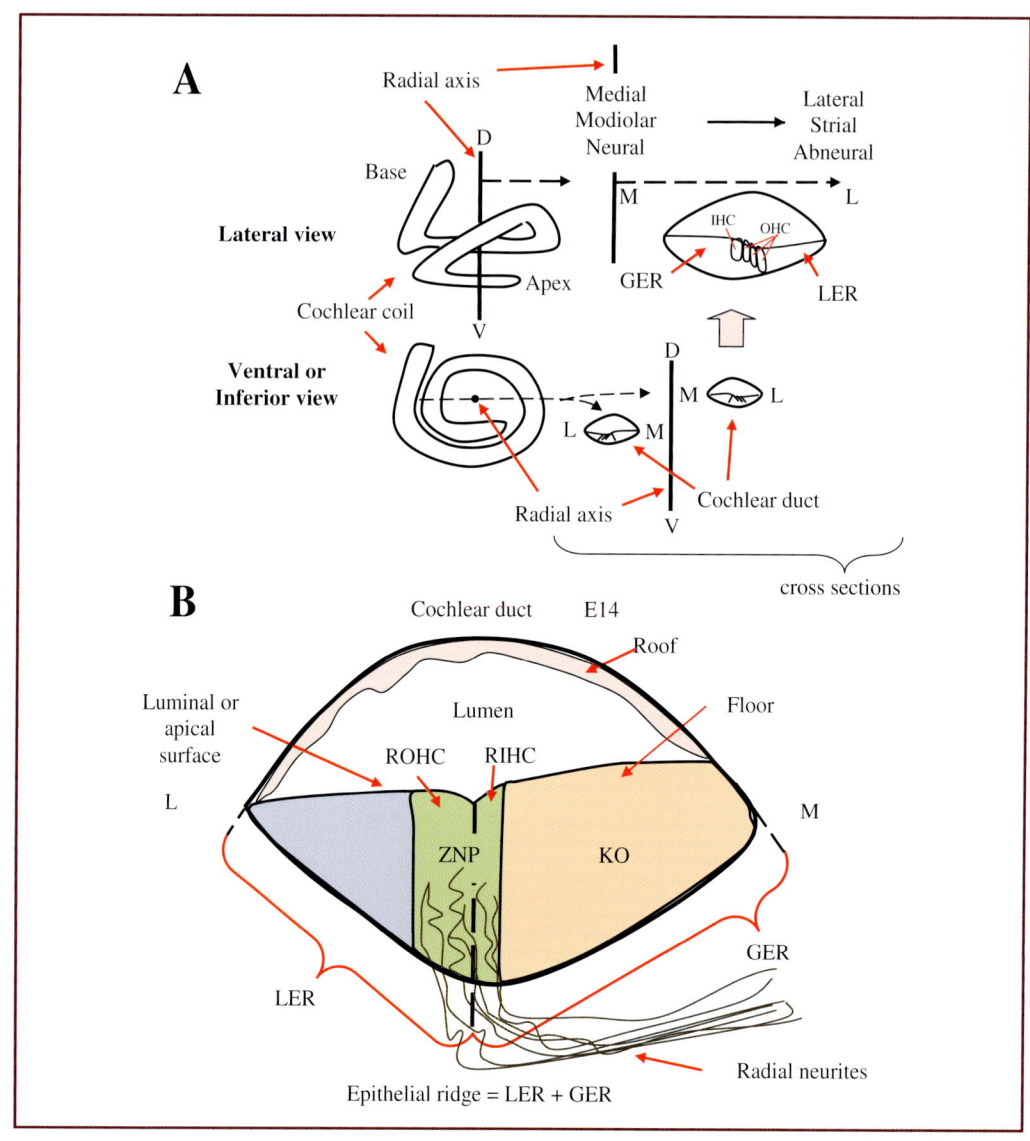

Figure 6–4. **A.** Schematic anatomical axes and landmarks of the developing mouse cochlea. Note the alignment of the principal central cochlear spiral axis with the dorsal (D) ventral (V) axis in the embryonic mouse (Morsli et al., 1998). In the human, the radial axis takes on a more mediolateral orientation especially in the later stages of formation (week 9, Sulik & Cotanche, 2004). Cross sections of the cochlear duct are represented to the right. The location of the cross sections is indicated by the dashed arrow over the inferior view of the cochlea to left. **B.** Schematic of the cochlear duct at about E14 in the mouse. According to Lim and Rueda (1992), the greater epithelial ridge (GER) incorporates the region of inner hair cells, whereas the lesser epithelial ridge (LER) incorporates the region where outer hair cells will form. Together they form the floor of the early cochlear duct also called generally the epithelial ridge. The zone of nonproliferation (ZNP) identifies the epithelial region of prospective inner and outer hair cells. Kolliker's organ (KO) has been defined a number of ways. Here, we have adopted the definition given by Kelley (2007) where Kolliker's organ includes only that portion of the GER that does not include prospective sensory hair cells. Lim and Anniko (1985) use primordial organ of Corti and Kolliker's organ interchangeably. Sulik and Cotanche (2004) use Kolliker's organ to refer to the entire epithelial ridge. RIHC = region of inner hair cells; ROHC = region of outer hair cells.

ridge changes as the duct elongates during development. This ridge of cells contains prosensory cells that have segregated ventrally from the sacculus (approximately E13.5 for mouse, week 6 for human). The ridge is marked by *Sox2* and *Lfng* and contains the presumptive organ of Corti. All cochlear hair cells and supporting cells will arise from this epithelial ridge.

Elongation and coiling of the cochlea continues with the leading edge of extension represented at the apical tip. Beginning at about E12.5 (late week 5), a central strip of the apical epithelial ridge begins to express the cyclin dependent kinase inhibitor *Cdkn1b*, (formerly p27^{Kip1}) (Figure 6–5) (Chen, Johnson, Zoghbi, & Segil, 2002; Chen & Segil, 1999; Lee, Liu, & Segil, 2006; also reviewed by Driver & Kelley, 2009). *Cdkn1b* causes cells to exit the cell cycle and terminate mitosis. Over a period of about a day in mice (E12.5 to E13.5), equivalent to a week in humans (late week 5 to week 6), a wave of *Cdkn1b* expression sweeps along a central strip of the epithelial ridge from the cochlear apex to base. This is followed shortly afterward by a wave of cells exiting the cell cycle, which also moves along the central strip from the cochlear apex to base forming a cellular zone of nonproliferation (ZNP) along the epithelial ridge (Figures 6-4B and 6-5) (Chen, 2002; Chen & Segil, 1999; Lee et al., 2006). These postmitotic cells form the undifferentiated prosensory domain of the organ of Corti, and its formation is completed before E14 (weeks 7 to 8 in humans) (Chen et al., 2002; Chen & Segil, 1999; Lee et al., 2006; Ruben, 1967). At this time, the ZNP is flanked on the modiolar (Kolliker's organ) and strial (lesser epithelial ridge) sides by epithelial cells that continue to show evidence of cell division (see Figures 6–4 and 6–5). However, it is the central ZNP that outlines the boundary of the future organ of Corti and therefore presumably contains the precursor cells specified to become sensory hair cells. Although the mechanism responsible for placing the ZNP strip at a particular position within the epithelial ridge is unknown, there is evidence that the modiolar edge (most medial) of the ZNP is lined with *Jag1* expression, and this Notch ligand may play some role in establishing the mediolateral position of the ZNP in the cochlear duct (Chen, 2002). *Gli3* also plays a role in establishing prosensory boundaries (Driver et al., 2008).

Exiting the cell cycle is a critical step for cochlear hair cell progenitors, and in the mammal this state is actively maintained following differentiation by many redundant cell cycle inhibitors. The quiescent state may be critical in mammals since an active cell cycle may be incompatible with a functional hair cell inasmuch as activation of the cell cycle in differentiated hair cells leads to cell death. Restricting the cell cycle also regulates the total number of hair cells formed. Interference with *Cdkn1b* results in supernumerary hair cells and continued proliferation accompanied by profound hearing loss (Chen & Segil, 1999). During hair cell differentiation, *Cdkn1b* is downregulated in hair cells but continues to be expressed in differentiating supporting cells.

ELONGATION AND REORGANIZATION OF THE COCHLEAR ZNP

By E14 (approximately week 8 in humans), the cochlear coil has reached 1.25 turns and the ZNP is in place (Chen & Segil, 1999). The region of the initial ventromedial cochlear extension in the otocyst continues to elongate during the coiling

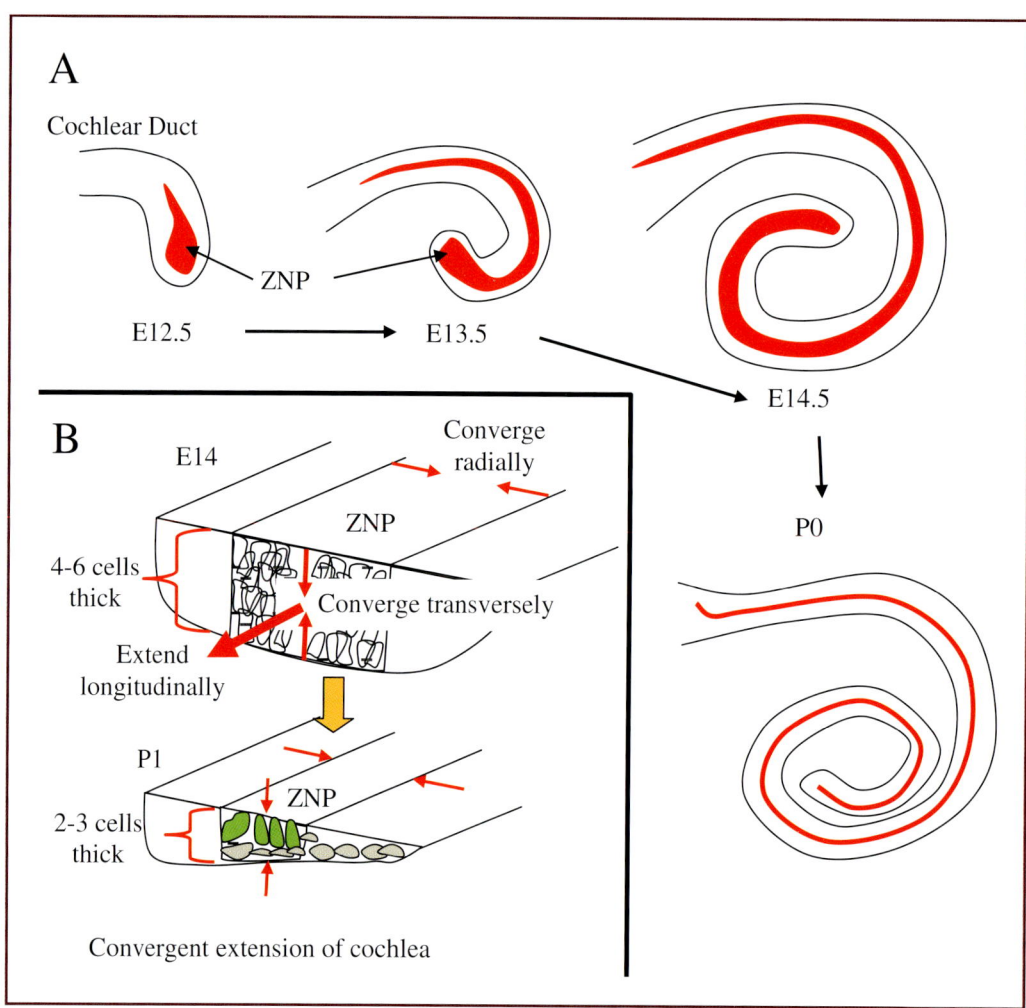

Figure 6–5. *A. Formation of the zone of non-proliferation (ZNP, red marker for* **Cdkn1b***) in the cochlear duct at four stages of development. The ZNP appears first in the apex and then sweeps towards the base.* ***B.*** *Convergent extension in the cochlear duct. One model explaining the continued longitudinal elongation of cochlear coils in the absence of cell proliferation is convergent extension. The schematic in B illustrates how two convergent migrations could contribute to the extension and elongation of the cochlea: (1) lateral or radial convergence, and (2) a transverse convergence or thinning of the epithelium). Green denotes differentiated hair cells.*

process, thus increasing the distance between the saccular macula and the first turn of the cochlea (Morsli et al., 1998). Elongation of the cochlear epithelium after E14 (weeks 7 to 8) must be accomplished without ZNP cell division since prosensory cells in the ZNP have exited the cell cycle. Completion of the adult number of coils is not accomplished until late embryonic ages (E17 to E19) for the mouse (weeks 9 to 10 for the human) (Chen & Segil, 1999; Morsli et al., 1998; Sher, 1971). Despite reaching the mature number of coils by approximately weeks

9 to 10 in the human, the cochlea is only about 20 mm long. It will continue to elongate to reach its mature length (approximately 35 mm) at approximately 16 weeks (Bredberg, 1968; Sulik & Cotanche, 2004). During cochlear elongation in both man and mouse, two processes are occurring simultaneously: (1) cellular reorganization and structural elongation of the ZNP and cochlear duct, and (2) differentiation of prosensory cells into hair cells and supporting cells (see below).

The final coiling and extension of the cochlea over the period from E14 to E18 (weeks 8 to 10) is sensitive to alterations in planar cell polarity (PCP) signaling, and it has been proposed that the process involves convergent extension in some form (see Chapter 4) (Jones & Chen, 2007; Kelley & Chen, 2007; McKenzie, Krupin, & Kelley, 2004; Montcouquiol et al., 2003; Wang et al., 2005; Wang, Guo, & Nathans, 2006). Mutations in or knockouts of core PCP genes or related signaling pathways prevent elongation and result in a shortened cochlea. In addition, inactivation of the intraflagellar transport protein 88 (*Ift88*) (Jones et al., 2008) also disrupts the planar alignment of outer hair cells (i.e., stereociliary bundles are disoriented) and produces shorter cochleas. However, *Ift88* inactivation does not interfere with the asymmetric distribution of core PCP gene products. These results suggest that *Ift88* inactivation uncouples PCP signals from the outer hair cells' intrinsic polarity. Intrinsic cell polarity is reflected in the orientation of stereociliary bundles. One measure of this orientation is the morphological polarization vector (MPV) itself, which describes the preferred direction for activating hair cells during the deflection of hair bundles (see Figures 6–8 and 6–9 later in this chapter, and Chapter 1). Interestingly, in *Ift88* null mutants, MPV orientation in inner hair cells is minimally affected, and in the vestibular epithelia the kinocilium is intact and MPV orientation is normal (Jones et al., 2008). Therefore, it appears that *Ift88* is critical for OHC stereociliary bundle orientation. Some of these effects are described in more detail below.

During elongation, the organ of Corti thins, going from an epithelium of about five cells thick to the normal two-cell thickness (hair cells and supporting cells, E14 to E19 or weeks 8 to 10 in humans) (see Figure 6–5B). This could be viewed as a transverse convergence that may be taken to mean a decrease in the distance between the luminal surface to the base of the ZNP. This convergence presumably leads to cochlear extension (longitudinal lengthening) as cells undergo intercalation. In addition, the mediolateral width of the ZNP decreases between E14.5 and P0 (approximately weeks 8 to 14) (McKenzie et al., 2004). The latter decrease in width could be viewed as a mediolateral or radial convergence (radial relative to the axis of the cochlear coil, see Figures 6–4A and 6–4B). This convergence presumably also leads to cochlear extension as cells intercalate mediolaterally. Although it is highly likely that convergent extension occurs in the cochlea, the exact process has not been clarified and no complete models of the process have been offered. Any such model would be required to explain both the radial and the transverse cellular intercalations during the differentiation of hair cells (see Figure 6–5B).

HAIR CELL DIFFERENTIATION

Before differentiation can begin, auditory and vestibular prosensory epithelial cells must complete the last or terminal cell division. This is often referred to as terminal mitosis. The postmitotic prosensory epithelial cells of the cochlear ZNP appear structurally like any other cells in the epithelial ridge. There are no superficial structural distinctions and no obvious evidence of mediolateral axial polarization. Prosensory cells must initiate differentiation before such distinctions will appear.

Differentiation of prosensory cells of the cristae, maculae, and cochlea all require the induction of an atonal homolog (*Atoh*) gene, specifically *Atoh1* (also known as mouse atonal homolog *Math1*), which is a proneural basic helix loop helix (bHLH) transcription factor (Bermingham et al., 1999). In the cochlea, prosensory cells exit the cell cycle and ultimately express *Atoh1* before differentiating into hair cells. *Atoh1* expression and differentiation of hair cells begins in the cristae and maculae about one day (approximately one week in humans) before similar levels are expressed in the cochlea (Bermingham et al., 1999; Chen et al., 2002; Woods, Montcouquiol, & Kelley, 2004).

Recent studies have indicated that prior to the appearance of Atoh1, there is a combined down-regulation of members of the family of "inhibitors of differentiation and DNA binding proteins" (Ids). These proteins are coded by a family of four genes (*Id1, Id2, Id3,* and *Id4*), although *Id4* is not expressed in the ear (Jones, Montcouquiol, Dabdoub, Woods, & Kelley, 2006). *Ids* contain the HLH motif but lack a basic DNA binding domain. *Ids* regulate bHLH transcription factors such as Atoh1 during development by competing for E-proteins. E-proteins are bHLH gene regulators that contain a functional basic DNA binding domain. Atoh1 must bind with E-proteins in order to bind DNA at selected genes and implement changes in transcription. When Ids bind E-proteins, the heterodimer produced is nonfunctional and cannot bind DNA. For every E-protein bound by Ids there is one less E-protein available for Atoh1. Thus, if Ids are at sufficient concentrations, the action of Atoh1 is inhibited due to the unavailability of E-proteins. Between E14 and E18 (weeks 8 and 10), Ids are significantly down-regulated in selected cells, thus disinhibiting the action of Atoh1 and permitting the onset of differentiation.

Atoh1 expression appears in the cochlea among postmitotic cells of the ZNP by approximately E13 (week 7) (Chen et al., 2002; Lanford, Shailam, Norton, Gridley, & Kelley, 2000; Woods et al., 2004;). At E14 (week 8), the epithelial ridge of the cochlear duct is composed of a sheet of cell columns 4 to 5 or more cells thick. In basal regions, the ZNP occupies a position almost midway between the modiolar (medial) and strial (lateral) sides of the duct and is some 20 to 40 microns wide (see Figures 6–4 and 6–5). During differentiation and elongation of the cochlear coil, the ZNP will be reduced to a bilayer of hair cells and supporting cells, ultimately containing a single row of inner hair cells and three rows of outer hair cells. The presence and distribution of Atoh1 appears somewhat different, depending on the method used to mark expression; however, based on the most sensitive methods, expression extends along the entire length of the cochlea by E14 (week 8), but is restricted to a relatively diffuse band of cells corresponding to the ZNP. Initially, it marks the entire depth and width of the ZNP columns, presumably including progenitors of both supporting and hair cells (Kelley, 2007; Lanford et al., 2000; Woods et al., 2004). Atoh1 appears transiently and relatively weakly in presumptive supporting cells, whereas in presumptive hair cells labeling intensifies with differentiation and continues throughout the embryonic period. By E15.5 (week 9), a pattern of discretely labeled individual cells surrounded by unmarked cells is evident in the most mature regions (basal cochlea), whereas in the less mature apical regions a more dispersed pattern is still present. A basal to apical gradient of maturation is evident through these periods. By P0 (13 to 15 weeks), *Atoh1* is expressed only in hair cells (see Figure 6–5B).

In addition to a general base to apex maturation gradient, differentiation and maturation of the cochlear organ of Corti occurs along a medial to lateral (inner hair cell to outer hair cell, IHC to OHC) gradient. Mediolateral differentiation closely follows the ongoing longitudinal differentiation. At about E15.5 (week 9), a specific hair cell marker, MyosinVIIa (Myo7a), first appears near the base in IHCs along the medial border of the ZNP. The appearance of Myo7a represents an early specific sign of stereociliary bundles. By this time, regions of the basal ZNP have reorganized to form two layers including the sensory hair cells and supporting cells. Differentiation progresses toward the cochlear hook and to the apex longitudinally and mediolaterally from IHCs to OHCs (Anniko, 1983b; Chen et al., 2002; Chen & Segil, 1999; Lim & Anniko, 1985; Morrison, Hodgetts, Gossler, Hrabe de Angelis, & Lewis, 1999; Sher, 1971). Apical regions do not exhibit the more mature bilayer configuration until about E17.5 (week 11). By E18 to E19 (weeks 11 to 14), the full complement of differentiated hair cells and supporting cells has appeared, and they are arranged in the adult configuration (one row of IHCs and three rows of OHCs). At this time, the marker *Cdkn1b* is fully downregulated and *Atoh1* fully upregulated specifically in hair cells.

PATTERNING OF HAIR CELLS AND SUPPORTING CELLS IN THE EPITHELIUM

During differentiation of the sensory epithelium, a remarkable pattern of hair cells and supporting cells emerges. Our understanding of the molecular signals orchestrating these events is more detailed for the cochlea than vestibular sensors, but in either case, there are many more questions to answer. One feature of both auditory and vestibular epithelia suggests a common organizing mechanism. The presence of orderly patterned mosaics of hair cells and supporting cells in all inner ear sensory epithelia implies a role for Notch-mediated lateral inhibition. In particular, all epithelia are arranged such that every hair cell is isolated from other hair cells by a ring of supporting cells or their processes. Lateral inhibition is one strategy that produces a "center on surround off" pattern, where "on" refers to a hair cell and "off" to a supporting cell.

Figures 6–6 and 6–7 schematically outline a *Notch1/Atoh1* signaling model adapted from Kelley (2007). In this model, the epithelia of the cochlear duct broadly express *Notch1* at E13 (week 7) and contain relatively high concentration levels of Ids (Jones et al., 2006; Lanford et al., 1999). Atoh1 appears at low to modest levels at this time and begins to mark out the ZNP throughout its depth and lateral extent, ultimately spanning the entire length of the cochlea (Chen et al., 2002; Lanford et al., 2000; Woods et al., 2004). Initially, the level of Atoh1 is similar in all cells. During the early stages, the action of Atoh1 remains repressed as its levels are insufficient to compete effectively with Ids for E-proteins (see Figure 6–6) (Jones et al., 2006). However, beginning about E15 (weeks 8 to 9), Ids levels

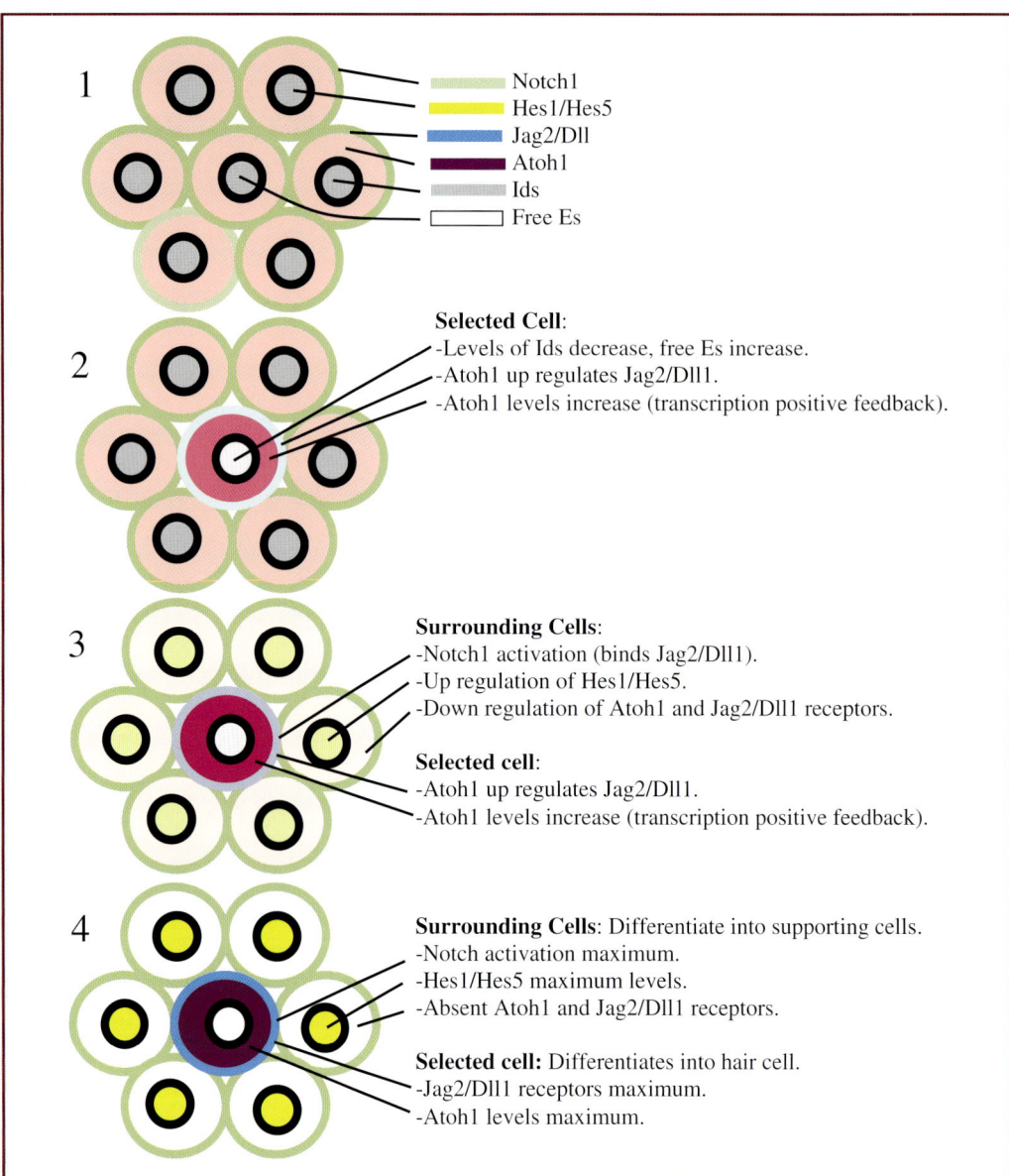

Figure 6–6. *Schematic representation of prosensory epithelial cells of the zone of nonproliferation (ZNP). Four stages (1–4) illustrate the progressive changes thought to accompany the determination of hair cell and supporting cell fates. (1) A group of prosensory cells have completed terminal mitosis and are equally capable of assuming a hair cell or supporting cell fate. High levels of Ids prevent low levels of Atoh1 from upregulating Jag2/Dll1 receptors. (2) Ids levels decrease in selected cells, permitting the binding of Atoh1 to E-proteins, upregulation of Jag2/Dll1 receptors and initiation of positive feedback regulation of* Atoh1 *transcription. (3) Notch1 receptors in adjacent cells are activated by binding Jag2/Dll1 receptors thus upregulating* Hes1/Hes5 *and in turn down-regulating* Atoh1. *Increasing Atoh1 levels in selected cell increases Jag2/Dll1 receptors and Atoh1 levels further. (4) Notch1 and Hes1/Hes5 levels reach a maximum in surrounding cells, whereas Atoh1 and Jag2/Dll1 receptors reach a maximum in selected cell. Thus a center "on" (hair cell) surround "off" (supporting cell) pattern of lateral inhibition is produced through* Notch1 *signaling. This arrangement ultimately represents the only stable configuration for cells in the ZNP.*

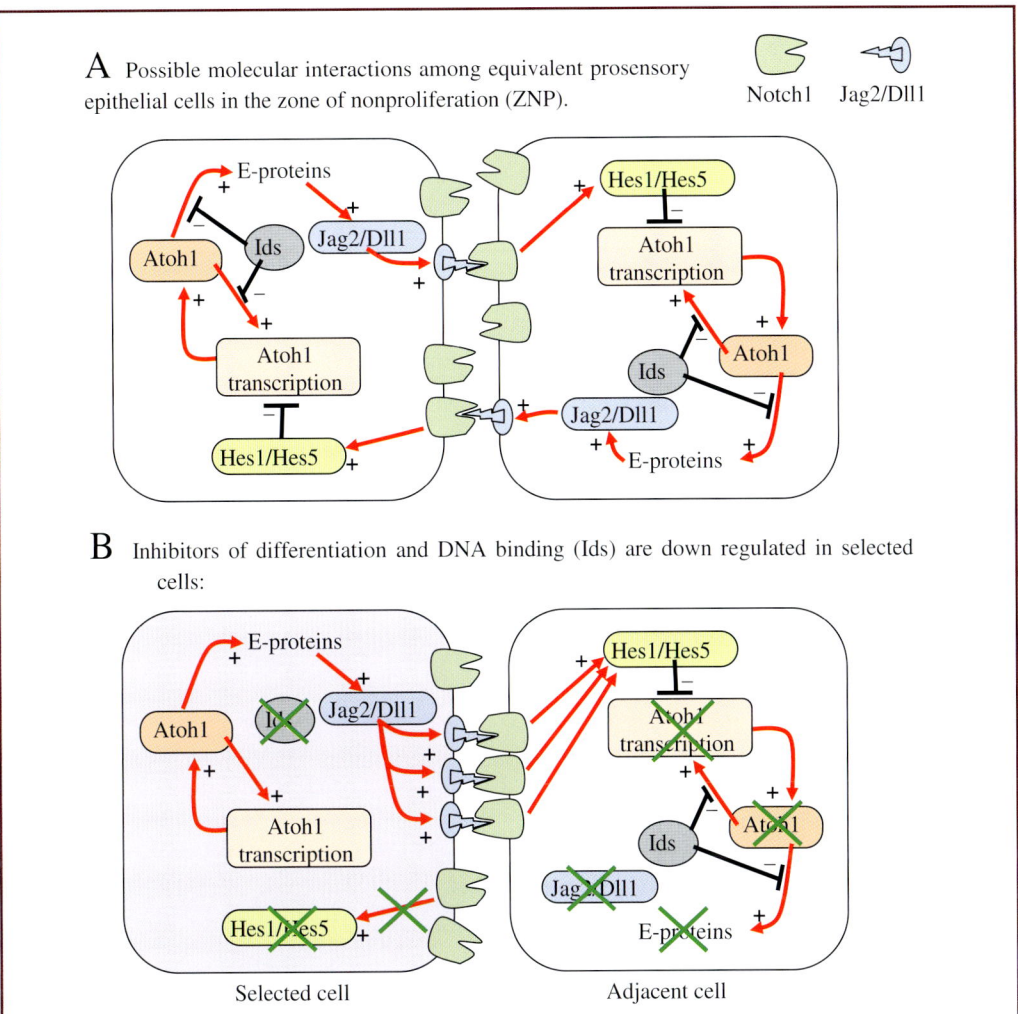

Figure 6–7. Notch signaling components during differentiation of hair cells and supporting cells in the cochlear zone of non-proliferation (ZNP). Notch1 is broadly expressed in all cells. **A.** Scheme representing the major possible molecular interactions between prosensory cells of the ZNP. All cells in the ZNP have completed terminal mitosis and presumably have equal capacity to become hair cells or supporting cells. Key processes include the transcription of Atoh1 (required for hair cell determination), upregulation of Jag2/Dll1 (requires E-proteins) and upregulation of Hes1/Hes5 (required to extinguish Atoh1 transcription in supporting cells). Ids are present in all cells initially and act to prevent full up-regulation of Atoh1 immediately following terminal mitosis. **B.** Down-regulation of Ids in selected cells increases E-protein levels thus upregulating Jag2/Dll1 receptors. It also releases the Atoh1/Atoh1 transcription positive feedback loop from inhibition, which further increases Atoh1 levels. In turn, Jag2/Dll1 receptor levels increase on the selected cell surface leading to the activation of additional Notch1 receptors in adjacent cells. This downregulates Atoh1 transcription decreasing Atoh1 levels in adjacent cells. Decreased Atoh1 levels in adjacent cells should eliminate Jag2/Dll1 surface ligands, which eliminates activated Notch1 receptors in the selected cell. The absence of activated Notch1 receptors down-regulates Hes1/Hes5 in the selected cell and prevents the downregulation of Atoh1 transcription. Atoh1 transcription is maximized in the selected cell. Ultimately, selected cells express Atoh1 maximally and differentiate into hair cells, whereas adjacent cells express Hes1/Hes5 maximally and differentiate into supporting cells.

in selected cells of the ZNP begin to drop. To help illustrate the molecular scheme further, we will select one cell (the center cell) of an otherwise indistinguishable group of seven prosensory cells to illustrate the effects of decreases in Ids (see Figures 6–6 and 6–7). As Ids levels fall in the selected cell, free E-protein concentrations increase, thus providing Atoh1 access to form heterodimers and up-regulate target genes (see Figure 6–7B). This drop in Ids levels also facilitates the transcription of *Atoh1* and fosters a positive feedback loop for *Atoh1* transcription (discussed below, Helms, Abney, Ben-Arie, Zoghbi, & Johnson, 2000). As a result, Atoh1 level increases as Ids levels decrease in selected cells (Jones et al., 2006).

Upregulation of Atoh1 target genes in the selected cell results in the appearance of Notch1 ligands, Delta1 (Dll1) and Jagged2 (Jag2). Once Jag2/Dll1 insert into the plasma membrane at the cell surface of the selected cell, there is an opportunity for Notch1 receptors on adjacent cells to bind with the newly inserted ligands (see Figure 6–7B). Binding of Jag2/Dll1 ligands to Notch1 receptor on adjacent cells activates Notch1 intracellular signaling leading to the upregulation of *Hes1* and *Hes5* in adjacent cells. Atoh1 signals present in adjacent cells are repressed by the upregulation of *Hes1* and *Hes5*, which in turn downregulates Atoh1 target genes. The downregulation of Atoh1 and its target genes in adjacent cells reduces any preexisting Jag2/Dll1 ligands. The reduction in ligands on adjacent cells reduces the number of activated Notch1 receptors on the selected cell. Reduced Notch1 signaling in the selected cell downregulates *Hes1* and *Hes5* and thus potentially upregulates Atoh1 target genes. Thus, upregulation of *Atoh1* in the selected cell leads to activation of more Notch1 receptors in adjacent cells, which potentially leads to an upregulation of Atoh1 levels in the selected cell. This is one potential positive feedback loop leading to ever-increasing levels of Atoh1 in the selected cell. However, this feedback cycle is limited by the size of the remaining Atoh1 pools and Notch1 ligands in adjacent cells, which steadily decrease during this process. Instead, this feedback process may serve to help spread or propagate the mosaic pattern through the epithelial field and to stabilize the configuration by preventing the upregulation of *Hes1/Hes5* in "on" cells.

As noted above, there is a second positive feedback loop mediated within the selected cell itself. Increasing levels of Atoh1 under conditions of reduced Ids results in a further increase in *Atoh1* transcription, thus producing even higher levels of Atoh1 in the selected cell (see Figure 6–7B). This positive feedback loop is recruited with the initial decrease in Ids and serves to initiate and strengthen a rise in Atoh1 and to increase the number of surface Notch1 ligands (Dll1/Jag2) in the selected cell. This activates more Notch1 receptors, further increasing Hes1 and Hes5 in adjacent cells. These processes drive Atoh1 levels to a maximum in the selected cell, whereas all adjacent cells are simultaneously driven to maximum activation of the Notch1 receptors and soaring levels of Hes1 and Hes5 (Figure 6–6). Dominated by *Atoh1* signaling, the selected cell is configured for differentiation to form a hair cell, whereas adjacent cells are dominated by *Hes1* and *Hes5* and initiate programs of differentiation to form supporting cells.

In due course across the entire ZNP, the molecular scheme can be imagined to produce a mosaic of patches, where each patch is composed of a hair cell surrounded by supporting cells (see Figure 6–6). This scheme is simplistic and incomplete, but it provides a basis for us to begin to understand how such mosaics can be shaped by molecular signaling and how the remarkable structures of the cochlea and vestibular system may be formed. Among the key questions regarding this model is, what factors mediate the reduction in Ids at the beginning of the *Atoh1*/*Notch1* process of lateral inhibition?

Continued differentiation, maturation, and survival of both hair cells and supporting cells require additional signals. For example, *Fgfr3* is likely involved in supporting cell differentiation (Kelley, 2007) whereas newly formed hair cells degenerate and lead to deafness when one of several factors are absent, including *Pou3f4* (also known as *Brn-3c*) (Xiang et al., 1997; Xiang, Gao, Hasson, & Shin, 1998), *Barhl1* (homeobox gene) (Li et al., 2002), and *Gfi1* (zinc finger) (Wallis et al., 2003; reviewed by Gao, 2003).

The process of sorting hair cells into repeating patterns occurs simultaneously with convergent extension of the cochlear duct and elongation of the coiling cochlea. By E19 (weeks 11 to 12), although not mature, the cochlear duct has the appropriate number of coils and its full complement of inner and outer hair cells. The onset of hearing will not begin until after birth at approximately P10 to P14 in the mouse (during the last trimester in the human).

ADDITIONAL INDICATORS OF HAIR CELL DIFFERENTIATION

Prior to differentiation, prosensory cells are homogenous epithelial cells interconnected to each other by gap junctions, among other junctions (Ginzberg & Gilula, 1979). Gap junctions are lost during the earliest period of hair cell differentiation (Bryant, Forge, & Richardson, 2005; Forge, Souter, & Denman-Johnson, 1997). Thus, one of the earliest events includes the isolation of hair cells from surrounding supporting cells by eliminating direct electrochemical coupling between them.

The appearance of a number of other gene products also signal the onset of hair cell differentiation including calretinin (also known as calbindin 2 [*Calb2*]) (Dechesne, Rabejac, & Desmadryl, 1994), calmodulin (Zheng & Gao, 1997), Pou4f3, and myosins VI and VIIa (Hasson et al., 1997; Xiang et al., 1998). More recently, the accumulation of actin filaments in the lateral walls was used to identify inner hair cells, which revealed a disorderly row of inner hair cells at E14.5 (week 8) in the region of the base (McKenzie et al., 2004). These gene products can be used as markers to identify hair cells among supporting cells in the absence of data regarding other outward structural changes.

STEREOCILIARY BUNDLES

The traditional definitive structural sign of hair cell differentiation is the elaboration of microvilli and formation of stereociliary bundles. Using scanning electron microscopy, vestibular hair cells can be definitively identified with the appearance of stereociliary bundles by E13.5 (week 7) (Bryant et al., 2005; Denman-Johnson & Forge, 1999; Deschesne, 1992; Mbiene & Sans, 1986). In contrast, the first morphological evidence of cochlear stereocilia begins to emerge between E14 and E15 (weeks 8 to 9) (Anniko, 1983b; Lim & Anniko, 1985; Pujol et al., 1998).

The earliest hair cells appear with microvilli or stereocilia encircling a single kinocilium in the center of the cell (Figure 6–8). The first sign of intrinsic hair cell polarization begins almost immediately with a movement of the kinocilium

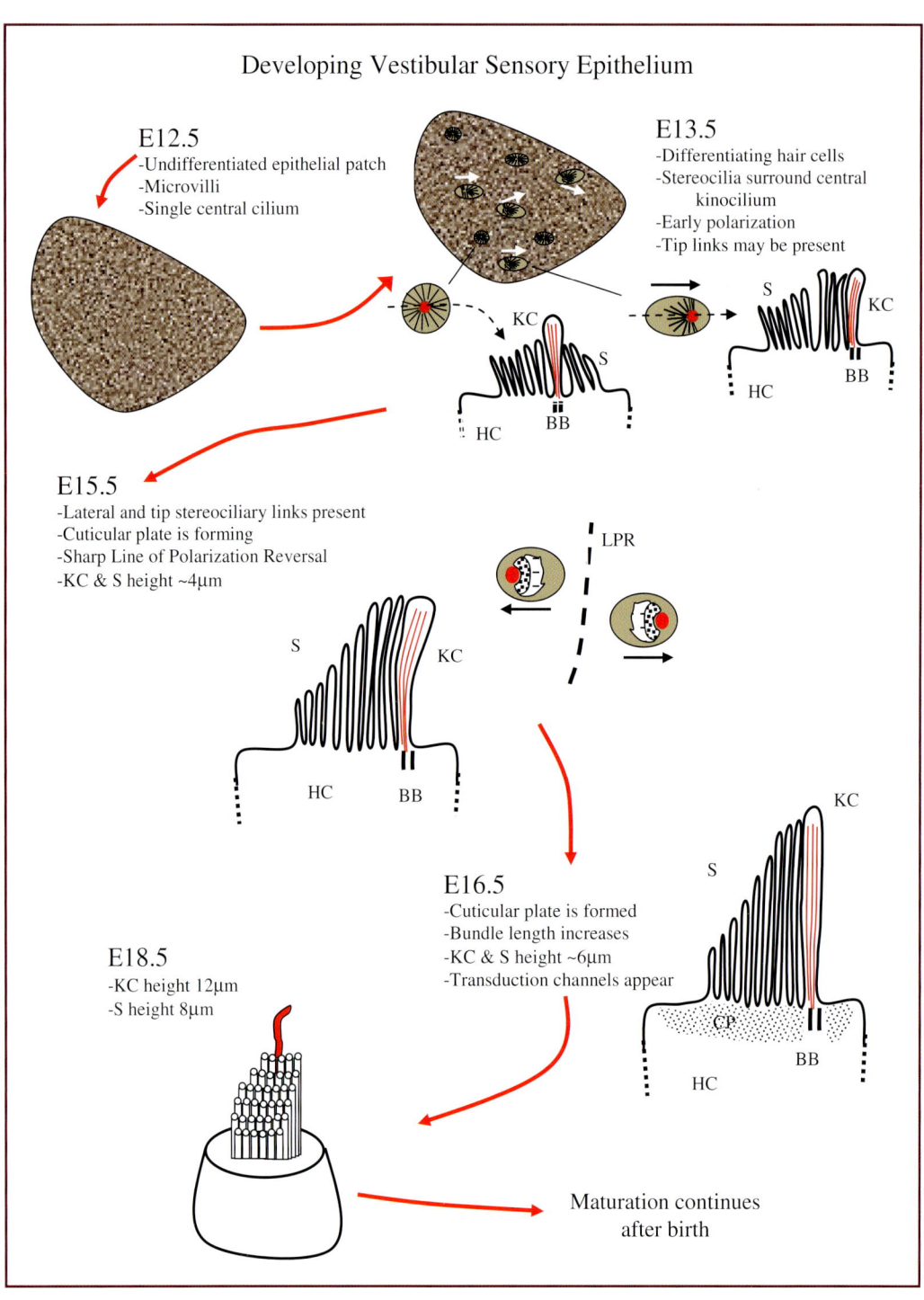

Figure 6–8. Development of vestibular stereociliary bundles. Ages and sizes represent mouse development. Early stages of development are similar for both auditory and vestibular hair cells although vestibular hair bundles begin appearing earlier than cochlear bundles. KC = kinocilium; S = stereocilia; HC = hair cell; BB = basal body; LPR = line of polarity reversal; CP = cuticular plate.

toward the periphery of the cell, thus taking an eccentric position in each hair cell. The direction of the movement reflects the emergence of the preferred direction of stimulation for each hair cell and hence the incipient morphological polarization vector (MPV). As noted above, the process associated with cilia formation and its direction of movement reflects the intrinsic cell polarization. Ablating the primary cilium (kinocilium) by preventing its formation using conditional inactivation of *Itf88* or other such tools prevents proper planar orientation of OHC stereociliary bundles (Jones et al., 2008). Disruption of PCP signaling pathways (*Vangl2, Celsr1*) has similar effects. In Looptail mouse mutants (*Vangl2/Ltap null*), auditory and vestibular hair cells are disoriented, indicating a role for Wnt-PCP signaling in establishing hair cell orientation (Montcouquiol et al., 2003, 2006; reviewed by Jones & Chen, 2007; Kelley & Chen, 2007; Rida & Chen, 2009).

As one might expect, the ontogeny of stereociliary bundles as well as other internal hair cell structures depends critically on the initiation of gene transcription for specific proteins. Many of these proteins have been identified because of natural mutation or genetic manipulation. Table 6–7 lists at least 18 genes that have been shown to be critical for stereocilia structure and function. There are also a number of important structures that serve to link stereocilia together as well as link the stereocilia to the kinocilium mechanically. There is of course the protein linkage to hair cell transduction channels called "tip links" that are obviously critical to sensory transduction. Cadherin 23 (Cdh23) and protocadherin 15 (Pcdh15) make up the tip links (Kazmierczak et al., 2007). Given the complex architecture of the stereociliary bundle (see Chapter 1), much more remains to be learned about its molecular and functional development.

Cochlear Stereociliary Bundles

A complex series of developmental steps leads to the growth and shaping of cochlear hair bundles for both inner and outer hair cells. Ultimately, in the mouse, the final maturation of inner ear stereociliary bundles occurs well after birth. In mammals, cochlear hair bundle formation begins first in the base (E14 to E15 in mouse, weeks 10 to 11 in humans) and proceeds over a considerable period of time to more apical locations (E17 to E18 in mouse, weeks 18 to 20+ in human) (Lim & Rueda, 1992). Final maturation of stereociliary bundles occurs first at the base and some two days later at the apex in the rat (Waguespack, Salles, Kachar, & Ricci, 2007). Moreover, bundles of inner hair cells (IHCs) mature faster than bundles of outer hair cells (OHCs), thus suggesting a second medio-lateral gradient for bundle maturation. For example, IHC bundles are mature at about P1, whereas OHC bundles mature between P5 and P14 in the mouse and hamster (Anniko, 1983a; Lim & Rueda, 1992; Kaltenbach, Falzarano, & Simpson,1994). Similar gradients are found in the human (Lavigne-Rebillard & Pujol, 1986). Bundles begin to emerge during week 11 and appear mature around week 20 for IHC and between weeks 22 to 25 for OHC. In the mature mammal, the rootlets of the hair bundle stereocilia are anchored in the apically located cuticular plate (Figures 6–8 and 6–9). The earliest evidence for an emerging cuticular plate was obtained using a specific immunological marker, which revealed labeling as early as E15 (week 11) in the base of the

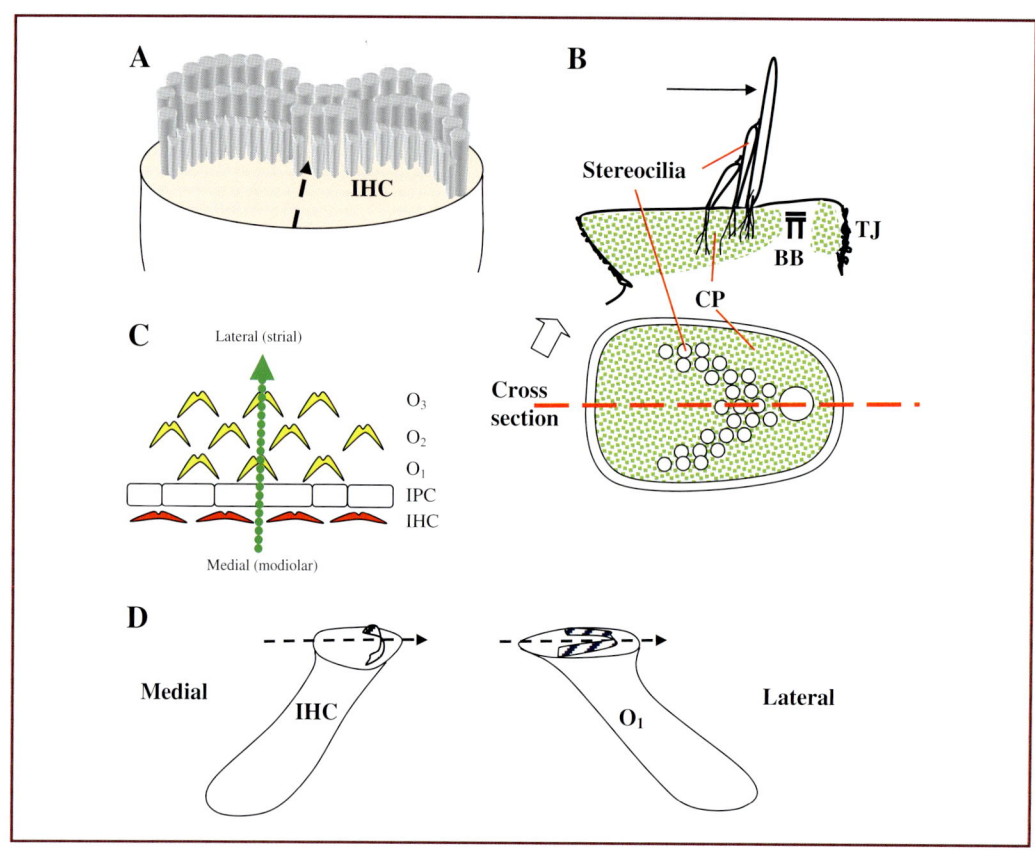

Figure 6–9. Schematic drawings of mature cochlear hair cells. Among the many differences between vestibular and cochlear hair cells are the general observations that cochlear hair cells ultimately lose the kinocilium during late development and adopt stereociliary bundle configurations with fewer rows of stereocilia (three to four), which mark out contours resembling "U"s, shallow crescents, "V"s or "W"s on the apical surface (compare with Figure 6–8). The basal body (BB) remains in place although the kinocilium disappears in the mature mammalian cochlear hair cell. Stereociliary bundles are securely anchored to the cuticular plate (CP) underlying the apical surface in both vestibular and cochlear hair cells. Arrows indicate the direction of maximal stimulation with bundle displacement. This direction is also termed the morphological polarization vector (MPV). In the cochlea this direction is radial from the center axis of the cochlear coil passing medial to lateral across the organ of Corti. The orientation of inner (IHC) and outer (OHC) hair cells of the organ of Corti is shown in relation to each other and to the MPV (C and D). **A.** Apex of inner hair cell. **B.** Medial to lateral cross section through stereocilia and basal body (BB). Schematic top view of apical surface of cochlear hair cell and line of cross section (red dashed). A specific monoclonal antibody can be used to mark the cuticular plate (here shown as green dotted texture). The fluorescent marker conspicuously outlines the position of the missing kinocilium, which appears as a small circle just in front of the stereociliary bundle overlying the basal body that is free of marker. TJ = tight junction. **C.** Schematic arrangement of hair cell stereociliary bundles showing the normal single row of inner hair cells (IHCs) and three rows of outer hair cells (O_1, O_2, O_3) as they are arranged on the apical surface of the sensory epithelium of the organ of Corti. IHCs occupy the most medial position toward the modiolus while OHCs are located laterally towards the stria vascularis side of the cochlear duct. Laterally directed deflections of hair cell bundles (green arrow) are in line with MPVs and thus activate (depolarize) hair cells. IPC = inner pillar cells. **D.** Side view orientations and MPVs (arrows) of an inner hair cell (IHC) and a first row outer hair cell (O_1) within the organ of Corti.

cochlea and a much later appearance in the apex at E18 (week 18) (Nishida, Rivolta, & Holley, 1998). Using conventional electron microscopy, the cuticular plate can be identified only as early as E17 to E18 (week 17 to 18) along with rootlets of stereocilia, which become more prominent during later development (Anniko, 1983b).

Figure 6–10 schematically illustrates the formation of the stereociliary bundles of inner and outer hair cells. Both begin as a circular tuft of stereocilia of similar height, which, apart from being more regularly spaced, look much like the surrounding microvilli. The tuft surrounds a central single cilium that will become the kinocilium. As development proceeds, IHC and OHC bundles take on different shapes. In both types, the kinocilium adopts an eccentric position within the cell located toward the strial side of the epithelium (lateral). The bundle grows in the form of a "V" or "W" shape where the rows on the outer edges of the tuft elongate at progressively faster rates. The row adjacent to the kinocilium becomes the tallest. This results in the familiar staircase pattern of increasing stereocilia row

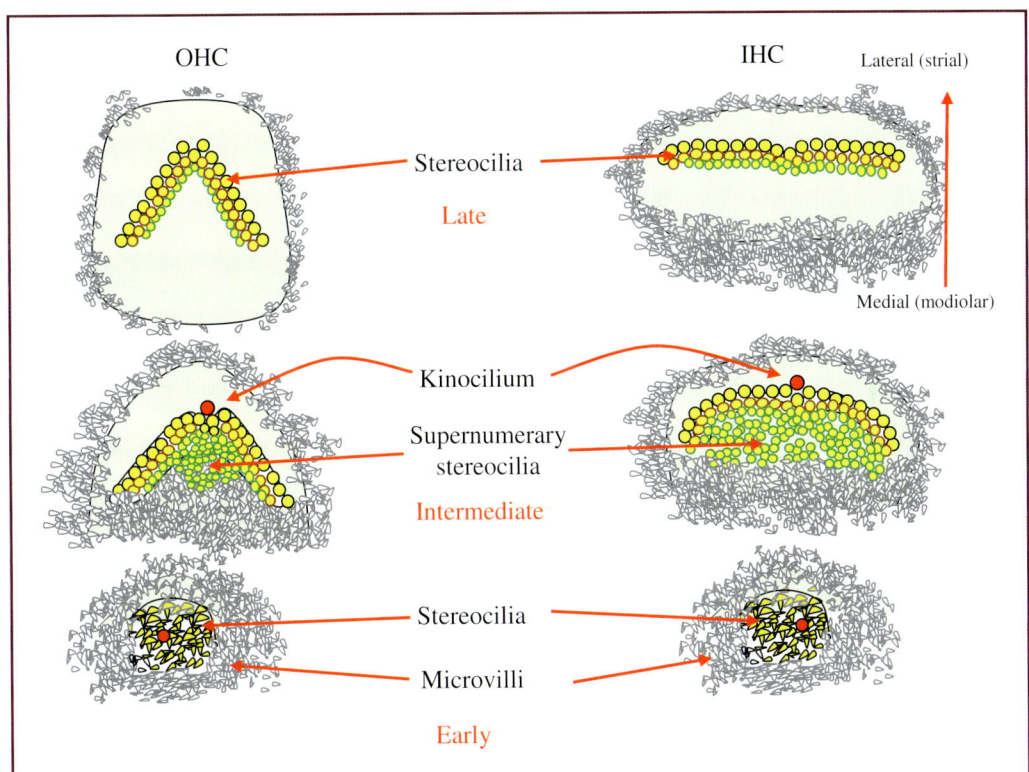

Figure 6–10. Development of cochlear stereociliary bundles. Early (depending on cochlear location): Appearance of apical surface when stereocilia begin to emerge. Intermediate: Intermediate stage of formation. Late: Final maturational changes. Different contours for IHC and OHC bundles emerge quickly where OHC bundles adopt a "V" or "W" shape and IHC bundles adopt a shallow crescent or straight linear arrangement. The kinocilium does not disappear until 7 to 14 days after birth depending on the species. In the human the kinocilium disappears during the last trimester.

height. Stereocilia cross-links appear with the emergence of the graded staircase pattern. IHCs quickly adopt a flatter "U" or crescent shape in appearance. In both IHCs and OHCs, the kinocilium remains on the lateral (strial) edge of the bundle as shown in Figure 6–10. During bundle growth OHCs refine the "V" shape, whereas IHCs rapidly adopt the shallow crescent or straight-line contour of the mature bundle. As bundles grow, the cuticular plate emerges. Stereociliary rootlets elongate, forming secure anchors, and supernumerary stereociliary bundles recede. The final configuration reveals the absence of a kinocilium and presence of three or four stereocilial rows forming near parallel lines on the apical surface of the IHCs and a "V" or "W" shape on OHCs (Anniko, 1983a; Anniko, 1983b; Kaltenbach, et al., 1994; Pujol et al., 1998).

As noted above, the kinocilium appears and establishes hair cell polarity. The kinocilium is anchored in the basal body. Recently, it has been recognized in the mammal that the initial direction of hair cell polarization in the cochlea may be slightly imprecise (Dabdoub et al., 2003; McKenzie et al., 2004). This is most apparent in the outer hair cells. However, MPV imprecision is corrected between E18 and P10 (second to third trimester) through some unknown mechanism during the period of hair cell bundle maturation. Realignment of the MPVs of cochlear hair cells is also a dramatic process in the bird (Cotanche & Corwin, 1991).

The kinocilium disappears in the cochlea by the end of the first postnatal week in the mouse (Anniko, 1983a) and during the last trimester of the human (Lavigne-Rebillard & Pujol, 1986). The position of the missing kinocilium is nonetheless marked by the basal body and the MPV of the cell is retained (see Figures 6–9 and 6–10).

Vestibular Stereociliary Bundles

Like the cochlea, each vestibular hair cell is surrounded by supporting cells and the decision made between hair cell and supporting cell fates in prosensory cells is likely mediated by *Notch1* signaling, as described above. However, there is an additional layer of organization. In the cochlea, the direction of polarization for the hair bundle is the same for all hair cells with respect to the axis of the cochlea (all point laterally from the central axis, see Figures 6–4A and 6–9). In vestibular maculae, hair bundle orientation is a function of position on the epithelial surface, and intrinsic vestibular hair cell polarization is marked by the MPV. Recall from Chapter 1 that MPVs of hair cells are arranged systematically over the surface of the macular epithelium and in the adult, both maculae show a line coursing through the middle regions of the epithelium (striolar regions), where MPV directions abruptly reverse. Across this line of polarity reversal, MPVs point in opposite directions. These MPV patterns arise during development and are thought to depend critically both on PCP signaling pathways and on intrinsic cell polarity.

Differentiation begins in vestibular hair cells well before it is initiated in cochlear hair cells. Vestibular afferent and efferent neurons reach sensory regions at about E12 (week 6). Markers for hair cell differentiation appear as early as E12.5 (week 6) in small numbers of cells in vestibular prosensory patches (Xiang et al., 1998). Within vestibular epithelia there are temporal and spatial gradients for the course of mitosis. In general, central regions (apex of crista and striolar regions of maculae) initiate and reach peak levels of mitosis well before peripheral regions (base cristae, edges of maculae). Terminal mitosis occurs in a similar order (Mbiene & Sans, 1986; Sans & Chat, 1982). Although these general trends may hold, for any given vestibular region there is a more or less continual eruption of new immature stereociliary bundles until late fetal periods.

Stereociliary bundles are not readily identified in vestibular hair cells before E13 (week 7), but as noted above they do begin to appear earlier than auditory hair cells between E12.5 and E14 (week 7) (Bryant et al., 2005; Dechesne, 1992; Denman-Johnson, & Forge, 1999; Forge et al., 1997; Mbiene, Favre, & Sans, 1984; Mbiene & Sans, 1986). Similar to the cochlea, the hair bundle generally is not polarized with its first appearance in a newly formed hair cell. The kinocilium is centrally located and surrounded by emerging stereocilia (see Figure 6–8). Although outward signs of polarization may not be apparent at this time, intracellular changes have already begun that clearly indicate proteins are being organized asymmetrically in the cell. For example, prickle-like-2 (Pk2, a member of the Wnt-PCP signaling pathway) and the homolog of frizzled 6, Fz6, are present and sequestered to different sides of presumptive hair cells (Deans et al., 2007). This reflects the early classical PCP asymmetry expected of hair cells. The direction of this polarization based on the markers Pk2 and Fz6 is reportedly the same for hair cells throughout the utricular epithelium, and thus, in general, is consistent with PCP organization. However, as noted, MPVs are not aligned in this manner in macular organs. For example, at the reversal line, MPVs are out of alignment with PCP by 180 degrees. This means that on one side of the reversal line Pk2 is localized near the kinocilium, whereas on the other side of the reversal line Pk2 is localized opposite the kinocilium. This suggests that the intrinsic polarity of a cell (e.g., MPV) may have different orientations in relation to the PCP organization, depending on the particular epithelium considered. Thus, normally, the relationship between the two may be different. These and other findings are remarkable as they suggest that there is some other yet unknown patterning signal that determines the orientation of the kinocilium and thus the MPV with respect to PCP markers. This unknown signal may mediate the linkage between PCP and intrinsic cellular polarity. Recent evidence may indicate some role for the homeobox transcription factor *Emx2*, which when knocked out (*Emx2$^{KO/KO}$*), eliminates the macular reversal line but has little effect on PCP in the maculae (Holley et al., 2010).

The progressive development of selected features of vestibular hair cell bundles over the period E12 to E18.5 in the mouse (weeks 7 to 15) is illustrated in Figure 6–8. Morphological polarization occurs rapidly as the kinocilium assumes an eccentric position on the apical surface of hair cells. By E13.5 (week 8), signs of morphological polarization are seen in large numbers of cells (see Figure 6–8). Moreover, already at this stage there is evidence of a planar organization of MPVs. MPV angles shift systematically as a function of position over the macular surface. A clear line of MPV reversal, however, is not seen. Thus, like auditory hair cells, planar organization of vestibular MPVs does not precisely match the mature organization initially. Reorientation of incipient MPVs is required. This happens quickly. By E15.5 (week 10) the striolar line of MPV reversal is sharp and clearly established (Denman-Johnson & Forge, 1999). The cuticular plate is first apparent in some macular hair cells using electron microscopy at about this time (E15.5, week 10), although traces of the cuticular plate are seen earlier using specific immunological markers (E13-E14) (Nishida et al., 1998). Once bundles appear in vestibular hair cells, development of the normal staircase form and maturation occurs rapidly (see Figure 6–8). By E15.5 (week 11), typical staircase shapes and numerous lateral and tip stereocilia links can be found (Anniko, 1983c; Mbiene & Sans, 1986; Forge et al., 1997). Bundle height increases progressively from E15 (e.g., utricle on average ~4 mm) until maturation after birth (e.g., stereocilia utricle ~12 mm) (Denman-Johnson & Forge, 1999). In the human, bundle height increases dramatically between weeks 10 and 11, and the bundles have achieved adult size by week 15 (Dechesne, 1992). Functional transduction channels make their appearance in the tips of stereocilia at about E16 (Geleoc & Holt, 2003), estimated to be week 12 for the human. Nascent hair cells continue to appear at the vestibular epithelial surface. Discrete samples in time give the impression of successive waves of additional immature bundles. Hair bundles at various stages of development continue to mature, and the overlying otoconial membrane continues to elaborate until relatively mature hair bundle forms dominate the surface at E18 (weeks 11 to 12) (Denman-Johnson & Forge, 1999).

INNERVATION OF THE INNER EAR

As reviewed in Chapter 1, there are three types of innervation to the inner ear. First, cochlear and vestibular receptors communicate information about environmental sounds or head motion, respectively, to the brain via primary sensory afferent neurons having cell bodies in the peripheral statoacoustic ganglion. Statoacoustic ganglion neurons arise from within the otocyst during development. Second, the brain can also modify peripheral sensory receptors by adjusting activity in efferent neurons that have cell bodies in the brainstem and axon terminals on the inner ear hair cell sensors or on primary afferent terminals. Efferent neurons arise from rhombomere 4 during development. Third, blood vessels of the inner ear are under the control of sympathetic neurons. These autonomic neurons originate from neural crest cells during development.

Development of the Statoacoustic Ganglion

The statoacoustic ganglion and its neural projections form over the period from E10.5 to E17 (weeks 4 to 9) as progenitor cells delaminate from the anteroventral wall of the otocyst. By E13 (week 7), cochlear and vestibular anlagen can be distinguished histologically as medial and lateral portions of the statoacoustic ganglion, respectively (Figure 6–11). Although distinguishable, these two portions remain as a contiguous collection of cells until about E18 (by week 9) when they actually separate physically into the spiral and vestibular ganglia (Sher, 1971; Sulik & Cotanche, 2004). The geniculate ganglion (seventh or facial cranial nerve) separates completely from the vestibular ganglion finally on P1 in mouse.

Most sensory neurons are born between E11 and E14 (weeks 4 and 8). On E12 (week 6), fibers from the statoacoustic ganglion can be seen entering the rostrolateral wall and projecting well into the epithelium near the luminal surface of the otocyst (see Figure 6–11) (Sher, 1971;

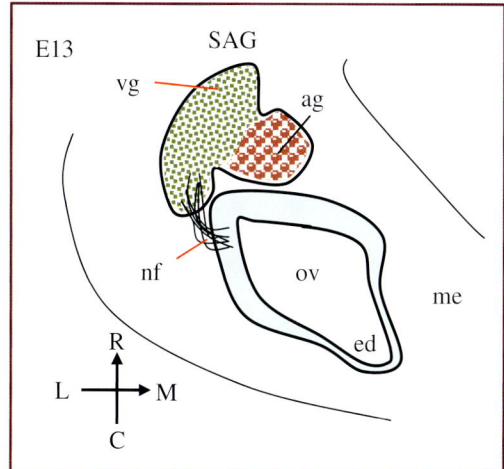

Figure 6–11. Schematic rendering of horizontal section through the otic vesicle (ov), statoacoustic ganglion (vestibular ganglion, vg and acoustic ganglion, ag) at E13 in the mouse. Neural fibers (nf) projecting into the lateral wall of the otic vesicle. Arrows designate orientations for rostro-caudal axis (R, C respectively) and mediolateral axis (M, L respectively). ed = endolymphatic duct, me = mesenchyme.

Van de Water, 1984). Efferent neurites arrive at about the same time (Bruce, Kingsley, Nichols & Fritzsch, 1997). The presence of auditory and vestibular afferents in epithelia is thought to precede slightly the arrival of efferent terminals and efferent neurites appear to follow afferent tracts during their growth (Bruce et al., 1997).

Neurotrophins regulate primary afferent innervation density. Neurotrophins are proteins secreted by target tissues that serve to prevent the natural cell death of path finding neurons (Davies, 1996; Lewin & Barde, 1996; Levi-Montalcini, 1987; Levi-Montalcini & Angeletti, 1968). Examples of neurotrophins include nerve growth factor (NGF), brain-derived neurotrophic factor (BDNF), and neurotrophin-3 (NT-3). These proteins bind with and act through high affinity receptors tyrosine kinase B (TrkB) and tyrosine kinase C (TrkC). Hypothetically, the amount of neurotrophin present ultimately determines the number of neural cells that survive to innervate target cells. This ability to determine whether developing neurons survive is known as a neurotrophic effect, and hence these proteins are called neurotrophins. Elevated amounts of neurotrophin can cause excessive growth of neurites whereas reduced levels of neurotrophin decrease neurite outgrowth and decrease survival of cells. The absence of neurotrophins can result in the loss of innervation entirely. Neurotrophins also have neurotropic effects, that is, they may serve to guide neurites along their growth paths (Fekete & Camparo, 2007; Fritzsch, Silas-Santiago, Bianchi, & Farinas, 1997; Levi-Montalcini, 1987).

The elaboration of afferent neurites appears to occur simultaneously with the delamination process in the otocyst. Two processes underlying afferent innervation of sensors have been emphasized (Fekete & Camparo, 2007; Fritzsch et al., 1997). In one model, neuroblasts send neurites back into sensory epithelia after delaminating and migrating out of the otocyst. This model requires a major guidance signal to aid neurites in their pathfinding. In the second model, the dendritic terminal endings of afferent neurites remain in the region of the original site of delamination, whereas the cell bodies migrate (translocate) to the mesenchyme rostromedial to the otocyst (see Figure 6–11). This leaves a ready-made dendritic path to target sensory regions. In this case, even though neurites are initially in the proximity of target sensory regions they must yet grow extensively and find their specific final sensory destinations. There is evidence for both models, and it is conceivable that both models operate to some extent depending on the sensory organ involved. The importance of each model may depend on the species in question (e.g., bird versus mammal). The molecular cues operating to guide neurite growth are not clear in the inner ear, although several candidates have been entertained (Fekete & Camparo, 2007; Fritzsch et al., 1997; Pauley, Matei, Beisel, & Fritzsch, 2005). The candidates include neurotrophins that may be involved in both guidance and survival of primary afferent dendrites. The neurotrophins BDNF and NT-3 are required for proper innervation patterns and maintenance of all inner ear ganglion cells. In the absence of BDNF and NT-3, all ganglion cells die before birth (Ernfors, Van de Water, Loring, & Jaenisch, 1995; Liebl, Tessarollo, Palko, & Parada, 1997; Silos-Santiago, Fagan, Garber, Fritzsch, & Barbacid, 1997).

Statoacoustic ganglion neurons must also form central projections. Neurites of the central axon must grow and terminate on cells within the cochlear nucleus, vestibular nuclei of the brainstem, and cerebellum. Guidance mechanisms for axons projecting to the CNS are independent of those responsible for peripheral afferent terminations (Pauley et al., 2005). The signals guiding central afferent projections are unknown.

By E13 (week 7), peripheral afferent projections extend to all presumptive vestibular sensory epithelia (cristae and maculae) as well as to the wall of the cochlear duct (Dechesne, 1992; Sher, 1971). Terminals for both auditory and vestibular afferent neurons at this stage are immature, and the final differentiation and refinement of these projections takes place over a prolonged period. In the mouse, final refinements in both modalities are made during postnatal periods. In the human, final refinements occur during the last trimester.

Vestibular Afferent Innervation

Vestibular afferent neurites arrive in the undifferentiated prosensory epithelium of the otocyst by E12 (week 6). Specialized synaptic contacts have been reported for the mouse vestibular epithelium as early as E15 (Mbiene et al., 1988). Early vestibular synaptic contacts form on postmitotic prosensory cells before or coincident with the onset of hair cell differentiation (Desmadryl, Dechesne, & Raymond, 1992). In the human, vestibular primary afferents arrive in the undifferentiated epithelium during weeks 6 to 7. They form numerous synaptic contacts on hair cells with emerging kinocilia by week 8. By 14 to 15 weeks, most hair bundles are relatively mature with only a few immature bundles appearing on the surface (Dechesne, 1992). Thus, it is likely that the earliest primary afferent contacts are made with immature "nonpolarized" hair cells. Through collateralization, contacts may be made with a mixture of nonpolarized and polarized hair cells. Descriptions of the surface of the vestibular epithelium at E13 to E15 (weeks 8 to 10) suggest that, at any given time, despite a clearly established striolar boundary, there are a wide variety of hair cell developmental stages coexisting in the same vicinity. Moreover, this developmental mosaic is generalized across the macula (Denman-Johnson & Forge, 1999). Hair cells with immature bundles appear between relatively more mature hair cells forming a complex mosaic of hair cell stages. It is reasonable to imagine that a sensory unit, defined as one primary afferent and all hair cells it innervates, at this stage also incorporates a mosaic of hair cell polarization stages and directions.

Thus, nascent sensory units are likely composed of hair cells that do not have a uniform hair bundle polarization status. Indeed, it would appear that hair cells at this stage may take on a wide range of immature features including hair cells with varying and just emerging polarization vectors, particularly those innervating regions of the line of polarity reversal.

It is important to note that during the period from E14 to E16 (weeks 7 to 8), otoconial growth begins and calcification rates are at their highest levels (Lim, 1984; Nakahara & Bevelander, 1979; Salamat, Ross, & Peacor, 1980; Veenhof, 1969). Presumably therefore, stimulus-dependent developmental processes in macular organs could become effective only during and after this period of otoconial formation. Although initial contacts may form early (~E12, week 7), they may not be functional synaptic contacts as most of the differentiation of vestibular neural dendrites occurs relatively late, from E18 to P9 (weeks 11 to 23) (Dechesne et al., 1994; Rusch, Lysakowski, & Eatock, 1998; Van de Water, Anniko, & Wersall, 1977). Similarly the final maturation of vestibular hair cells (membrane conductances), especially type I hair cells occurs during the first and second postnatal weeks for mice (week 23+ for humans) (Dechesne, 1992).

By E17 in the mouse crista (week 8), primary afferent dendrites penetrate the basal lamina and send a single process passing through lower layers to the superficial apical layer where they branch to produce several undifferentiated collaterals (Desmadryl et al., 1992). The terminals are initially restricted in their extent and density but by E20 (week 12) they begin to ramify considerably, covering distances of 30 to 50 microns in diameter. At E18 (week 8), clear evidence of synaptic contacts between afferents and hair cells is present. At these early stages the afferent terminals are still too immature to be formally classified (e.g., calyx only versus dimorph or bouton types). Soon thereafter (E18 to E20) (week 12), the first evidence of incomplete calyces as well as bouton terminals can be seen (Dechesne et al., 1994; Rusch et al., 1998; Van De Water et al., 1977). On P0 (weeks 12 to 13), there is a slight improvement where incomplete calyces, boutons, and type I and type II hair cells can be distinguished but are not mature. By approximately P5 (week 20), all three dendritic types are present and are distributed in their normal proportions over the crista. Although most features of the mature cristae are present, fine structure and function of the vestibular system in general continues to mature over several weeks after birth. These features are discussed in Chapter 7.

Cochlear Afferent Innervation:

Cochlear primary afferents also differentiate prior to hair cells. As noted above, cochlear and vestibular afferent fibers reach undifferentiated hair cell regions as early as E12 (week 6). In the region of the early cochlear duct, afferent endings penetrate the outer wall of the otocyst and enter the undifferentiated sensory epithelium. The afferent endings extend through the 4 to 6 layers of the epithelium to reach the most luminal epithelial layers of the otocyst (Sher, 1971). As the prosensory epithelium exits mitosis and initiates differentiation over the next several days (weeks), afferent fibers form contacts on hair cells that are associated with membrane swelling and presynaptic specializations (12-week-old human fetus) (Pujol et al., 1998). These afferent terminations begin as hair cells initiate differentiation first at IHCs (recall that IHCs form before

OHCs in any given cochlear region) and then extend to the OHCs. Although the precise timing and nature of early events remains to be clarified, it is clear that in most mammals primary afferent synapses form on IHCs and OHCs well before the onset of auditory function.

By E14.5 (weeks 7 to 8), afferents form bundles of radial fibers in the base and midcochlear regions, and by E15 (week 9) afferent neurons in the apex exit mitosis, representing the last spiral ganglion cells to do so (Bruce et al., 1997). Radial bundles appear in the apex by E18.5 (week 10). This basal to apex maturational gradient is observed generally. Afferent fibers reach the OHCs early in the base and only after E18.5 (week 10) in the apex.

Well before hearing begins (P0 to-P1 for the mouse, weeks 12 to 15 in the human), the configuration of afferent terminals is still immature and varies considerably along the length of the cochlea (Bruce et al., 1997; Echteler, 1992; Perkins & Morest, 1975; Pujol et al., 1998; Simmons, 1994; Simmons, Manson-Gieseke, Hendrix, Morris, & Williams, 1991). Spiral ganglion cells and fibers within the spiral lamina are unmyelinated at these ages. In some cases, fibers exhibit growth cones and filopodia, further indicating their immature status and that some endings have not acquired a hair cell target (early prehearing period, Figure 6–12). There is considerable sprouting and branching of dendrites during this period. Afferent terminals form ribbon synapses on both IHCs and OHCs. Interestingly, about 70% of the ribbons are found in OHCs whereas 30% are in IHCs, suggesting that afferent terminals are widely and abundantly distributed between OHCs and IHCs during this early prehearing period (Sobkowicz, Rose, Scott, & Slapnick, 1982). However, it is not clear what afferent types are represented for contacts on IHCs and OHCs during this time (e.g., type I, radial afferent, versus type II, spiral afferent, terminals).

The majority of fibers project radially and elaborate short branches that contact multiple (e.g., two to eight) IHCs only (early prehearing period) (see Figure 6–12). These are likely nascent type I afferent fibers. Radial fibers briefly innervating both IHCs and OHCs have also been described, and contacts by these fibers on OHCs disappear soon after birth for the mouse (Echteler, 1992). Innervation of both IHCs and OHCs by outer spiral afferents has been noted as an early transient configuration by many investigators (Perkins & Morest, 1975; Sobkowicz, 1992). Afferent fibers innervating only OHC have also been described at birth in animal models. Many of these fibers are reminiscent of outer spiral fibers and evidence extensive branching with terminal endings typical of outer spiral fibers (type II), although they lack an appreciable spiraling path at these young ages (Simmons, 1994). Longer spiraling trajectories appear later, the lengths of which increase progressively as hearing onset is approached. At later prehearing stages, spiral afferent fibers reach lengths between 400 and 1,000 μm depending on species (transitional and late prehearing periods) (see Figure 6–12). These long, spiraling afferent fibers course for considerable distances well below the OHCs and rise to make synaptic contacts in the final terminal OHC fields.

In addition, during the first week after birth in rodents (weeks 14 to 16 in humans) there is a large increase in the number of short afferent radial fibers projecting directly and primarily to the first row of outer hair cells. The appearance of these new afferent terminals is an immediate precursor to a transition in innervation patterns (transitional prehearing

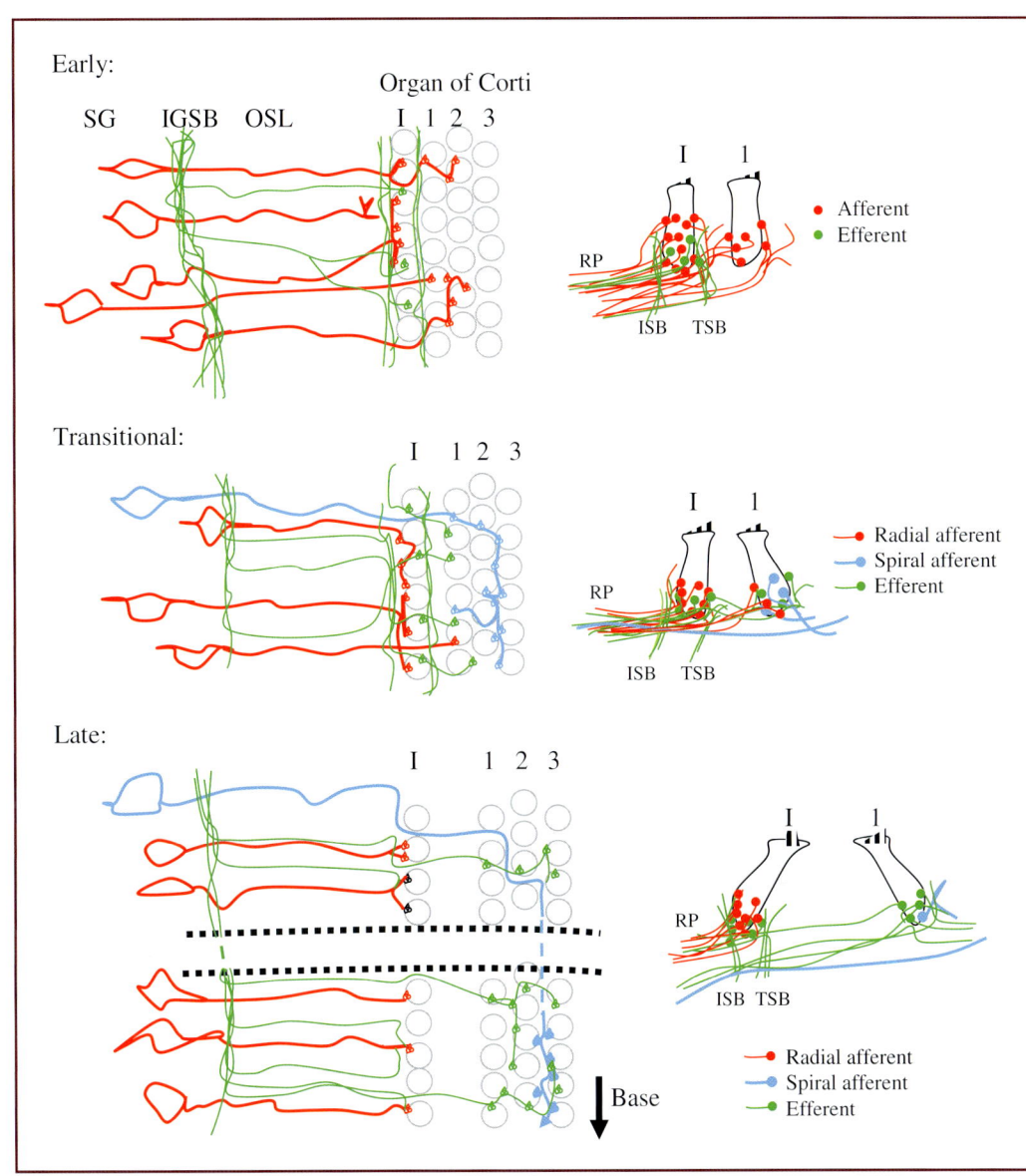

Figure 6–12. Development of mammalian cochlear afferent and efferent innervation. Schematic representation of early, transitional, and late prehearing periods of development. The mature pattern of innervation appears gradually during the first and second neonatal weeks in many mammals (e.g., mice, rats, cats, gerbils and hamsters). We estimate that this period corresponds approximately to weeks 12 to 22 in the human. In general, maturation of the base precedes that of the apex for both afferent and efferent innervation and afferents mature before efferents. IGSB = intraganglionic spiral bundle, SG = spiral ganglion; OSL = osseous spiral lamina; I = inner hair cell; 1, 2, 3 = first, second and third row of outer hair cells.

period) (see Figure 6–12). These outer radial fiber terminals are estimated to represent more than half of all afferent OHC synaptic contacts during their peak numerical expression, where the balance is represented by outer spiral afferent fibers. These outer radial afferent fiber terminal contacts on OHCs disappear abruptly days later as afferent contacts on OHC decrease and approach a mature configuration (transitional prehearing) (see Figure 6–12) (Simmons, 1994). A similar transition period is seen in the developing bird auditory papilla, where a major remodeling of afferent terminal endings and synaptic contacts occurs during prehearing periods following initial synaptogensis (approximately E11 to E17) (Fermin & Cohen, 1984a, 1984b; Rebillard & Pujol, 1983; Whitehead & Morest, 1985a, 1985b).

Although immature during the neonatal prehearing period, the afferent synaptic contacts on IHCs are likely functional (Beutner & Moser, 2001; Glowatzki & Fuchs, 2002; Knipper et al., 1997; Luo et al., 1995; Sobkowicz et al., 1982). Indeed, it is during this period of development, prior to the onset of hearing, when rhythmic spontaneous discharge (endogenous signaling) of cochlear afferents has been reported (Jones, Jones, & Paggett, 2001; Jones, Leake, Snyder, Stakhovskaya, & Bonham, 2007). Such discharge patterns are thought to play a role in guiding the refinements of central auditory tonotopic pathways. This period is discussed in more detail in Chapter 7. It is also during this period (one to two weeks after birth, depending on the species) when radial afferents eliminate terminal branches and retain a single synaptic contact on an IHC, leaving each IHC with numerous (e.g., 10 to 30) unbranched primary afferent terminal synapses.

The changes in afferent innervation patterns that occur during neonatal prehearing periods are illustrated in Figure 6–12 and represent a major reorganization and refinement of afferent terminal configurations. The refinement of afferent termination patterns is also accompanied by programmed cell death (apoptosis) among spiral ganglion neurons (e.g., Echteler, Magardino, & Rontal, 2005). Refinement continues as afferent projections form nearly mature terminal distributions just before the onset of hearing (late prehearing period, see Figure 6–12). Reorganization of efferent innervation patterns also takes place during this prehearing endogenous signaling period.

Development of Efferent Innervation

The first efferent axons arrive in the prosensory regions of the mouse otocyst at about E12 (week 6) (Fritzsch & Nichols, 1993; Pujol et al., 1998). These neural processes have origins (cell bodies) in rhombomere 4 of the hindbrain (future brainstem) and appear in the otocyst before hair cell differentiation (Bruce et al., 1997). Both ipsilateral and contralateral cell bodies give rise to these early projections. There is some evidence that the earliest arrivals in the cochlea are medial olivocochlear fibers (reviewed by Simmons, 2002). Little is known about the nature of terminal contacts made by efferents during these very early embryonic periods. On either side of the cochlear and vestibular ganglia, efferent fibers tend to grow along the afferent tracts, and their appearance in time follows just behind the leading edge of afferent projections (Bruce et al., 1997; Bruce, Christensen, & Warr, 2000). Efferents travel in the vestibular nerve until they reach the vestibulocochlear anastomosis, at which point cochlear efferents segregate from vestibular efferents and enter the spiral ganglion, forming the intraganglionic spiral bundle (IGSB). The IGSB follows the spiral ganglion through the course of the cochlear coil. Along the course of the IGSB, efferent fibers turn and run radially toward the sensory epithelium along the paths of afferent radial fibers (see Figures 1–11, 1–13, and 6–12) (Ginzberg & Morest, 1983; Perkins & Morest, 1975; Simmons, Manson-Gieseke, Hendrix, & McCarter, 1990; Spoendlin, 1966). By E14 (week 7), efferents penetrate the sensory epithelium of the emerging organ of Corti, and by E15 to E16 (week 8), efferents are found in the proximity of the greater epithelial ridge and IHCs (Bruce et al., 1997; Simmons, 2002; Sobkowicz, 1992). By E17 (weeks 11 to 12), radial efferent fibers contact IHCs, turn, and run spirally just medial and lateral beneath IHCs. These spiraling tracts ultimately form the inner and tunnel spiral bundles found in the mature cochlea (ISB & TSB) (see Figure 6–12) (Bruce et al., 1997; Fritzsch, Barbacid, & Silos-Santiago, 1998; Simmons et al., 1990).

At the time of birth, most efferents make direct axosomatic synaptic contacts on IHCs (early prehearing period) (see Figure 6–12). The exact origin of the efferent fibers forming early synapses on IHCs during this period has not been definitively established; they could be medial or lateral olivocochlear fibers. There has been some debate regarding this issue (Pujol et al., 1998; Simmons, 2002; Simmons, Mansdorf, & Kim, 1996). However, growing evidence suggests that medial olivocochlear efferent terminals make up the largest proportion of synapses on IHCs during this period (Cole & Robertson, 1992; Simmons, 2002; Simmons, Bertolotto, Kim, Raji-Kubba, & Mansdorf, 1998; Simmons et al., 1996). The high proportion of efferent axosomatic synaptic contacts on IHC at this stage is a transient configuration since it disappears with maturation. At the time of birth in the mouse, rat, hamster, and cat, there are also considerable numbers of afferent contacts on OHC, whereas there are relatively few if any efferent synapses (early prehearing period) (see Figure 6–12). Axodendritic efferent contacts begin to appear in the postnatal period in these mammals. In the human, this occurs well before birth where efferent axodendritic contacts have been demonstrated as early as week 14 (Pujol et al., 1998).

During the immediate postnatal period over several days in the mouse (weeks 15 to 20 in the human), efferent neurites begin to extend terminals and contact OHCs, and many radially projecting efferents make contacts on both IHCs and OHCs (transitional prehearing period) (see Figure 6–12) (Bruce et al., 1997; Simmons et al., 1990). Moreover, increasing numbers of efferent axodendritic synapses are found at the IHCs. This period is considered an intermediate time of transition. Between the time that efferents first synapse in large numbers on IHCs and the onset of cochlear function there is a striking reorganization of both efferent and afferent innervation in the cochlea. During this final period of refinement and maturation, IHC efferent axosomatic synapses decrease as the number of efferent synaptic contacts on OHCs increases. Ultimately this transitional phase gives way to a more adult-like configuration (late prehearing period) (see Figure 6–12) where axodendritic efferent synapses prevail on afferents innervating the IHC (lateral OCB) and axosomatic synapses are dominant on OHCs (medial OCB). Mature efferent synaptic contacts do not fully appear until after the onset of hearing, which corresponds approximately to 12 to 14 days after birth for the mouse, cat, and gerbil, and weeks 20 to 25 in the human fetus (Pujol et al., 1998).

Table 6–2. Genes Involved in Otic Placode Commitment and Specification

Age	Event	Gene	Locus	Human Disorder
3	Induction of preplacodal region	Fgf3	11q13	Congenital deafness, inner ear agenesis, microtia, and microdontia[1]
		Fgf10	5p13-p12	Lacrimo-auriculo-dento-digital (LADD) syndrome[2]
	Preotic field	Pax8	2q12-q14	Congenital Hypothyroidism[3]
		Pax2	10q24.3-q25.1	Renal Coloboma Syndrome[4]
	Otic placode from epidermis	Fgf19	11q13.1	
		Wnt8c		

Loci from the Online Mendelian Inheritance in Man (OMIM) Web site (http://www.ncbi.nlm.nih.gov/sites/entrez?cmd=search&db=omim)

[1]Tekin et al. (2007, 2008); [2]Rohmann et al. (2006); [3]Macchia et al. (1998); [4]Sanyanusin et al. (1995).

Table 6–3. Genes Expressed in the Otic Vesicle

Age	Event	Gene	Locus	Human Disorder
4	Otic Vesicle	Six1	14q23	Branchio-Oto (BO) Syndrome[5]
		Jag1	20p12	Alagille Syndrome[6]
		Lfng	7p22	Spondylocostal Dysostosis 3[7]
		Sox2	3q26.3-q27	
		Eya1	8q13.3	Branchial Oto-Renal (BOR) Syndrome[8]
		Tbx1	22q11.2	Velocardiofacial/DiGeorge Syndromes[9]
		Fgfr2	10q26	Crouzon Syndrome,[10] Apert Syndrome,[11] Pfeiffer Syndrome[12]
		Pax2	10q24.3-q25.1	Renal Coloboma Syndrome[4]
		Hmx3	10q36.13	
		Dlx5	7q22	
		Dach1	13q22	
		Dach2	Xq13.1-q21.33	

Loci from the Online Mendelian Inheritance in Man (OMIM) Web site (http://www.ncbi.nlm.nih.gov/sites/entrez?cmd=search&db=omim)

[5]Ruf et al. (2004); [6]Oda et al. (1997); [7]Sparrow et al. (2006); [8]Abdelhak et al. (1997); [9]Yagi et al. (2003), Merscher et al. (2001); [10]Reardon et al. (1994); [11]Wilkie et al. (1995); [12]Teebi et al. (2002).

Table 6–4. Genes for Delamination and Specification of Prosensory Domain

Age	Event	Genes	Locus	Human Disorder
4	Delamination and specification of neural domain	NeuroD1	2q32	Type II Diabetes[13]
		Ngn1	5q23-q31	
		Gata3	10p15	Hypoparathyroidism, Sensorineural Deafness, and Renal Disease (HDR) Syndrome[14]
		Eya1	8q13.3	BOR Syndrome[8]
		Fgfr2	10q26	Crouzon Syndrome,[10] Apert Syndrome,[11] Pfeiffer Syndrome[12]
	Specification of prosensory domain	Lfng	7p22	Spondylocostal Dysostosis 3[7]
		Jag1	20p12	Alagille Syndrome[6]
		Bmp4	14q22-q23	Anophthalmia/Microphthalmia, Cleft Lip and Palate[15]
		Sox2	3q26.3-q27	
		Gli3*	7p13	Pallister Hall Syndrome[16]

Loci from the Online Mendelian Inheritance in Man (OMIM) Web site (http://www.ncbi.nlm.nih.gov/sites/entrez?cmd=search&db=omim)

*Restricts the boundary of the prosensory domain.

[13]Malecki et al. (1999); [14]Van Esch et al. (2000); [15]Bakrania et al. (2008), Suzuki et al. (2009); [16]Kang, Graham, Olney, & Biesecker (1997).

Table 6–5. Genes for Prosensory Patches and Terminal Mitosis

Age	Event	Gene	Locus	Human Disorder
5	Prosensory Patches-Cristae	Bmp4	14q22–q23	Anophthalmia/Microphthalmia, Cleft Lip and Palate[15]
		Lfng	20p12	Spondylocostal Dysostosis 3[7]
		Pax2	10q24.3-q25.1	Renal Coloboma Syndrome[4]
		Fgf3	11q13	Congenital deafness, inner ear agenesis, microtia, and microdontia[1]
		Fgf10	5p13-p12	LADD Syndrome[2]
		Hmx3	10q26.13	
		Otx1	2p13	
	Sensory Patches-Maculae	Lfng	20p12	Spondylocostal Dysostosis 3[7]
		Pax2	10q24.3-q25.1	Renal Coloboma Syndrome[4]
		Fgf3	11q13	Congenital deafness, inner ear agenesis, microtia, and microdontia[1]
		Lmx1a	1q22-q23	
6	Sensory Patch-Cochlea (1/2 turn)	Jag1	20p12	Alagille Syndrome[6]
		Sox2	3q26.3-q27	
		Lmx1a	1q22-q23	
5–8	Terminal Mitosis-ZNP	Cdkn1b	12p13	Multiple Endocrine Neoplasia, Type IV[17]

Loci from the Online Mendelian Inheritance in Man (OMIM) Web site (http://www.ncbi.nlm.nih.gov/sites/entrez?cmd=search&db=omim)

[17]Pellegata et al. (2006).

Table 6–6. Genes Involved in Hair Cell Differentiation and Patterning of the Epithelia

Age	Event	Gene	Locus	Human Disorder
6 7	Onset of Hair Cell Differentiation— Cristae/Maculae Cochlea (1 full turn)	Atoh1 Calb2	4q22 16q22.1	
8–13	Hair Cell/Supporting Cell patterning	Notch1 Dll1 Jag2 Hes1 Hes5 Atoh1	9q34.3 6q27 14q32 3q28-q29 1q36.31 4q22	Aortic Valve Disease[18]
7 9	First appearance of stereocilia— Vestibular Cochlear (1.25 turns)	Myo6 Myo7a	6q13 11q13.5	DFNA22,[19] DFNB37[20] Usher 1B Syndrome,[21] DFNB2,[22] DFNA11[23]

Loci from the Online Mendelian Inheritance in Man (OMIM) Web site (http://www.ncbi.nlm.nih.gov/sites/entrez?cmd=search&db=omim)

[18]Garg et al. (2005); [19]Malchionda et al. (2001), Hilgert et al. (2008), Saangard et al. (2008); [20]Ahmed et al. (2003); [21]Weil et al. (1995), Weston et al. (1996); [22]Weil et al. (1997); [23]Liu et al. (1997).

Table 6–7. Genes Expressed in the Stereociliary Bundles (listed in no particular order). The genes listed encompass both vestibular and cochlear stereocilia.

Age	Event	Gene	Locus	Human Disorder
	Development of stereociliary bundles	Myo6	6q13	DFNA22,[19] DFNB37[20]
		Myo7a	11q13.5	Usher 1B Syndrome,[21] DFNB2,[22] DFNA11[23]
8–15	Vestibular Cristae	Ush1c	11p15.1	Usher 1C Syndrome,[24] DFNB18[25]
10–15	Maculae	Cdh23	10q21-q22	Usher 1D Syndrome,[26] DFNB12[27]
	Cochlear			
10–20	IHC	Pcdh15	10q21-q22	Usher 1F Syndrome,[28,29] DFNB23[29]
12–24	OHC			
		Sans	17q24-q25	Usher 1G Syndrome[30]
		Ush2a	1q41	Usher 2A Syndrome[31]
		Gpr98	5q14.3	Usher 2C Syndrome[32]
		Whrn	9q32–q34	Usher 2D Syndrome,[33] DFNB31[34]
		Espn	1p36.3-p36.1	DFNB36[35]
		Triobp	22q13.1	DFNB28[36]
		Myo3a	10p11.1	DFNB30[37]
		Myo15a	17p11.2	DFNB3[38]
		Myh9	22q11.2	May-Hegglin Anomaly, Fechtner Syndrome, Sebastian Syndrome,[39] DFNA17[40]
		Rdx	11q23	DFNB24[41]
		Lhfpl5	6p21.3	DFNB66/67[42]
		Ptprq	12q21.2	DFNB84[43]
		Actg1	17q25.3	DFNA20[44]

Loci from the Online Mendelian Inheritance in Man (OMIM) Web site (http://www.ncbi.nlm.nih.gov/sites/entrez?cmd=search&db=omim)

[24]Verpy et al. (2000); Bitner-Glindzicz et al. (2000); [25]Ahmed et al. (2002); [26]Bolz et al. (2001); [27]Bork et al., (2001); [28]Alagramam et al. (2001); [29]Ahmed et al. (2001); [30]Weil et al. (2003); [31]Eudy et al. (1998); [32]Westin et al. (2004); [33]Ebermann et al. (2007); [34]Mburu et al. (2003); [35]Naz et al. (2004); [36]Riazuddin et al. (2006); Shahin et al. (2006); [37]Walsh et al. (2002); [38]Wang et al. (1998); [39]May-Hegglin/Fechtner Syndrome Consortium (2000); [40]Lalwani et al. (2000); [41]Khan et al. (2007); [42]Shabbir et al. (2006); Kalay et al. (2006); [43]Schraders et al. (2010); [44]Zhu et al. (2003).

Table 6–8. Some of the Genes Involved in Overlying Membranes and Pillar Cells

Age	Event	Gene	Locus	Human Disorder
	Membranes—			
6–8	Cupular	Otog	11p14.3	
8–23	Otoconial	Nox3	6q25.1	
		Noxo1	16p13	
		Cyba	16q24	Chronic Granulomatosis Disease (CGD)[45]
		Otop1	4p16.2	
		Oc90	8q24	
		Tecta	11q22-q24	DFNA12/8[47], DFNB12[48]
		Tectb		
		Otog	11p14.3	
9–20+	Tectorial	Otoa	16p12.2	DFNB22[46]
		Tecta	11q22-q24	DFNA12/8,[47] DFNB12[48]
		Tectb		
		Col2a1	12q13.11-q13.2	Stickler Syndrome Type 1[49]
11–24	Pillar Cells	Fgf3	11q13	Congenital deafness, inner ear agenesis, microtia, and microdontia[1]

Loci from the Online Mendelian Inheritance in Man (OMIM) Web site (http://www.ncbi.nlm.nih.gov/sites/entrez?cmd=search&db=omim)

[45]Dinauer et al. (1990); [46]Zwaenepoel et al. (2002); [47]Verhoeven et al. (1998); [48]Mustapha et al. (1999); [49]Brown et al. (1992).

Table 6–9. Genes Involved in Nonsensory Cells

Age	Event	Gene	Locus	Human Disorder
	Supporting Cells	Tak1	3p25	
		Gjb2	13q11-q12	DFNB1A,[50] DFNA3A,[50] Vohlwinkel Syndrome,[51] Keratitis-ichthyosis-deafness syndrome (KID)[52]
		Gjb6	13q12	DFNB1B,[53] DFNA3B[54]
		Gjb3	1p35.1	DFNA2B[55]
		Slc12a6	15q13-q14	
		Slc12a7	5p15.3	
		Slc26a4	7q31	Pendred Syndrome,[56] Enlarged Vestibular Aqueduct[57]
		Cldn14	21q22.3	DFNB29[58]
		Esrrb	14q24.3	DFNB35[59]
		Myh14	19q13.33	DFNA4[60]
	Stria Vascularis/ Spiral Ligament	Gjb1	Xq13.1	Charcot-Marie-Tooth Disease[61]
		Gjb2	13q11-q12	DFNB1A,[50] DFNA3A,[50] Vohlwinkel Syndrome,[51] Keratitis-ichthyosis-deafness syndrome (KID)[52]
		Gjb3	1p35.1	DFNA2B[55]
		Gjb6	13q12	DFNB1B,[53] DFNA3B[54]
		Gja1	6q21-q23.2	Oculodentaldigital Dysplasia[62]
		Bsnd	1p31	Bartter Syndrome Type 4A[63]
		Coch	14q12-q13	DFNA9[64]
		Myh14	19q13.33	DFNA4[60]
		Tak1	3p25	
		Kcne1	21q22.1-q22.2	Jervell and Lange-Nielsen Syndrome 2,[65] Long QT Syndrome 5[66]
		Kcnq1	11p15.5	Jervell and Lange Nielsen Syndrome 1,[67] Long QT Syndrome 1[68]
		Kcnj10	1q23.2	SeSAME Syndrome[69]
		Slc12a2	5q23.3	
		Cldn11	3q26.2-q26.3	

Loci from the Online Mendelian Inheritance in Man (OMIM) Web site (http://www.ncbi.nlm.nih.gov/sites/entrez?cmd=search&db=omim)

[50]Kelsell et al. (1997); [51]Maestrini et al. (1999); [52]Richard et al. (2002); [53]delCastillo et al. (2002); [54]Grifa et al. (1999); [55]Xie et al. (1998); [56]Everett et al. (1997); [57]Li et al. (1998); [58]Wilcox et al. (2001); [59]Collin et al. (2008); [60]Donaudy et al. (2004); [61]Bergoffen et al. (1993); [62]Paznekas et al. (2003); [63]Birkenhager et al. (2001); [64]Robertson et al. (1998); [65]Tyson et al. (1997); Schulze-Bahr et al. (1997); [66]Splawski et al. (1997); [67]Neyroud et al. (1999); [68]Wang et al. (1996); [69]Scholl et al. (2009).

Table 6–10. Germ Layer Derivatives of Inner Ear Structures

Ectoderm	Neural Crest Mesenchyme	Mesodermal Mesenchyme
Membranous Labyrinth	Melanocytes	Bony Labyrinth
Hair Cells	Stria Vascularis Basal Cells	Basilar Membrane
Stria Vascularis Marginal Cells	Stria Vascularis Intermediate Cells	Stria Vascularis Basal Cells
Supporting Cells		Spiral Limbus
Reissner's Membrane Scala Media side		Reissner's Membrane Scala Vestibuli side
		Spiral Ligament
		Blood Vessels and Capillaries

GENERAL REFERENCES

Carlson, B. M. (2004). *Human embryology and developmental biology* (3rd ed.). Philadelphia, PA: Mosby.

Kelley, M. W., Wu, D., Popper, A. N., & Fay, R. R. (2005). *Development of the inner ear*. Springer Handbook of Auditory Research. New York, NY: Springer-Verlag.

Romand, R., & Varela-Nieto, I. (2003). *Development of auditory and vestibular systems 3, Molecular development of the inner ear* (Series Ed.), Current Topics in Developmental Biology: Vol. 57. Amsterdam, The Netherlands: Elsevier Academic Press.

Rubel, E. W., Popper, A. N., & Fay, R. R. (1998). *Development of the auditory system*. Springer Handbook of Auditory Research. New York, NY: Springer-Verlag.

LITERATURE CITED

Abdelhak, S., Kalatzis, V., Heilig, R., Compain, S., Samson, D., Vincent, C., . . . Petit, C. (1997). A human homologue of the Drosophila eyes absent gene underlies branchio-oto-renal (BOR) syndrome and identifies a novel gene family. *Nature Genetics, 15,* 157–164.

Ahmed, Z. M., Morell, R. J., Riazuddin, S., Gropman, A., Shaukat, S., Ahmad, M. M., . . . Wilcox, E. R.. (2003). Mutations of MYO6 are associated with recessive deafness, DFNB37. *American Journal of Human Genetics, 72,* 1315–1322.

Ahmed, Z. M., Riazuddin, S., Bernstein, S. L., Ahmed, Z., Khan, S., Griffith, A. J., . . . Wilcox, E. R. (2001). Mutations of the protocadherin gene PCDH15 cause Usher syndrome type 1F. *American Journal of Human Genetics, 69,* 25–34.

Ahmed, Z. M., Smith, T. N., Riazuddin, S., Makishima, T., Ghosh, M., Bokhari, S., . . . Wilcox, E. R. (2002). Nonsyndromic recessive deafness DFNB18 and Usher syndrome type IC are allelic mutations of USHIC. *Human Genetics, 110,* 527–531.

Alagramam, K. N., Yuan, H., Kuehn, M. H., Murcia, C. L., Wayne, S., Srisailpathy, C. R. S., . . . Smith, R. J. (2001). Mutations in the novel protocadherin PCDH15 cause Usher syndrome type 1F. *Human Molecular Genetics, 10,* 1709–1718.

Anniko, M. (1983a). Postnatal maturation of sensory hairs in the mouse. *Anatomy and Embryology, 166*, 355–368.

Anniko, M. (1983b). Cytodifferentiation of cochlear hair cells. *American Journal of Otolaryngology, 4*, 375–388

Anniko, M. (1983c). Embryonic development of vestibular sense organs and their innervation. In Romand, R. (Ed.), *Development of auditory and vestibular systems* (pp. 375–423). New York, NY: Guilford Press.

Baker, C. V. H., & Bronner-Fraser, M. (2001). Vertebrate cranial placodes. I. Embryonic induction. *Developmental Biology, 232*, 1–61.

Bakrania, P., Efthymiou, M., Klein, J. C., Salt, A., Bunyan, D. J., Wyatt, A., . . . Ragge, N. K. (2008). Mutations in BMP4 cause eye, brain, and digit developmental anomalies: Overlap between the BMP4 and hedgehog signaling pathways. *American Journal of Human Genetics, 82*, 304–319.

Bergoffen, J., Scherer, S. S., Wang, S., Oronzi Scott, M., Bone, L. J., Paul, D. L., . . . Fischbeck, K. H. (1993). Connexin mutations in X-linked Charcot-Marie-Tooth disease. *Science, 262*, 2039–2042.

Bermingham, N. A., Hassan, B. A., Price, S. D., Vollrath, M. A., Ben-Arie, N., Eatock, R. A., . . . Zoghbi, H. Y.(1999). *Math1*: An essential gene for the generation of inner ear hair cells. *Science, 284*, 1837–1841.

Beutner, D., & Moser, T. (2001). Presynaptic function of mouse cochlear inner hair cells during development of hearing. *Journal of Neuroscience, 21*(13), 4593–4599.

Birkenhager, R., Otto, E., Schurmann, M. J., Vollmer, M., Ruf, E. M., Maier-Lutz, I., . . . Hildebrandt, F. (2001). Mutation of BSND causes Bartter syndrome with sensorineural deafness and kidney failure. *Nature Genetics, 29*, 310–314.

Bok, J., Chang, W., & Wu, D. K. (2007). Patterning and morphogenesis of the vertebrate inner ear. *International Journal of Developmental Biology, 51*, 521–533.

Bolz, H., von Brederlow, B., Ramirez, A., Bryda, E. C., Kutsche, K., Nothwang, H. G., . . . Kubisch, C. (2001). Mutation of CDH23, encoding a new member of the cadherin gene family, causes Usher syndrome type 1D. *Nature Genetics, 27*, 108–112.

Bork, J. M., Peters, L. M., Riazuddin, S., Bernstein, S. L., Ahmed, Z. M., Ness, S. L., . . . Morell, R. J. (2001). Usher syndrome 1D and nonsyndromic autosomal recessive deafness DFNB12 are caused by allelic mutations of the novel cadherin-like gene CDH23. *American Journal of Human Genetics, 68*, 26–37.

Bredberg, G. (1968). Cellular pattern and nerve supply of the human organ of Corti [Supplemental material]. *Acta Oto-Laryngologica, 236*, 1–135.

Brown, D. M., Nichols, B. E., Weingeist, T. A., Sheffield, V. C., Kimura, A. E., & Stone, E. M. (1992). Procollagen II gene mutation in Stickler syndrome. *Archives in Ophthalmology, 110*, 1589–1593.

Bruce, L. L., Christensen, M. A., & Warr, B. (2000). Postnatal development of efferent synapses in the rat cochlea. *Journal of Comparative Neurology, 423*, 532–548.

Bruce, L. L., Kingsley, J., Nichols, D. H., & Fritzsch, B. (1997). The development of vestibulocochlear efferents and cochlear afferents in mice. *International Journal of Developmental Neuroscience, 15*(4-5), 671–692.

Bryant, J. E., Forge, A., & Richardson, G. P. (2005). The differentiation of hair cells. In M. W. Kelley, D. Wu, A. A. Popper, & R. R. Fay (Eds.), *Development of the inner ear* (pp. 158–203). New York, NY: Springer-Verlag.

Cantos, R., Cole, L. K., Acampora, D., Simeone, A., & Wu, D. K. (2000). Patterning of the mammalian cochlea. *Proceedings of the National Academy of Sciences, 97*(22), 11707–11713.

Chen, P. (2002). The role of math1 in inner ear development: Uncoupling the establishment of the sensory primordium from hair cell fate determination. *Development, 129*, 2495–2505.

Chen, P., & Segil, N. (1999). P27kip1 links cell proliferation to morphogenesis in the organ of Corti. *Development, 126*, 1581–1590.

Chen, P., Johnson, J. E., Zoghbi, H.Y., & Segil, N. (2002). The role of *Math1* in inner ear development: Uncoupling the establishment of the sensory primordium from hair cell fate determination. *Development*, *129*, 2495–2505.

Cole, K. S., & Robertson, D. (1992). Early efferent innervation of the rat cochlea studied with a carbocyanine dye. *Brain Research*, *575*, 223–230.

Collin, R. W. J., Kalay, E., Tariq, M., Peters, T., van der Zwaag, B., Venselaar, H., . . . Kremer, H. (2008). Mutations of ESRRB encoding estrogen-related receptor beta cause autosomal-recessive nonsyndromic hearing impairment DFNB35. *American Journal of Human Genetics, 82*, 125–138.

Cotanche, D., & Corwin, J. (1991). Stereociliary bundles during hair cell development and regeneration in the chick cochlea. *Hearing Research, 52*, 379–402

Dabdoub, A., Donohue, M. J., Brennan, A., Wolf, V., Montcouquiol, M., Sassoon, D.A., . . . Kelley, M. W. (2003). Wnt signaling mediates reorientation of outer hair cell stereociliary bundles in the mammalian cochlea. *Development, 130*, 2375–2384.

Davies, A. M. (1996). The neurotrophic hypothesis: Where does it stand? *Philosophical Transactions of the Royal Society of London B, 351*, 389–394.

Deans, M. R., Antic, D., Suyama, K., Scott, M. P., Axelrod, J. D., & Goodrich, L. V. (2007). Asymmetric distribution of prickle-like 2 reveals an early underlying polarization of vestibular sensory epithelia in the inner ear. *Journal of Neuroscience, 27*(12), 3139–3147.

Dechesne, C. J. (1992). The development of vestibular sensory organs in human. In R. Romand (Ed.), *Development of auditory and vestibular systems 2* (pp. 419–447). Amsterdam, The Netherlands: Elsevier.

Dechesne, C. J., Rabejac, D., & Desmadryl, G. (1994). Development of calretinin immunoreactivity in the mouse inner ear. *Journal of Comparative Neurology, 346*, 517–529.

del Castillo, I., Villamar, M., Moreno-Pelayo, M. A., del Castillo, F. J., Alvarez, A., Telleria, D., . . . Moreno, F. (2002). A deletion involving the connexin 30 gene in nonsyndromic hearing impairment. *New England Journal of Medicine, 346*, 243–249.

Denman-Johnson, K., & Forge, A. (1999). Establishment of hair bundle polarity and orientation in the developing vestibular system of the mouse. *Journal of Neurocytology, 28*, 821–835.

Desmadryl, G., Dechesne, C., & Raymond, J. (1992). Recent aspects of development of the vestibular sense organs and their innervation. In R. Romand (Ed.), *Development of auditory and vestibular systems 2* (pp. 461–487). Amsterdam, The Netherlands: Elsevier.

Dinauer, M. C., Pierce, E. A., Bruns, G. A. P., Curnutte, J. T., & Orkin, S. H. (1990). Human neutrophil cytochrome b light chain (p22-phox): Gene structure, chromosomal location, and mutations in cytochrome-negative autosomal recessive chronic granulomatous disease. *Journal of Clinical Investigation, 86*, 1729–1737.

Donaudy, F., Snoeckx, R., Pfister, M., Zenner, H. P., Blin, N., Di Stazio, M., . . . Savoia, A. (2004). Nonmuscle myosin heavy-chain gene MYH14 is expressed in cochlea and mutated in patients affected by autosomal dominant hearing impairment (DFNA4). *American Journal of Human Genetics, 74*, 770–776.

Driver, E. C., & Kelley, M. W. (2009). Specification of cell fate in the mammalian cochlea. *Birth Defects Research* (Part C), *87*, 212–221.

Driver, E. C., Pryor, S. P., Hill, P., Turner, J., Ruther, U., Biesecker, L. G., . . . Kelley, M. W. (2008). Hedgehog signaling regulates sensory cell formation and auditory function in mice and humans. *Journal of Neuroscience, 28*(29), 7350–7358.

Ebermann, I., Scholl, H. P. N., Issa, P. C., Becirovic, E., Lamprecht, J., Jurklies, B., . . . Bolz, H. (2007). A novel gene for Usher syndrome type 2: Mutations in the long isoform of whirlin are associated with retinitis pigmentosa and sensorineural hearing loss. *Human Genetics*, *121*, 203–211.

Echteler, S. M. (1992). Developmental segregation of the afferent projections to mammalian auditory hair cells. *Proceedings of the National Academy of Sciences, 89*, 6324–6327.

Echteler, S. M., Magardino, T., & Rontal, M. (2005). Spatiotemporal patterns of neuronal programmed cell death during postnatal development of the gerbil cochlea. *Developmental Brain Research, 157,* 192–200.

Ernfors, P., Van de Water, T., Loring, J., & Jaenisch, R. (1995). Complementary roles of BDNF and NT-3 in vestibular and auditory development. *Neuron, 14,* 1153–1164.

Eudy, J. D., Weston, M. D., Yao, S., Hoover, D. M., Rehm, H. L., Ma-Edmonds, M., . . . Sumegi, A. (1998). Mutation of a gene encoding a protein with extracellular matrix motifs in Usher syndrome type IIa. *Science, 280,* 1753–1757.

Everett, L. A., Glaser, B., Beck, J. C., Idol, J. R., Buchs, A., Heyman, M., . . . Green, E. D. (1997). Pendred syndrome is caused by mutations in a putative sulphate transporter gene (PDS). *Nature Genetics, 17,* 411–422.

Fekete, D. M., & Camparo, A. M. (2007). Axon guidance in the inner ear. *International Journal of Developmental Biology, 51,* 549–556.

Fekete, D. M., & Wu, D. K. (2002). Revisiting cell fate specification in the inner ear. *Current Opinion in Neurobiology, 12,* 35–42.

Fermin, C. D. & Cohen, G. M. (1984a). Developmental gradients in the embryonic chick's basilar papilla. *Acta Oto-Laryngolica, 97,* 39–51.

Fermin, C.D. & Cohen, G.M. (1984b). Development of the embryonic chick's statoacoustic ganglion. *Acta Oto-Laryngolica, 98,* 42–52.

Forge, A., Souter, M., & Denman-Johnson, K. (1997). Structural development of sensory cells in the ear. *Seminars in Cell and Developmental Biology, 8,* 225–237.

Fritzsch, B., Beisel, K. W., Jones, K., Farinas, I., Maklad, A., Lee, J., & Reichardt, L. F. (2002). Development and evolution of inner ear sensory epithelia and their innervations. *Journal of Neurobiology, 53,* 143–156.

Fritzsch, B., & Nichols, D. H. (1993). DiI reveals a prenatal arrival of efferents at the differentiating otocyst of mice. *Hearing Research, 65,* 51–60.

Fritzsch, B., Silas-Santiago, I., Bianchi, L. M., & Farinas, I. (1997). The role of neurotrophic factors in regulating the development of inner ear innervation. *Trends in Neuroscience, 20*(4), 159–164.

Fritzsch, G., Barbacid, M., & Silos-Santiago, I. (1998). The combined effects of TrkB and TrkC on the innervation of the inner ear. *International Journal of Developmental Neuroscience, 16*(6), 493–505.

Gao, W. Q. (2003). Hair cell development in higher vertebrates. In R. Romand, & I. Varela-Nieto (Eds.), *Development of auditory and vestibular systems 3: Molecular development of the inner ear* (pp. 294–321). San Diego, CA: Elsevier Academic Press.

Garg, V., Muth, A. N., Ransom, J. F., Schluterman, M. K., Barnes, R., King, I. N., . . . Srivastava, D. (2005). Mutations in NOTCH1 cause aortic valve disease. *Nature, 437,* 270–274.

Geleoc, G. S. G., & Holt, J. R. (2003). Developmental acquisition of sensory transduction in hair cells of the mouse inner ear. *Nature Neuroscience, 6,* 1019–1020.

Ginzberg, R. D., & Gilula, N. B. (1979). Modulation of cell junctions during differentiation of the chicken otocyst sensory epithelium. *Developmental Biology, 68,* 110–129.

Ginzberg, R. D. & Morest, D. K. (1983). A study of cochlear innervation in the young cat with the Golgi method. *Hearing Research, 10,* 227–246.

Glowatzki, E., & Fuchs, P.A. (2002). Transmitter release and the hair cell ribbon synapse. *Nature Neuroscience, 5*(2), 147–154.

Grifa, A., Wagner, C. A., D'Ambrosio, L., Melchionda, S., Bernardi, F., Lopez-Bigas, N., . . . Gasparini, P. (1999). Mutations in GJB6 cause nonsyndromic autosomal dominant deafness at DFNA3 locus. *Nature Genetics, 23,* 16–18.

Groves, A.K., (2005). The induction of the otic placode. In M. W. Kelley, D. K. Wu, A. N. Popper, & R. R. Fay (Eds.), *Development of the ear* (pp. 10–42). New York, NY: Springer-Verlag.

Hasson, T., Gillespie, P. G., Garcia, J. A., MacDonald, R. B., Zhao, Y. D., Yee, A. G., . . . Corey, D. P. (1997). Unconventional myosins in inner-ear sensory epithelia. *Journal of Cell Biology, 137,* 1287–1307.

Helms, A. W., Abney, A. L., Ben-Arie, N., Zoghbi, H.Y., & Johnson, J. E. (2000). Autoregulation and multiple enhancers control *Math1* expression in the developing nervous system. *Development, 127*, 1185–1196.

Hilgert, N., Topsakal, V., van Dinther, J., Offeciers, E., Van de Heyning, P., & Van Camp, G. (2008). A splice-site mutation and overexpression of MYO6 cause a similar phenotype in two families with autosomal dominant hearing loss. *European Journal of Human Genetics, 16*, 593–602.

Holley, M., Rhodes, C., Kneebone, A. Herde, M. K., Fleming, M., & Steel, K. P. (2010). Emx2 and early hair cell development in the mouse inner ear. *Developmental Biology.* doi:10.1016/j.ydbio.2010.02.004

Jeffery, N., & Spoor, F. (2004). Prenatal growth and development of the modern human labyrinth. *Journal of Anatomy, 204*, 71–92.

Jones, C., & Chen, P. (2007). Planar cell polarity signaling in vertebrates. *BioEssays, 29*, 120, 132.

Jones, C., Roper, V., Foucher, I., Qian, D., Banizs, B., Petit, C., . . . Chen, P. (2008). Ciliary proteins link basal body polarization to planar cell polarity regulation. *Nature Genetics, 40*(1), 69–77.

Jones, J. M., Montcouquiol, M., Dabdoub, A., Woods, C., & Kelley, M. W. (2006). Inhibitors of differentiation and DNA binding (Ids) regulate Math1 and hair cell formation during the development of the organ of Corti. *Journal of Neuroscience, 26*(2), 550–558.

Jones, T. A., Leake, P. A., Snyder, R. L., Stakhovskaya, O., & Bonham, B. (2007). Spontaneous discharge patterns in cochlear spiral ganglion cells prior to the onset of hearing in cats. *Journal of Neurophysiology, 98*, 1898–1908.

Jones, T. A., Jones, S. M., & Paggett, K. M. (2001). Primordial rhythmic bursting in embryonic cochlear ganglion cells. *Journal of Neuroscience, 21*(20), 8129–8135.

Kalay, E., Li, Y., Uzumcu, A., Uyguner, O., Collin, R. W., Caylan, R., . . . Wollnik, B. (2006). Mutations in the lipoma HMGIC fusion partner-like 5 (LHFPL5) gene cause autosomal recessive nonsyndromic hearing loss. *Human Mutations, 27*, 633–639.

Kaltenbach, J. A., Falzarano, P. R., & Simpson, T. H. (1994). Postnatal development of the hamster cochlea. II. Growth and differentiation of stereocilia bundles. *Journal of Comparative Neurology, 350*, 187–198.

Kang, S., Graham, J. M., Jr., Olney, A. H., & Biesecker, L. G. (1997). GLI3 frameshift mutations cause autosomal dominant Pallister-Hall syndrome. *Nature Genetics, 15*, 266–268.

Kazmierczak, P., Sakaguchi, H., Tokita, J., Wilson-Kubalek, E. M., Milligan, R. A., Müller, U., & Kachar, B. (2007). Cadherin 23 and protocadherin 15 interact to form tip link filaments in sensory hair cells. *Nature, 449*(7158), 87–91.

Kelley, M. W. (2007). Cellular commitment and differentiation in the organ of Corti. *International Journal of Developmental Biology, 51*, 571–583.

Kelley, M. W., & Chen, P. (2007). Shaping the mammalian sensory organ by the planar cell polarity pathway. *International Journal of Developmental Biology, 51*, 535–547.

Kelsell, D. P., Dunlop, J., Stevens, H. P., Lench, N. J., Liang, J. N., Parry, G., . . . Leigh, I. M. (1997). Connexin 26 mutations in hereditary non-syndromic sensorineural deafness. *Nature, 387*, 80–83.

Khan, S. Y., Ahmed, Z. M., Shabbir, M. I., Kitajiri, S., Kalsoom, S., Tasneem, S., . . . Riazuddin, S. (2007). Mutations of the RDX gene cause nonsyndromic hearing loss at the DFNB24 locus. *Human Mutations, 28*, 417–423.

Knipper, M., Kopschall, I., Rohbock, K., Kopke, A. K. E., Bonk, I., & Zenner, H. P. (1997). Transient expression of NMDA receptors during arrangement of AMPA-receptor-expressing fibers in the developing inner ear. *Cell and Tissue Research, 287*, 23–41.

Lalwani, A. K., Goldstein, J. A., Kelley, M. J., Luxford, W., Castelein, C. M., & Mhatre, A. N. (2000). Human nonsyndromic hereditary deafness DFNA17 is due to a mutation in nonmuscle myosin MYH9. *American Journal of Human Genetics, 67*, 1121–1128.

Lanford, P. J., Lan, Y., Jiang, R., Lindsell, C., Weinmaster, G., Gridley, T., . . . Kelley, M. W. (1999). Notch signaling pathway mediates hair cell development in mammalian cochlea. *Nature Genetics, 21,* 289–292.

Lanford, P. J., Shailam, R., Norton, C. R., Gridley, T., & Kelley, M. W. (2000). Expression of *Math1* and HES5 in the cochlea of wild-type and *Jag2* mutant mice. *Journal of the Association for Research in Otolaryngology, 1,* 161–171.

Lavigne-Rebillard, M., & Pujol, R. (1986). Development of the auditory hair cell surface in human fetuses. A scanning electron microscopy study. *Anatomy and Embryology, 174,* 369–377.

Lee, Y. S., Liu, F., & Segil, N. (2006). A morphogenetic wave of p27kip1 transcription directs cell cycle exit during organ of Corti development. *Development, 133,* 2817–2826.

Levi-Montalcini, R. (1987). The nerve growth factor: 35 years later. *Science, 237,* 1154–1162.

Levi-Montalcini, R., & Angeletti, P. U. (1968). Nerve growth factor. *Physiological Reviews, 48*(3), 534–569.

Lewin, G. R., & Barde, Y. A. (1996). Physiology of the neurotrophins. *Annual Review of Neuroscience, 19,* 289–317.

Li, S., Price, S. M., Cahill, H., Ryugo, D., Shen, M. M., & Xiang, M. (2002). Hearing loss caused by progressive degeneration of cochlear hair cells in mice deficient for *Barhl1* homeobox gene. *Development, 129,* 3523–3532.

Li, X. C., Everett, L. A., Lalwani, A. K., Desmukh, D., Friedman, T. B., Green, E. D., & Wilcox, E. R. (1998). A mutation in PDS causes nonsyndromic recessive deafness. *Nature Genetics, 18,* 215–217.

Liebl, D. J., Tessarollo, L., Palko, M. E., & Parada, L. F. (1997). Absence of sensory neurons before target innervation in brain derived neurotrophic factor-, neurotrophin 3-, and Trk-C deficient embryonic mice. *Journal of Neuroscience, 17*(23), 9113–9121.

Lim, D. J. (1984). The development and structure of otoconia. In I. Friedman & J. Ballantyne (Eds.), *Ultrastructural atlas of the inner ear* (pp. 245–269). London, UK: Butterworth.

Lim, D. J., & Anniko, M. (1985). Developmental morphology of the mouse inner ear *Acta Oto-Laryngologica Supplement, 422,* 1–69.

Lim, D. J. & Rueda, J. (1992). Structural development of the cochlea. In R. Romand (Ed.), *Development of the auditory and vestibular systems 2* (pp. 33–58). Amsterdam, The Netherlands: Elsevier Science.

Liu, X. Z.,Walsh, J., Tamagawa, Y., Kitamura, K., Nishizawa, M., Steel, K. P., & Brown, S. D. M. (1997). Autosomal dominant non-syndromic deafness caused by a mutation in the myosin VIIA gene. *Nature Genetics, 17,* 268–269.

Luo, L., Brumm, D., & Ryan, A. F. (1995). Distribution of non-NMDA gluatamate receptor mRNAs in the developing rat cochlea. *Journal of Comparative Neurology, 361,* 372–382.

Ma, Q., Anderson, D. J., & Fritzsch, B. (2000). Neurogenin1 null mutant ears develop fewer, morphologically normal hair cells in smaller sensory epithelia devoid of innervation. *Journal of the Association for Research in Otolaryngology, 1,* 129-143.

Macchia, P. E., Lapi, P., Krude, H., Pirro, M. T., Missero, C., Chiovato, L., . . . DiLauro, R. (1998). PAX8 mutations associated with congenital hypothyroidism caused by thyroid dysgenesis. *Nature Genetics, 19,* 83–86.

Maestrini, E., Korge, B. P., Ocana-Sierra, J., Calzolari, E., Cambiaghi, S., Scudder, P. M., . . . Munro, C. S.. (1999). A missense mutation in connexin26, D66H, causes mutilating keratoderma with sensorineural deafness (Vohwinkel's syndrome) in three unrelated families. *Human Molecular Genetics, 8,* 1237–1243.

Malecki, M. T., Jhala, U. S., Antonellis, A., Fields, L., Doria, A., Orban, T., . . . Krolewski, A. S. (1999). Mutations in NEUROD1 are associated with the development of type 2 diabetes mellitus. *Nature Genetics, 23,* 323–328.

May-Hegglin/Fechtner Syndrome Consortium. (2000). Mutations in MYH9 result in the May-Hegglin anomaly, and Fechtner and Sebastian syndromes. *Nature Genetics, 26*, 103–105.

Mbiene, J. P., Favre, D., & Sans, A. (1984). The pattern of ciliary development in fetal mouse vestibular receptors. *Anatomy and Embryology, 170*, 229–238.

Mbiene, J. P., Favre, D., & Sans, A. (1988). Early innervation of hair cells in the vestibular epithelia of mouse embryos: SEM and TEM study. *Anatomy and Embryology, 177*, 331–340.

Mbiene, J. P., & Sans, A. (1986). Differentiation and maturation of the sensory hair bundles in the fetal and postnatal vestibular receptors of the mouse: A scanning microscopy study. *Journal of Comparative Neurology, 254*, 271–278.

Mburu, P., Mustapha, M., Varela, A., Weil, D., El-Amraoui, A., Holme, R. H., . . . Brown, S. D. (2003). Defects in whirlin, a PDZ domain molecule involved in stereocilia elongation, cause deafness in the whirler mouse and families with DFNB31. *Nature Genetics, 34*, 421–428.

McKenzie, E., Krupin, A., & Kelley, M. W. (2004). Cellular growth and rearrangement during the development of the mammalian organ of Corti. *Developmental Dynamics, 229*, 802–812.

Melchionda, S., Ahituv, N., Bisceglia, L., Sobe, T., Glaser, F., Rabionet, R., . . . Gasparini, P. (2001). MYO6, the human homologue of the gene responsible for deafness in Snell's Waltzer mice, is mutated in autosomal dominant nonsyndromic hearing loss. *American Journal of Human Genetics, 69*, 635–640.

Merscher, S., Funke, B., Epstein, J. A., Heyer, J., Puech, A., Lu, M. M., . . . Kucherlapati, R. (2001). TBX1 is responsible for cardiovascular defects in velo-cardio-facial/DiGeorge syndrome. *Cell, 104*, 619–629.

Montandon, P., Gacek, R. R., & Kimura, R. S. (1970). Crista neglecta in the cat and human. *Annals of Otology Rhinology and Laryngology, 79*, 105–112.

Montcouquiol, M., Rachel, R. A., Lanford, P. J., Copeland, N. G., Jenkins, N. A., & Kelley, M. W. (2003). Identification of Vangl2 and Scrb1 as planar polarity genes in mammals. *Nature, 423*, 173–177.

Montcouquiol, M., Sans, N., Huss, D., Kach, J., Dickman, J., D., Forge, A., . . . Kelley, M. W. (2006). Asymmetric localization of Vangl2 and Fz3 indicate novel mechanisms for planar cell polarity in mammals. *Journal of Neuroscience, 26*(19), 5265–5275.

Morrison, A., Hodgetts, C., Gossler, A., Hrabe de Angelis, M., & Lewis, J. (1999). Expression of Delta1 and Serrate1 (Jagged1) in the mouse inner ear. *Mechanisms of Development, 84*, 169–172.

Morsli, H., Choo, D., Ryan, A., Johnson, R., & Wu, D. K. (1998). Development of the mouse inner ear and origin of its sensory organs. *Journal of Neuroscience, 18*(9), 3327–3335.

Mustapha, M., Weil, D., Chardenoux, S., Elias, S., El-Zir, E., Beckmann, J. S., . . . Petit, C. (1999). An alpha-tectorin gene defect causes a newly identified autosomal recessive form of sensorineural pre-lingual non-syndromic deafness, DFNB21. *Human Molecular Genetics, 8*, 409–412.

Nakahara, H., & Bevelander, G. (1979). An electron microscope study of crystal calcium carbonate formation in the mouse otolith. *Anatomical Record, 193*, 233–242.

Naz, S., Griffith, A. J., Riazuddin, S., Hampton, L. L., Battey, J. F., Jr., Khan, S. N., . . . Friedman, T. B. (2004). Mutations of ESPN cause autosomal recessive deafness and vestibular dysfunction. *Journal of Medical Genetics, 41*, 591–595.

Neyroud, N., Richard, P., Vignier, N., Donger, C., Denjoy, I., Demay, L., . . . Guicheney, P. (1999). Genomic organization of the KCNQ1 K+ channel gene and identification of C-terminal mutations in the long-QT syndrome. *Circulation Research, 84*, 290–297.

Nichols, D. H., Pauley, S., Jahan, I., Beisel, K. W., Millen, K. J., & Fritzsch, B. (2008). *Lmx1a* is required for segregation of sensory epithelia and normal ear histogenesis and morphogenesis. *Cell and Tissue Research, 334*(3), 339–358.

Nishida, Y., Rivolta, M. N., & Holley, M. C. (1998). Timed markers for the differentiation of the cuticular plate and stereocilia in hair cells from the mouse inner ear. *Journal of Comparative Neurology, 395*, 18–28.

Oda, T., Elkahloun, A. G., Pike, B. L., Okajima, K., Krantz, I. D., Genin, A., . . . Chandrasekharappa, S. C. (1997). Mutations in the human Jagged1 gene are responsible for Alagille syndrome. *Nature Genetics, 16*, 235–242.

Ohyama, T., Groves, A. K., & Martin, K. (2007). The first steps toward hearing: Mechanisms of otic placode induction. *International Journal of Developmental Biology, 51*, 463–472.

Ozaki, H., Nakamura, N., Funahashi, J., Ikeda, K., Yamada, G., Tokano, H., . . . Kawakami, K. (2004). *Six1* controls patterning of the mouse otic vesicle. *Development, 131*, 551–562.

Pauley, S., Matei, V., Beisel, K. W., & Fritzsch, B. (2005). Wiring the ear to the brain: The molecular basis of neurosensory development, differentiation, and survival. In M. W. Kelley, D. K. Wu, A. N. Popper, & R. R. Fay (Eds.), *Development of the inner ear* (pp. 85–121). New York, NY: Springer-Verlag.

Paznekas, W. A., Boyadjiev, S. A., Shapiro, R. E., Daniels, O., Wollnik, B., Keegan, C. E., Innis, J. W., . . . Jabs, E. W. (2003). Connexin 43 (GJA1) mutations cause the pleiotropic phenotype of oculodentodigital dysplasia. *American Journal of Human Genetics, 72*, 408–418.

Pellegata, N. S., Quintanilla-Martinez, L., Siggelkow, H., Samson, E., Bink, K., Hofler, H., . . . Atkinson, M. J. (2006). Germ-line mutations in p27(Kip1) cause a multiple endocrine neoplasia syndrome in rats and humans. *Proceedings of the National Academy of Sciences, 103*, 15558–15563.

Perkins, R. E., & Morest, D. K. (1975). A study of cochlear innervation patterns in cats and rats with the Golgi method and Nomarski optics. *Journal of Comparative Neurology, 163*, 129–158.

Pujol, R., Lavigne-Rebillard, M., & Lenoir, M. (1998). Development of sensory and neural structures in the mammalian cochlea. In E. W. Rubel, A. N. Popper, & R. R. Fay (Eds.), *Development of the auditory system* (pp. 146–192). New York, NY: Springer-Verlag.

Reardon, W., Winter, R. M., Rutland, P., Pulleyn, L. J., Jones, B. M., & Malcolm, S. (1994). Mutations in the fibroblast growth factor receptor 2 gene cause Crouzon syndrome. *Nature Genetics, 8*, 98–103.

Rebillard, M., & Pujol, R. (1983). Innervation of the chicken basilar papilla during development. *Acta Oto-Laryngolica, 96*, 379–388.

Riazuddin, S., Khan, S. N., Ahmed, Z. M., Ghosh, M., Caution, K., Nazli, . . . Friedman, T. B. (2006). Mutations in TRIOBP, which encodes a putative cytoskeletal-organizing protein, are associated with nonsyndromic recessive deafness. *American Journal of Human Genetics, 78*, 137–142.

Richard, G., Rouan, F., Willoughby, C. E., Brown, N., Chung, P., Ryynanen, M., . . . Russell, L. (2002). Missense mutations in GJB2 encoding connexin-26 cause the ectodermal dysplasia keratitis-ichthyosis-deafness syndrome. *American Journal of Human Genetics, 70*, 1341–1348.

Rida, P. C. G., & Chen, P. (2009). Line up and listen: Planar cell polarity regulation in the mammalian inner ear. *Seminars in Cell and Developmental Biology, 20*, 978–985.

Robertson, N. G., Lu, L., Heller, S., Merchant, S. N., Eavey, R. D., McKenna, M., . . . Seidman, J. G. (1998). Mutations in a novel cochlear gene cause DFNA9, a human nonsyndromic deafness with vestibular dysfunction. *Nature Genetics, 20*, 299–303.

Rohmann, E., Brunner, H. G., Kayserili, H., Uyguner, O., Nurnberg, G., Lew, E. D., . . . Wollnik, B.. (2006). Mutations in different components of FGF signaling in LADD syndrome. *Nature Genetics, 38*, 414–417.

Romand, R. (2003). The roles of retinoic acid during inner ear development. In R. Romand & I. Varela-Nieto (Eds.), *Development of auditory and vestibular systems 3: Molecular development of the inner ear* (pp. 261–292). Amsterdam, The Netherlands: Elsevier Academic Press.

Romand, R., Dollé, P., & Hashino, E. (2006). Retinoid signaling in inner ear development. *Journal of Neurobiology, 66*(7), 687–704.

Ruben, R. J. (1967). Development of the inner ear of the mouse: A radioautographic study of terminal mitoses [Supplemental Material]. *Acta Oto-Laryngologica, 220*, 5–44.

Ruf, R. G., Xu, P. X., Silvius, D., Otto, E. A., Beekmann, F., Muerb, U. T., . . . Hildebrandt, F. (2004). SIX1 mutations cause branchio-oto-renal syndrome by disruption of EYA1-SIX1-DNA complexes. *Proceedings of the National Academy of Sciences, 101,* 8090–8095.

Rusch, A., Lysakowski, A., & Eatock, R. A. (1998). Postnatal development of type I and type II hair cells in the mouse utricle: Acquisition of voltage-gated conductances and differentiated morphology. *Journal of Neuroscience, 18,* 7487–7501.

Salamat, M. S., Ross, M. D., & Peacor, D. R. (1980). Otoconia formation in the fetal rat. *Annals of Otology, Rhinology, and Laryngology, 89,* 229–238.

Sanggaard, K. M., Kjaer, K. W., Eiberg, H., Nurnberg, G., Nurnberg, P., Hoffman, K., . . . Tranebjaerg, L. (2008). A novel nonsense mutation in MYO6 is associated with progressive nonsyndromic hearing loss in a Danish DFNA22 family. *American Journal of Medical Genetics, 146A,* 1017–1025.

Sans, A., & Chat, M. (1982). Analysis of temporal and spatial patterns of rat vestibular hair cell differentiation by tritiated thymidine radiography. *Journal of Comparative Neurology, 206,* 1–8.

Sans, A., & Deschesne, C. (1985). Early development of vestibular receptors in human embryos: An electron microscopic study [Supplemental Material]. *Acta Oto-Laryngologica, 423,* 51–58.

Sanyanusin, P., Schimmenti, L. A., McNoe, L. A., Ward, T. A., Pierpont, M. E. M., Sullivan, M. J., . . . Eccles, M. R. (1995). Mutation of the PAX2 gene in a family with optic nerve colobomas, renal anomalies and vesico-ureteral reflux. *Nature Genetics, 9,* 358–364.

Schneider-Maunoury, S., & Pujades, C. (2007). Hindbrain signals in otic regionalization: walk on the wild side. *International Journal of Developmental Biology, 51,* 495–506.

Scholl, U. I., Choi, M., Liu, T., Ramaekers, V. T., Hausler, M. G., Grimmer, J., . . . Lifton, R. P. (2009). Seizures, sensorineural deafness, ataxia, mental retardation, and electrolyte imbalance (SeSAME syndrome) caused by mutations in KCNJ10. *Proceedings of the National Academy of Sciences, 106,* 5842–5847.

Schraders, M., Oostrik, J., Huygen, P. L. M., Strom, T. M., van Wijk, E., Kunst, H. P. M., . . . Kremer, H.(2010). Mutations in PTPRQ are a cause of autosomal-recessive non-syndromic hearing impairment DFNB84 and associated with vestibular dysfunction. *American Journal of Human Genetics, 86,* 604–610.

Schulze-Bahr, E., Wang, Q., Wedekind, H., Haverkamp, W., Chen, Q., Sun, Y., . . . Isbrant, D. (1997). KCNE1 mutations cause Jervell and Lange-Nielsen syndrome. *Nature Genetics, 17,* 267–268.

Shabbir, M. I., Ahmed, Z. M., Khan, S. Y., Riazuddin, S., Waryah, A. M., Khan, S. N., . . . Riazuddin, S.. (2006). Mutations of human TMHS cause recessively inherited nonsyndromic hearing loss. *Journal of Medical Genetics, 43,* 634–640.

Shahin, H., Walsh, T., Sobe, T., Sa'ed, J. A., Rayan, A. A., Lynch, E. D., . . . Kanaan, M. (2006). Mutations in a novel isoform of TRIOBP that encodes a filamentous-acting binding protein are responsible for DFNB28 recessive nonsyndromic hearing loss. *American Journal of Human Genetics, 78,* 144–152.

Sher, A.E. (1971). The embryonic and postnatal development of the inner ear of the mouse. *Acta Oto-Laryngologica Supplement, 285,* 1–77.

Silos-Santiago, I., Fagan, A. M., Garber, M., Fritzsch, B., & Barbacid, M. (1997). Severe sensory deficits but normal CNS development in newborn mice lacking TrkB and TrkC tyrosine protein kinase receptors. *European Journal of Neuroscience, 9,* 2045–2056.

Simmons, D. D. (1994). A transient afferent innervation of outer hair cells in the postnatal cochlea. *NeuroReport, 5*(11), 1309–1312.

Simmons, D. D. (2002). Development of the inner ear efferent system across vertebrate species. *Journal of Neurobiology, 53,* 228, 250.

Simmons, D. D., Bertolotto, C., Kim, J., Raji-Kubba, J., & Mansdorf, N. B. (1998). Choline acetyltransferase expression during a putative developmental waiting period. *Journal of Comparative Neurology, 397,* 281–295.

Simmons, D. D., Mansdorf, N. B., & Kim, J. H. (1996). Olivocochlear innervations of inner and outer hair cells during postnatal maturation: Evidence for a waiting period. *Journal of Comparative Neurology, 370,* 551–562.

Simmons, D. D., Manson-Gieseke, L., Hendrix, T. W., & McCarter, S. (1990). Reconstruction of efferent fibers in the postnatal hamster cochlea. *Hearing Research, 49,* 127–140.

Simmons, D. D., Manson-Gieseke, L., Hendrix, T. W., Morris, K., & Williams, S. J. (1991). Postnatal maturation of spiral ganglion neurons: A horseradish peroxidase study. *Hearing Research, 55,* 81–91.

Sobkowicz, H. M. (1992). The development of innervations in the organ of Corti. In R. Romand (Ed.), *Development of auditory and vestibular systems 2* (pp. 59–100). Amsterdam, The Netherlands: Elsevier.

Sobkowicz, H. M., Rose, J. E., Scott, G. E., & Slapnick, S. M. (1982). Ribbon synapses in the developing intact and cultured organ of Corti in the mouse. *Journal of Neuroscience, 2*(7), 942–957.

Sparrow, D. B., Chapman, G., Wouters, M. A., Whittock, N. V., Ellard, S., Fatkin, D., . . . Dunwoodie, S. L. (2006). Mutation of the lunatic fringe gene in humans causes spondylocostal dysostosis with a severe vertebral phenotype. *American Journal of Human Genetics, 78,* 28–37.

Splawski, I., Tristani-Firouzi, M., Lehmann, M. H., Sanguinetti, M. C., & Keating, M. T. (1997). Mutations in the hminK gene cause long QT syndrome and suppress I(Ks) function. *Nature Genetics, 17,* 338–340.

Spoendlin, H. (1966). *The organization of the cochlear receptor.* Basal, Switzerland: Karger.

Sulik, K. K., & Cotanche, D.A. (2004). Embryology of the ear. In H. V. Toriello, W. Reardon, & R. J. Gorlin (Eds.), *Hereditary hearing loss and its syndromes* (2nd ed., pp. 17–36). New York, NY: Oxford University Press.

Suzuki, S., Marazita, M. L., Cooper, M. E., Miwa, N., Hing, A., Jugessur, A., . . . Murray, J. C. (2009). Mutations in BMP4 are associated with subepithelial, microform, and overt cleft lip. *American Journal of Human Genetics, 84,* 406–411.

Teebi, A. S., Kennedy, S., Chun, K., & Ray, P. N. (2002). Severe and mild phenotypes in Pfeiffer syndrome with splice acceptor mutations in exon IIIc of FGFR2. *American Journal of Medical Genetics, 107,* 43–47.

Tekin, M., Hismi, B. O., Fitoz, S., Ozdag, H., Cengiz, F. B., Sirmci, A., . . . Akar, N. (2007). Homozygous mutations in fibroblast growth factor 3 are associated with a new form of syndromic deafness characterized by inner ear agenesis, microtia, and microdontia. *American Journal of Human Genetics, 80,* 338–344.

Tekin, M., Ozturkmen, A. H., Fitoz, S., Birnbaum, S., Cengiz, F. B., Sennaroglu, L., . . . Duman, D. (2008). Homozygous FGF3 mutations result in congenital deafness with inner ear agenesis, microtia, and microdontia. *Clinical Genetics, 73,* 554–565.

Tyson, J., Tranebjaerg, L., Bellman, S., Wren, C., Taylor, J. F. N., Bathen, J., . . . Bitner-Glindzicz, M. (1997). IsK and KvLQT1: Mutation in either of the two subunits of the slow component of the delayed rectifier potassium channel can cause Jervell and Lange-Nielsen syndrome. *Human Molecular Genetics, 61,* 2179–2185.

Van De Water, T. R. (1984). Developmental mechanisms of mammalian inner ear formation. In C. Berlin (Ed.), *Hearing science* (pp. 49–170). San Diego, CA: College-Hill Press.

Van de Water, T. R., Anniko, M., & Wersall, J. (1977). Embryonic development of the sensory cells in macula utriculae of mouse. In M. Portmann & J. M. Aran (Eds.), *Inner ear biology VIVth Workshop* (Vol. 68, pp. 25–35), Paris, France: INSERM.

Van Esch, H., Groenen, P., Nesbit, M. A., Schuffenhauer, S., Lichtner, P., Vanderlinden, G., . . . Devriendt, K. (2000). GATA3 haploinsufficiency causes human HDR syndrome. *Nature, 406,* 419–422.

Veenhof, V. B. (1969). *The development of statoconia in mice.* Amsterdam-London, UK: N.V. North-Holland.

Verhoeven, K., Van Laer, L., Kirschhofer, K., Legan, P. K., Hughes, D. C., Schatteman, I., . . . Van Camp, G. (1998). Mutations in the human alpha-tectorin gene cause autosomal dominant non-syndromic hearing impairment. *Nature Genetics*, *19*, 60–62.

Verpy, E., Leibovici, M., Zwaenepoel, I., Liu, X. Z., Gal, A., Salem, N., . . . Petit, C. (2000). A defect in harmonin, a PDZ domain-containing protein expressed in the inner ear sensory hair cells, underlies Usher syndrome type 1C. *Nature Genetics*, *26*, 51–55.

Waguespack, J., Salles, F. T., Kachar, B., & Ricci, A. (2007). Stepwise morphological and functional maturation of mechanotransduction in rat outer hair cells. *Journal of Neuroscience*, *27*, 13890–13902.

Wallis, D., Hamblen, M., Zhou, Y., Venken, K. J. T., Schumacher, A., Grimes, H. L., . . . Bellen, H. J. (2003). The zinc finger transcription factor, Gfi1, is required for inner ear hair cell differentiation and survival. *Development*, *130*, 221–232.

Wang, A., Liang, Y., Fridell, R. A., Probst, F. J., Wilcox, E. R., Touchman, J. W., . . . Friedman, T. B. (1998). Association of unconventional myosin MYO15 mutations with human nonsyndromic deafness DFNB3. *Science*, *280*, 1447–1451.

Wang, J., Mark, S., Zhang, X., Qian, D., Yoo, S. J., Radde-Gallwitz, K., . . . Chen P. (2005). Regulation of polarized extension and planar cell polarity in the cochlea by the vertebrate PCP pathway. *Nature Genetics*, *37*(9), 980–985.

Wang, Q., Curren, M. E., Splawski, I., Burn, T. C., Millholland, J. M., VanRaay, T. J., . . . Keating, M. T. (1996). Positional cloning of a novel potassium channel gene: KVLQT1 mutations cause cardiac arrhythmias. *Nature Genetics*, *12*, 17–23.

Wang, Y., Guo, N., & Nathans, J. (2006). The role of *Frizzled3* and *Frizzled6* in neural tube closure and in the planar polarity of inner-ear sensory hair cells. *Journal of Neuroscience*, *26*(8), 2147–2156.

Weil, D., Blanchard, S., Kaplan, J., Guilford, P., Gibson, F., Walsh, J., . . . Petit, C. (1995). Defective myosin VIIA gene responsible for Usher syndrome type 1B. *Nature*, *374*, 60–61.

Weil, D., Kussel, P., Blanchard, S., Levy, G., Levi-Acobas, F., Drira, M., . . . Petit, C. (1997). The autosomal recessive isolated deafness, DFNB2, and the Usher 1B syndrome are allelic defects of the myosin-VIIA gene. *Nature Genetics*, *16*, 191–193.

Weil, D., El-Amraoui, A., Masmoudi, S., Mustapha, M., Kikkawa, Y., Laine, S., . . . Petit, C. (2003). Usher syndrome type IG (USH1G) is caused by mutations in the gene encoding SANS, a protein that associates with the USH1C protein, harmonin. *Human Molecular Genetics*, *12*, 463–471.

Weston, M. D., Kelley, P. M., Overbeck, L. D., Wagenaar, M., Orten, D. J., Hasson, T., . . . Kimberling, W. J. (1996). Myosin VIIA mutation screening in 189 Usher syndrome type 1 patients. *American Journal of Human Genetics*, *59*, 1074–1083.

Weston, M. D., Luijendijk, M. W. J., Humphrey, K. D., Moller, C., & Kimberling, W. J. (2004). Mutations in the VLGR1 gene implicate G-protein signaling in the pathogenesis of Usher syndrome type II. *American Journal of Human Genetics*, *74*, 357–366.

Whitehead, M. C., & Morest, D. K. (1985a). The development of innervation patterns in the avian cochlea. *Neuroscience*, *14*, 255–276.

Whitehead, M. C., & Morest, D. K. (1985b). The growth of cochlear fibers and the formation of their synaptic endings in the avian inner ear: A study with the electron microscope. *Neuroscience*, *14*, 277–300.

Wilcox, E. R., Burton, Q. L., Naz, S., Riazuddin, S., Smith, T. N., Ploplis, B., . . . Friedman, T. B. (2001). Mutations in the gene encoding tight junction claudin-14 cause recessive deafness DFNB29. *Cell*, *104*, 165–172.

Wilkie, A. O. M., Slaney, S. F., Oldridge, M., Poole, M. D., Ashworth, G. J., Hockley, A. D., . . . Reardon, W. (1995). Apert syndrome results from localized mutations of FGFR2 and is allelic with Crouzon syndrome. *Nature Genetics*, *9*, 165–172.

Woods, C., Moncouquiol, M., & Kelley, M. W. (2004). Math1 regulates the development of the sensory epithelium in the mammalian cochlea. *Nature Neuroscience*, *7*, 1310–1318.

Xia, J., Liu, C., Tang, B., Pan, Q., Huang, L., Dai, H., . . . Huang, J. C. (1998). Mutations in the gene encoding gap junction protein beta-3 associated with autosomal dominant hearing impairment. *Nature Genetics, 20,* 370–373.

Xiang, M., Gan, L., Li, D., Chen, Z. Y., Zhou, L., O'Malley, B. W., . . . Nathans, J. (1997). Essential role of POU domain factor Brn-3c in auditory and vestibular hair cell development. *Proceedings of the National Academy of Sciences, 94,* 9445–9450.

Xiang, M., Gao, W. Q., Hasson, T., & Shin, J. J. (1998). Requirement for Brn-3c in maturation and survival, but not in fate determination of inner ear hair cells. *Development, 125,* 3935–3946.

Yagi, H., Furutani, Y., Hamada, H., Sasaki, T., Asakawa, S., Minoshima, S., . . . Matsuoka, R. (2003). Role of TBX1 in human del22q11.2 syndrome. *Lancet, 362,* 1366–1373.

Zheng, J. L., & Gao, W. L. (1997). Analysis of rat vestibular hair cell development and regeneration using calretinin as an early marker. *Journal of Neuroscience, 17*(21), 8270–8282.

Zhu, M., Yang, T., Wei, S., DeWan, A. T., Morell, R. J., Elfenbein, J. L., . . . Friderici, K. H. (2003). Mutations in the gamma-actin gene (ACTG1) are associated with dominant progressive deafness (DFNA20/26). *American Journal of Human Genetics, 73,* 1082–1091.

Zwaenepoel, I., Mustapha, M., Leibovici, M., Verpy, E., Goodyear, R., Liu, X. Z., . . . Petit, C. (2002). Otoancorin, an inner ear protein restricted to the interface between the apical surface of sensory epithelia and their overlying acellular gels, is defective in autosomal recessive deafness DFNB22. *Proceedings of the National Academy of Sciences, 99,* 6240–6245.

7

Emergence of Inner Ear Function

KEY POINTS

1. The human is precocious with regard to auditory and vestibular functional development; however, final maturation of function occurs after birth.
2. Membranes overlying the inner ear sensory receptors are first present between 8 to 10 weeks of age before hair cells have fully differentiated. The overlying membrane is critical for adequate stimulation of the sensory hair cells.
3. Hair cells become active upon the appearance of ion channels and membrane currents. Newly differentiated hair cells lack most voltage-gated membrane channels present in mature cells and the membrane currents poorly follow the applied stimulus.
4. In vestibular hair cells, an orderly appearance of different ion channels shapes the receptor potential during development. There is no evidence of Ca^{2+} dependent spiking in vestibular hair cells.
5. Vestibular primary afferents display an irregular discharge during development and show no evidence of rhythmic bursting. Spontaneous discharge rates increase and the number of neurons displaying regular discharge patterns increases.
6. Prior to the onset of hearing, the cochlea undergoes a prehearing endogenous signaling period when immature inner hair cells generate spontaneous action potentials, which drive neural activity in immature neurons. These spontaneous waves of hair cell activity are initiated by adenosine triphosphate (ATP), which induces calcium currents generating hair cell action potentials.

> 7. Late in the prehearing period, endogenous ATP-initiated inner hair cell activity decreases, neural reorganization occurs, and synaptic machinery is refined.
> 8. With the onset of hearing to ambient sound levels, frequency selectivity and sensitivity improve coinciding with increases in endolymphatic potential, outer hair cell motility, and changes in basilar membrane stiffness.

INTRODUCTION

Although still immature, the human can respond to auditory and vestibular stimuli at birth. Therefore, a functional inner ear emerges in the human fetus and, for this reason, the human is considered to be precocial. However, full maturation of the auditory and vestibular systems occurs over several years. In contrast, many nonhuman mammals (e.g., mice, rats, hamsters, cats, etc.) are deaf at birth and thus are considered altricial (or altricous). Being born deaf, they must acquire the ability to hear during the first few days or weeks after birth. Vestibular function is also immature in these neonatal, nonhuman mammals and matures during subsequent weeks.

In order to hear a sound, the sound must reach the organ of Corti, hair cells must transduce the sound into membrane currents and release neurotransmitter, and postsynaptic spiral ganglion neurons must respond to the neurotransmitter and transmit discharges to the central nervous system (CNS). Once in the CNS, the signals must be processed and relayed to the auditory sensory regions of the neocortex (temporal lobe) where perception can take place. A similar chain of functional events can be drawn for the transduction, processing, and perception of head motion stimuli by the vestibular system. We focus here on the emergence of inner ear function, which covers adequate stimulation, the peripheral processes of prehearing endogenous signaling, stimulus transduction, and encoding of information in the primary afferents of the vestibular and auditory nerves.

Adequate Stimulation

Vestibular and auditory systems are examples of special senses that rely on elaborate ancillary structures to preferentially select, from among numerous potential environmental stimuli, only a few particular mechanical events that serve as adequate stimuli. For example, both vestibular and auditory hair cells are mechanoreceptors responding to displacement of the hair bundle along the axis of polarization. However, the most effective stimulus causing such shearing motion naturally is considerably different for the two sensors. In the vestibular system, normal head motion is effective in stimulating vestibular hair cells but not cochlear hair cells. Low levels of airborne sound effectively stimulate cochlear but not vestibular hair cells. Special structures are largely responsible for these functional attributes rather than substantial differences in the receptors themselves (although there are differences to be sure).

For cochlear hair cells, the outer and middle ears collect and amplify low level ambient sound and convert sound pressures into mechanical vibration of the stapes footplate in the oval window. The vast majority of vibrations introduced to the inner ear by the stapes footplate are transmitted across the cochlear partition along the lowest impedance paths to the round window. Vibration of the cochlear partition stimulates cochlear receptors via a shearing action between the basilar membrane and the tectorial membrane overlying the sensory receptors. In addition, active processes within the cochlea channel the energy traveling along the cochlear partition and amplify basilar membrane motion. Without these mechanisms, low level ambient sounds would be inaudible.

In the vestibular system, there are dense otoconia that are packed into the otoconial membrane lying immediately over the stereociliary bundles in the maculae. These dense otoconial crystals, being fixed in the otoconial matrix, introduce a shearing force on macular hair cell bundles when placed under a linear acceleration field. Such a field occurs in association with head movement or in the presence of gravity, thus leading to the displacement of the otoconial membrane and stimulation of macular sensors. Without dense otoconia, macular hair cells are not stimulated (Jones, Erway, Bergstrom, Schimenti, & Jones,1999; Jones, Jones, & Hoffman, 2008). Stimulation of ampullar organs depends on the semicircular canals and the inherent inertia of the endolymphatic fluids within the membranous labyrinth. During head rotation, pressure gradients develop across the cupula within the canal lumen due to the inertia of endolymphatic fluid. These pressure gradients distort the cupula and produce shearing forces at the surface of the sensory epithelium where hair cell bundles are displaced and hair cells are activated or inhibited. Without the canals and their fluid-filled patent lumen, the ampullae would be insensitive to head rotation.

Because normal function is strictly linked to ancillary structures, we must consider them when evaluating the emergence of function during ontogeny. The formation of the outer, middle, and inner ear divisions are described in some detail in Chapters 5 and 6. We will integrate information about key developmental changes in ancillary structures with our discussion of functional development.

Studies and Conventions

Detailed knowledge regarding the development of vestibular and auditory function comes primarily from the study of animals. Much, if not all, of the postnatal developmental changes observed in mammalian models occur prior to birth in the human. As in Chapter 6, we will indicate human ages estimated or found to correspond to those of animal models and the reader is referred to Figures 7–1 and 7–2 for a comparison of developmental time frames for chicken, mouse, and human. For the chicken, equivalent days of incubation (E) are given based on Hamburger and Hamiton (1951) staging. For the mouse, embryonic (E) days of gestation are equivalent to days postconception (dpc) where day E0.5 is the morning that the vaginal plug is found (Kaufman, 1992). Days postnatal are designated with a "P" where P0 is the day of birth or hatch. For the human, weeks of gestation are indicated.

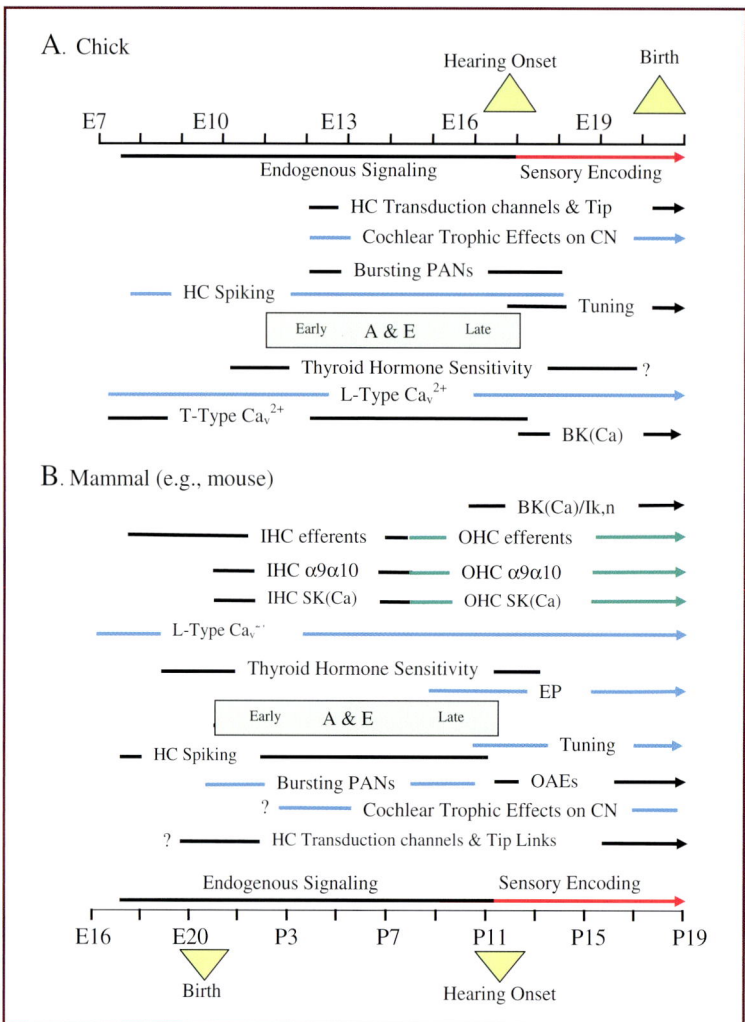

Figure 7–1. *Emergence of hearing. Time lines for various stages of auditory development in the chick (**A**) and mammal (**B**). Findings in the literature have been generalized for the mammal and recast in the time frame of the mouse. Key milestones are represented schematically according to the approximate corresponding ages during "Endogenous Signaling" and "Sensory Encoding" periods of development. HC = hair cell. IHC = inner hair cell. OHC = outer hair cell. CN= cochlear nucleus. PANs = primary afferent neurons. Tuning = the appearance, maturation and maintenance of frequency selectivity. "A & E" = period of afferent and efferent reorganization, which is associated with the "early" and "late" prehearing periods. L-Type Ca_v^{2+} = appearance and time course of L-type voltage sensitive Ca^{2+} channels ($Ca_v1.3$). T-Type Ca_v^{2+} = appearance and time course of T-type voltage sensitive Ca^{2+} channels ($Ca_v3.1$). BK(Ca) = appearance and maintenance of calcium activated big K+ channels in hair cells. BK(Ca)/Ik,n = the appearance and maintenance of BK(Ca) and Ik,n channels in IHC and OHC. IHC efferents = the appearance and time course of axosomatic efferent synapses on IHCs. OHC efferents = the appearance and time course of axosomatic efferent synapses on OHCs. IHC α9α10 = the appearance and time course of nicotinic acetylcholine receptors on IHC. OHC α9α10 = the appearance and time course of nicotinic acetylcholine receptors on OHC. IHC SK(Ca) = the appearance and time course of Ca^{2+}-activated small K+ channels in IHCs. OHC SK(Ca) = the appearance and time course of Ca^{2+}-activated small K+ channels in OHCs. EP = endocochlear potential. OAEs = Otoacoustic emissions.*

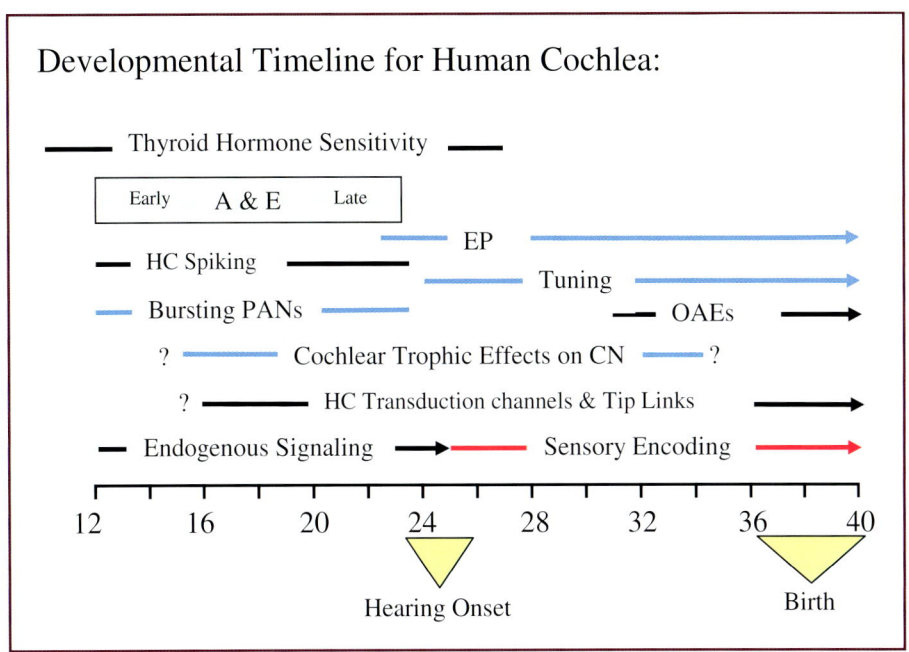

Figure 7–2. Emergence of hearing. Time lines for various stages of auditory development in the human. Key milestones are represented schematically according to the approximate corresponding ages during "Endogenous Signaling" and "Sensory Encoding" periods of development. In most instances, time lines had to be estimated based on human anatomy and functional measures in animals models. Compare with Figure 7–1. "A & E" = period of afferent and efferent reorganization, which is associated with the "early" and "late" prehearing periods. EP = endocochlear potential. OAEs = Otoacoustic emissions. PANs = primary afferent neurons. Tuning = the appearance, maturation and maintenance of frequency selectivity. CN = cochlear nucleus. HC = hair cell.

Much of the research on early development is done in vitro, that is, conducted on cells or tissues that are removed from the animal and placed in artificial physiological solutions. In many cases, we must assume that the processes described from in vitro studies are similar to those that function in vivo, or in the intact animal.

Defining the Onset of Hearing

In general, hearing onset is taken to be the earliest ages where responses to airborne sound can be obtained across a range of frequencies at stimulus levels below 80 to 90 dB SPL. This criterion helps identify when the cochlea begins to be influenced appreciably by natural environmental sound for most species.

DEVELOPMENT OF VESTIBULAR FUNCTION

In mammals, movements of the mother or the fetus could serve as natural stimuli before birth. However, the actual time of onset for vestibular function (i.e., an adequate stimulus producing a neural response) during embryonic development in mammals and birds has not been determined. Nonetheless, we can estimate an earliest age based on known requirements for a functioning vestibular apparatus. Mechanically, gravity receptors require otoconia and associated membrane whereas sensors in ampullae require fully patent fluid-filled semicircular ducts and a functional cupula for the coupling of inertial forces to the stereociliary bundles of hair cells in the crista.

In mice, the otoconial membrane and otoconia form from E14 to E16 and continue to elaborate at least until birth (Anniko, 1983; Lim 1984; Nakahara & Bevelander, 1979; Salamat, Ross, & Peacor, 1980; Veenhof, 1969). In the human this occurs by week 10 (Dechesne, 1992). The semicircular canals form initially as epithelial tubes (semicircular ducts) beginning about E12.5 (Lim & Anniko, 1985; Sher, 1971). The ducts generally are well formed and patent by about E15.5 (week 7). Between E12.5 and E14.5 (weeks 6 to 7), the cross-sectional shape of the ducts enlarges from a slit-like form to a much wider oval and the three ampullae are present by E13.5 (week 7). The cristae are emerging during this same period. The incipient cupula forms a thin membrane extending from the crista to the ampullar roof by E14.5 (Anniko, 1983; Lim & Anniko, 1985). In the human, the cupula is present by week 8 (Dechesne 1992). By birth in the mouse (weeks 10 to 12), the crista and cupula are well formed and semicircular ducts have the characteristic circular cross section (Lim & Anniko, 1985). Thus, it is conceivable that natural stimulation of ampullar and macular sensors could begin well before birth between E15.5 and P0 in rodents and at 9 to 12 weeks in the human (Dechesne, 1992). Are vestibular receptors capable of responding to natural stimulation prior to birth? The answer appears to be yes.

Transduction Channels and Associated Membrane Currents

Hair cell transduction channels are believed to be located in the tips or upper walls of stereocilia (Hudspeth, 1982). The opening and closing of transduction channels are thought to be controlled by tip links tethered to adjacent stereocilia (Hudspeth, 1985; Pickles, Comis, & Osborne, 1984). Functional transduction channels appear first in vestibular hair cells. In the mouse, they appear suddenly over a period of approximately 24 hours between E16 and E17 (Géléoc & Holt, 2003, estimated to be week 12 in the human). Tip links may appear one or two days earlier in the mouse (Denman-Johnson & Forge, 1999; Nayak, Ratnayaka, Goodyear, & Richardson, 2007). This timing parallels that of the appearance of otoconia in the maculae (Anniko, 1983; Lim, 1984; Nakahara & Bevelander, 1979; Salamat et al., 1980; Veenhof, 1969;) and the canal apparatus for the ampullae noted above. In the bird, hair cells are capable of responding to injected transduction currents very early (by E10, Masetto, Perin, Malusà, Lucca, & Valli, 2000). However, it is not clear when transduction channels are actually functional in the bird. E10 is two days before avian cochlear hair cells show evidence of open transduction channels (Si, Brodie, Gillespie, Vazquez, & Yamoah, 2003) and when short hair bundles in the cochlea are present with tip links (Tilney, Tilney, Saunders, & DeRosier, 1986). Given that immature otoconia are present beginning at about E5 to E6 in the bird, it is likely that transduction channels appear some time after the formation of otoconia (Dickman, Huss, & Lowe, 2004; Fermin & Igarashi, 1985, 1986; Kido, 1997; Kido et al., 1993). Thus, natural mechanical stimuli are likely present as vestibular hair cell receptors acquire the ability to transduce them into embryonioc receptor potentials.

Hair Cell Response to Transduction Currents

The vestibular hair cell response (i.e., the receptor potential) to transduction currents depends on the nature of ionic channels located within basolateral portions of the hair cell membrane. When gated open, these basolateral ion channels permit the movement of particular ions across the membrane, thus contributing to current flow. Collectively, the ability to conduct currents (i.e., the property of conductance, symbolized as "g") depends on the number of channels open for each type of channel. Each channel type is named according to the dominant ion species it conducts. Currents associated with particular channels are designated with an "I" and a subscript indicating the specific ion channel or current (e.g., I_K). Conductance associated with the current I_K for example is designated as g_K. There are a variety of voltage-gated channels that are important in shaping the receptor potential. These include K^+, Na^+, and Ca^{2+} channels, (Eatock & Hurley, 2003). Table 7–1 lists various currents that have been measured in hair cells or primary afferents of the inner ear along with their associated membrane channels. Table 7–2 lists some of the genes for these channels and the human genetic disorders that can occur from mutations in the genes. Several of these human disorders result in heart arrhythmias, but also can include hearing loss or dizziness/imbalance.

The mature hair cell receptor potential reproduces the shape of an applied depolarizing current. The receptor potential follows the input signal reliably so that the gating of channels at the basolateral surface and the release of neurotransmitter is synchronized to the stimulus input. The first hair cells born are not equipped with the adult complement of channels. Undifferentiated otocyst cells and new hair cells have few if any voltage-gated K^+ membrane channels (Correia, Rennie, & Koo, 2001; Eatock & Hurley, 2003; Sokolowski, Stahl, & Fuchs, 1993). Specific channels appear at different developmental stages in hair cells and each channel can impart different characteristics to the hair cells' response to transduction currents. By E10 in the chick, hair cell responses to simulated transduction currents show that early responses are slow and follow current profiles poorly. With the acquisition of new conductances, the hair cell's ability to follow stimuli improves. In the bird, mature-like vestibular hair cell membrane responses and a full complement of channels are present just before hatching (Masetto et al., 2000), whereas in the altricial mammal the mature configurations of channels and more mature responses emerge in the late embryo and neonate (E18 to P4; Géléoc, Risner, & Holt, 2004; Masetto et al., 2000;

Table 7–1. Some of the Ion Channels That Have Been Identified in Inner Ear Sensory Cells or Primary Afferents and Their Associated Currents

Channel	Current	Function	Representative Reference
$K_v\beta1.1, \alpha1.4$	I_A	Voltage Gated Transiently Activating Potassium Channel	Correia et al., 2008
BK, IK, SK	$I_{K(Ca)}$ ($I_{K,f}$)	Calcium Activated Potassium Channels	Li et al., 2009; Schweizer et al., 2009
$K_v3.1, 3.2, 3.3$	I_K	Voltage Gated Potassium Delayed Rectifier	Chen and Davis, 2006
$K_v7.4$, Erg1	$I_{K,L}, I_{Kn}$	Voltage Gated Potassium Rectifier	Rennie and Correia, 1994; Hurley et al., 2006
$K_{ir}2.1$	I_{Kir}	Voltage Gated Potassium Inward Rectifier	Ruan et al., 2008; Zampini et al., 2008
HCN1, 2, 4	I_h (I_H, I_f)	Hyperpolarization-Activated Potassium-Sodium Channel	Bakondi et al., 2009; Horwitz et al., 2010
$Ca_v1.3$	I_{Ca}	High Voltage Activated L-type Calcium Channel	Zampini et al., 2010
$Ca_v2.1$	I_{Ca}	High Voltage Activated P/Q-type Calcium Channel	Eatock et al., 2008
$Ca_v2.2$	I_{Ca}	High Voltage Activated N-type Calcium Channel	Eatock et al., 2008
$Ca_v2.3$	I_{Ca}	High Voltage Activated R-type Calcium Channel	Eatock et al., 2008
$Ca_v3.1, 3.2, 3.3$	I_{Ca}	Low Voltage Activated T-type Calcium Channel	Eatock et al., 2008
$Na_v1.5, 1.6, 1.7$	I_{Na}	Voltage Gated Sodium Channel	Rusznák and Szücs, 2009
$Na_v1.8, 1.9$	I_{Na}	Voltage Gated Sodium Channel	Eatock et al., 2008

Note: The assembly of subunits that form a given channel may vary, thereby altering the channel's function (e.g., from transiently activated to delayed rectifier). Representative references provide more details about the channels.

Rüsch & Eatock, 1996; Rüsch, Lysakowski, & Eatock, 1998). Figure 7–3 illustrates how a vestibular hair cell receptor response to a depolarizing current step changes with the acquisition of new channels during development. The depolarizing current is used to simulate depolarizing transduction currents. A relatively mature response (bottom trace) does not appear until the $g_{K,L}$ channels are present. The voltage

Table 7–2. Genes Associated With the Channels Listed in Table 7–1, Their Loci, and Human Disorders Associated With Mutations in the Listed Genes

Channel	Gene	Locus	Human Disorder
BK	Kcnma1	10q22.3	
	Kcnmb1	5q34	
SK2	Kcnn2	5q21.2-q22.1	
$K_v3.1$	Kcnc1	11p15	
$K_v7.4$	Kcnq4	1p34	DFNA2[1]
Erg1	Kcnh2	7q35-q36	Long QT Syndrome Type 2[2]
$K_{ir}2.1$	Kcnj2	17q23.1-24.2	Anderson Syndrome[3]
HCN1	Hcn1	5p12	
HCN2	Hcn2	19p13.3	
HCN4	Hcn4	15q24-q25	Sick Sinus Syndrome[4]
$Ca_v1.3$	Cacna1d	3p14.3	SANDD Syndrome[5]
$Ca_v2.1$	Cacna1a	19p13	Familial Hemiplegic Migraine, Episodic Ataxia[6]
$Ca_v2.2$	Cacna1b	9q34	
$Ca_v2.3$	Cacna1e	1q25-1q31	
$Ca_v3.1$	Cacna1g	17q22	
$Ca_v3.2$	Cacna1h	16p13.3	Susceptibility to Childhood Epilepsy[7]
$Ca_v3.3$	Cacna1i	22q13.1	
$Na_v1.5$	Scn5a	3p21	Long QT Syndrome Type 3, Brugada Syndrome[8]
$Na_v1.6$	Scn8a	12q13	Cerebellar Atrophy, Ataxia, and Mental Retardation[9]
$Na_v1.7$	Scn9a	2q24.3	Erythermalgia[10]
$Na_v1.8$	Scna10	3p24.2-p22	
$Na_v1.9$	Scn11a	3p24-p21	

SANDD = SinoAtrial Node Dysfunction and Deafness.

[1]Kubisch et al. (1999), Coucke et al. (1999); [2]Curran et al. (1995); [3]Plaster et al. (2001); [4]Nof et al. (2007); [5]Baig et al. (2009); [6]Ophoff et al. (1996); [7]Chen et al. (2003); [8]Chen et al. (1998), Milanesi, Baruscotti, Gnecchi-Ruscone, & DiFrancesco (2006); [9]Trudeau, Dalton, Day, Ranum, & Meisler (2006); [10]Yang et al. (2004).

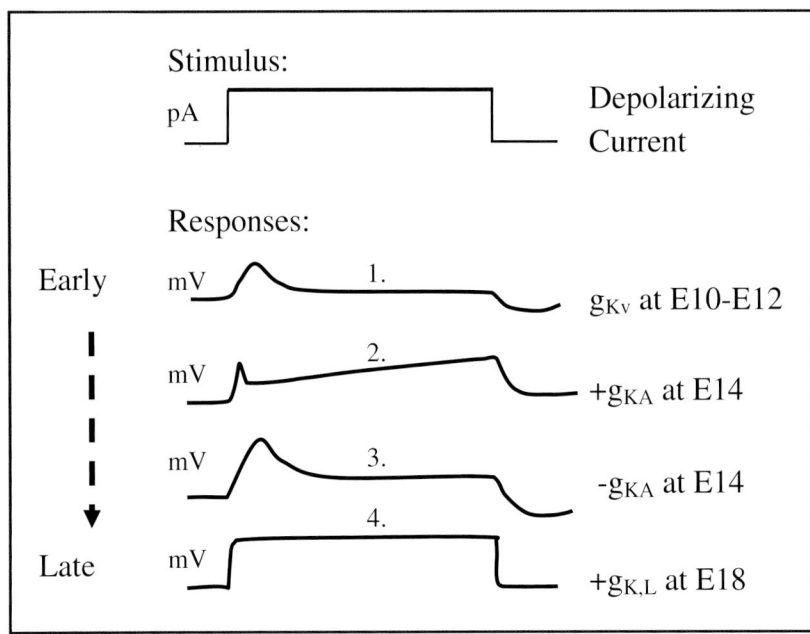

Figure 7–3. The effects of acquiring selected K+ channels on the hair cell receptor response. Top tracing shows the hypothetical step depolarizing current applied. This stimulus (levels reflected in picoamps, pA) simulates transduction currents. The four traces below the stimulus trace (1-4) are schematic representations of the receptor potential response. During development, the shape of the receptor potential changes with the addition or removal of each channel. The first response tracing represents a young age where the hair cell has acquired only the early delayed rectifier, g_{Kv} (trace 1). Additional channels appear successively over time (e.g., g_{KA}, $g_{K(Ca)}$ and $g_{K,L}$ traces 2 & 4). Traces 2 and 3 illustrate the kind of change in membrane response produced by electrically inactivating g_{KA} ($-g_{KA}$) before presenting the stimulus (trace 3). Corresponding ages in the chick are shown to the right of traces. Note how ultimately the response follows the step depolarizing currents closely as the cell acquires $g_{K,L}$ (trace 4) and matures. Hair cell membrane voltage responses (mV) are schematic representations of data reported by Masetto et al. (2000) and Chen and Eatock (2000) with permission.

response of the hair cell is the resulting receptor potential, which modulates transmitter release from the hair cell. Note the changes in the shape of the receptor potential as new ion channels are added at different ages of development.

A number of studies have examined the temporal sequence of channel acquisition (Eatock & Rüsch, 1997; Géléoc et al., 2004; Géléoc & Holt 2003; Hurley et al., 2006; Li, Meredith, & Rennie, 2010; Masetto et al., 2003; Rüsch, Lysakowski, et al., 1998; Sokolowski et al., 1993). We have tabulated the sequence of acquisition of several membrane ion channels for the chick and mouse in Table 7–3 based on the work of several laboratories (i.e., Géléoc et al., 2004; Masetto et al., 2000; Rüsch,

Table 7–3. Age of Appearance in Chicken and Mouse for Various Currents Recorded From Vestibular Hair Cells or Primary Afferents

Current	Chick	Mouse
I_{Kv}	E10	E14
I_{Ca}	E10	E15–birth*
I_{KA}	E12	?
$I_{K(Ca)}$	E14	P12–P23**
I_h	E16	P3
$I_{K,L}$	E17	E18
I_{Kir}	E19	E15

Sources: Information from Masetto et al. (2000) and Géléoc et al. (2004) unless otherwise noted.

See Table 7–1 for descriptions of ion currents. E = Chick: equivalent day of incubation; Mouse: Embryonic day or days postconception. P = postnatal day.

*Expression in vestibular primary afferents beginning at E15 and decreasing by birth. Changes in the density of the different I_{Ca} types (L, P/Q, N, R, T) also occurred from E15 to birth (Chambard, Chabbert, Sans, & Desmadryl, 1999).

**Based on expression of BK channels in the rat (Schweizer, Savin, Luu, Sultemeier, & Hoffman, 2009), which first appear at P12 and diminish by P23.

Eatock, et al. 1998;). The genes coding for these channel proteins are upregulated at particular times and for some (e.g., g_{Na}) they may be downregulated again in the mature animal. One striking difference to be noted regarding the acquisition sequence in vestibular hair cells versus that for auditory hair cells (described below) is the fact that there are no descriptions of Ca^{2+}-based hair cell spiking during vestibular development. Although the reason for this has not been explored, one might speculate that the early appearance of the fast I_{KA} and $I_{K(Ca)}$ (BK) channels in vestibular hair cells may prevent the development of hair cell spiking. In the cochlea, BK channels appear late in the maturation of inner hair cells (IHCs), and prior to their appearance, the IHCs are generating spontaneous Ca^{2+} spiking. As we will see in the development of hearing, IHC spiking disappears as BK channels develop.

Behavioral Response to Head Motion

In the precocial chicken, there is no question that the vestibular system is virtually mature at hatch as chicks quickly learn to walk bipedally within hours. In the human, it also is likely that the peripheral vestibular system is relatively mature at birth. However, it will be at least a year before any walking is done. Considerable maturation is required in central motor control circuitry as well as in the skeleto-motor system. To evaluate the development of vestibular responses to head motion, it is useful to study the behavior of altricial mammals (rats, cats, mice, gerbils, etc.). In such animals, during a span of 2 or 3 weeks, the vestibular apparatus goes from poorly functioning to functionally mature and this occurs with observable behaviors. When a mature animal lying face up is dropped from a reasonable height onto a soft sponge base, it quickly will turn to right itself and land on its feet. This is the air-righting reflex and it can be used to assess the combined maturity of the vestibular and motor control systems. This reflex is absent at birth in rats when the vestibular system is still immature. It appears first between P10 and P15 (e.g., Hard & Larsson 1975; Laouris, Kalli-Laouri, & Schwartze, 1990). In contrast, head position compensation for turntable rotation appears as early as P4 (Parrad & Cotteraau, 1977). Thus, the vestibular epithelium is immature in the newborn rodent, but behavioral testing leaves open to question whether the behavioral immaturities reflect the functional status of peripheral or central components or both. Human studies evaluating eye movements in response to rotational stimuli suggest some aspects of compensatory eye movements are mature at 6 to 10 months of age (e.g., Cioni, Favilla, Ghelarducci, & La Noce, 1984; Cyr, Brookhouser, Valente, & Moore, 1985; Viener-Wacher, Toupet, & Narcy, 1996) although other aspects continue to mature up to late childhood or adolescence (e.g., Cyr et al., 1985; Herman, Maulucci, & Stuyck, 1982; Valente, 2007; Viener-Wacher et al., 1996).

Studies of standing balance function generally demonstrate continued maturation of balance into late childhood (e.g., Casselbrant et al., 2010; Charpiot, Tringali, Ionescu, Vital-Durand, & Ferber-Viart, 2010; Valente, 2007;). It is likely that behavioral maturation is due in large part to central refinements after birth.

Primary Afferent Function

When recording normal mature individual vestibular primary afferent neurons in the absence of head movement, the neurons are not silent but rather discharge spontaneously (Figure 7–4). Discharge rate (spikes/sec) remains relatively constant unless the head is moved thus stimulating hair cells and neurons innervating them. Spontaneous vestibular discharge patterns are of two types: regular and irregular. Just how regular or irregular the neural discharge is depends in part on the nature of the dendritic synaptic termination and on the nature of the membrane channels resident in the neuron (Eatock, Xue, & Kalluri, 2008; Iwasaki, Chihara, Komuta, Ito, & Sahara, 2008; Kalluri, Xue, & Eatock, 2010). The regularity can vary widely and this variety can be quantified using a single number called the coefficient of variation (CV = standard deviation/mean spike interval). The value of the CV varies generally between 0 and 1. The CV approaches 1.0 as the discharge pattern becomes more irregular whereas CV approaches 0.0 as the discharge becomes more regular. Figure 7–4 illustrates the discharge patterns of regular and irregular vestibular afferents in mature (Figure 7–4A) and developing (Figure 7–4B, P7 neonate) mice. These recordings were made in vivo from primary afferent neurons of the superior vestibular nerve. The discharge pattern

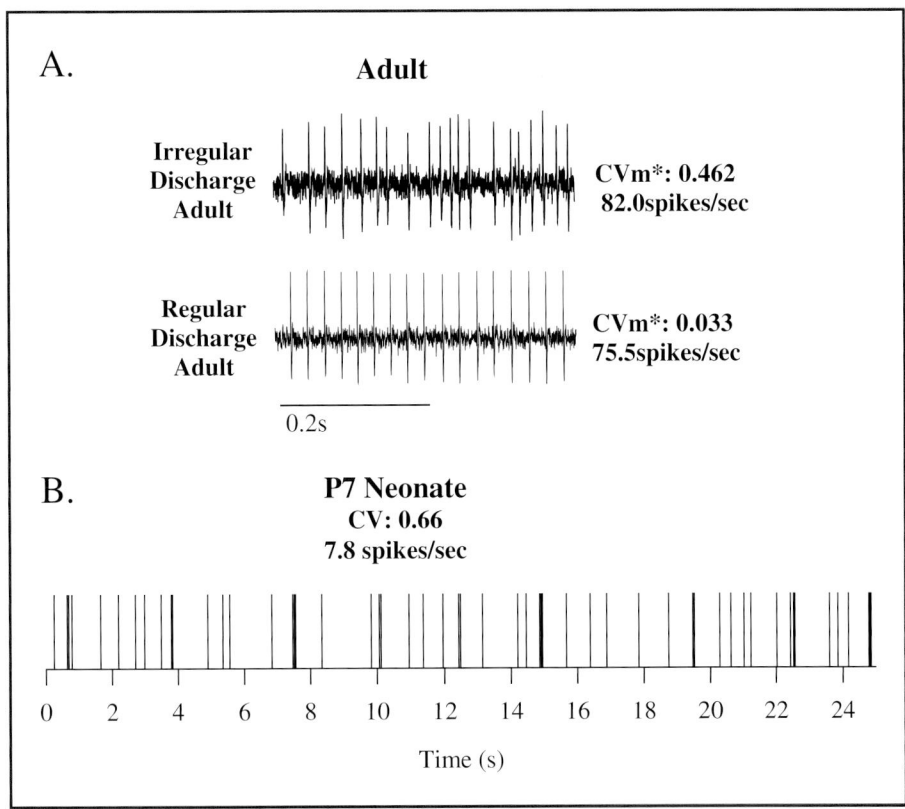

Figure 7–4. Spontaneous discharge activity of vestibular primary afferent neurons in mice. **A.** Adult mice: Each voltage spike represents an individual action potential. No stimulus is presented. These cells were chosen for illustration because they have similar high discharge rates. Two types of activity patterns are recognized. Irregular discharge (top tracing), which is characterized by irregular spacing between spikes. A regular spike discharge pattern tends to show regular spacing between spikes and a low CV. Modified from Jones et al. (2008). **B.** Spontaneous spike train of primary afferent neuron recorded from the superior vestibular nerve in the neonatal mouse at P7. Note the time scale difference in B. This record was made over a period of approximately 24 seconds. Each vertical "spike" represents the onset time of the neural spike discharge during this portion of the recording. Discharge rate is slow and an irregular firing pattern (with high CV) typical for neonatal vestibular neurons is apparent. Unpublished data.

of the neonate reflects a relatively low discharge rate with irregular discharge timing (CV = 0.66).

Vestibular primary afferents recorded in a mouse inner ear explant preparation (i.e., labyrinth and ganglia are removed and maintained in a physiological solution) are also spontaneously active (Desmadryl, Raymond, & Sans, 1986). Discharge patterns measured on different postnatal days showed that mean spontaneous discharge rates were low initially (P0, 5 to 10 spikes/sec) and all neurons displayed irregular activity. Remarkably, regular discharge patterns were found at P1 and older. Beginning between P6 and P8 discharge rates increased dramatically (>80 spikes/sec), and the proportion of regular fibers increased as well. In a similar preparation in the chicken, Galicia, Cotes, and Galindo (2010) reported irregular spontaneous discharge rates on the order of 40 spikes/sec and CVs above 1.0 in recordings as early as E15. Many of the patterns found in the in vitro preparation were similar to those reported for in vivo studies (discussed below). In vitro studies provide many practical advantages while raising the question of whether neurons behave the same when studied in their natural environment, that is, in vivo.

Spontaneous discharge patterns of horizontal canal neurons have been recorded in vivo in the rat from ages P1 to P20 (Curthoys, 1983). Discharge rates were low at the youngest ages (<10 spikes/sec, P1 to P3) and all neurons exhibited irregular spontaneous discharge patterns. The first regular fibers were seen on P4. An example of an irregularly discharging primary afferent at P7 is shown in Figure 7–4B. Discharge rates and the proportion of regular fibers increased substantially after P10 (30 to 40 spikes/sec). Nonetheless, rates for regular cells were still somewhat below the adult values at P20. In vivo recordings in the chick embryo at E19 showed that regular discharge patterns were present, and on average the rate for embryos was 22 spikes/sec versus 60 spikes/sec for posthatch animals (Jones & Jones, 2000b). Note that discharge rates were somewhat lower in the in vivo recordings compared to explants. This may reflect the effects of anesthesia used in in vivo preparations or conditions associated with in vitro preparations. In summary, in the rodent at birth, spontaneous activity is immature, evidencing low discharge rates and irregular firing patterns. Maturation progresses over a period of weeks.

One role of the horizontal canal is to detect head turning and to activate vestibular neurons to send a signal to the brainstem that produces eye movements that compensate for head motion and maintain gaze on a stationary visual target. This is known as the vestibulo-ocular reflex (VOR). One question to ask is, when are primary afferents capable of delivering a compensatory signal comparable to that of the mature system? Curthoys (1983) evaluated this question by measuring horizontal canal primary afferent neurons during head rotation in rats at ages from P1 to P20. At P1 neural responses were sluggish and highly variable. By P6 to P8, the neural response gain approached that of the adult. This suggests that, for some stimuli, the neuroepithelium at P8 is capable of generating signals comparable to those of an adult.

The basic structural elements of the macula are in place in the chick before hatch (by E16). The question is, when does macular function emerge? Little information is available regarding macular functional development. Recordings of vestibular compound action potentials (VsEPs) have been made in embryos and hatchling chicks (E18 to P22). Jones and Jones (2000a) revealed that responses could be obtained as early as E19 (1 to 2 days before hatch). Therefore, the onset of macular function occurs at least by E19 in the chick and likely earlier. Response threshold decreased rapidly to approach adult values within days of hatch. Other response characteristics also matured systematically over similar periods (response latency shortened and amplitudes increased). These findings and the fact that chicks are able to walk and run within hours of hatching show that, in the chick, the vestibular periphery matures early and is functional even before hatch.

We are only meagerly informed about the functional development of the vestibular system. Considerable work needs to be done to fill in the gaps in knowledge. Of course, that means there is much to be discovered and opportunity abounds!

DEVELOPMENT OF AUDITORY FUNCTION

The auditory periphery plays two fundamentally different functional roles during development in mammals and birds (see Figures 7–1 and 7–2). We are obviously most familiar with the sensory role of the mature cochlea in hearing. However, during development, the cochlea also plays a role as a signal generator. During late development before hearing begins, the cochlea is not silent but rather is surprisingly active. This intrinsic activity is independent of ambient sound or other extrinsic stimuli. It is characteristically rhythmic or repetitive, often intermittent, appearing as transient bursts of activity separated by long silent periods. This is an endogenous signaling period, the activity of which is thought to be important in shaping the final transition to a mature cochlea. In addition, the discharge of hair cells acts in turn to pattern the discharge of primary afferent neurons. Patterned primary afferent discharge provides signals that likely play a role in the refinement of cochleotopic projections through central auditory relays and may help prepare central circuits for their mature role of processing information about ambient sound. Intrinsic bursting cochlear activity ceases with the onset of hearing, when the cochlea becomes a sensor and receptors begin to signal the characteristics of ambient airborne sound. We next consider events leading up to and during this prehearing period and then characterize the emergence of auditory sensory function.

Prehearing Endogenous Signaling Period

Transduction Channels and Associated Membrane Currents

The ability to transduce mechanical stimuli into biological signals requires transduction channels and stereociliary bundle tip links. Functional channels and tip links appear in development well before hearing begins. In the cochlea, stereocilia tip links are present at least by the first few days after birth in rodents (Furness, Richardson, & Russell, 1989; Souter, Nevill, & Forge, 1995; Zine & Romand, 1996), by E12 in birds (Pickles, von Perger, Rouse, & Brix, 1991), both of these ages corresponding to week 15 to 18 in humans. Levigne-Rebillard, and Pujol (1986) described cross links and tip links in human fetuses at 20 to 22 weeks. Transduction currents in IHCs and outer hair cells (OHCs) (or tall and short hair cells in birds) although immature are already present during these periods in animal models (e.g., by E12 in the chick) (Géléoc, Lennan, Richardson, & Kros, 1997; Kros, Rüsch, & Richardson, 1992; Si et al., 2003). Hair cell receptor potentials produced in response to mechanical displacement of the hair bundle are also clearly present in hair cells harvested from rodents at P0 and cultured for several days (e.g., Russell, Cody, & Richardson, 1986; Russell, Richardson, & Cody, 1986). This early appearance of transduction currents is well before most primary afferent neurons begin to respond to acoustic stimuli, including stimuli that are directly applied to the oval window (Jones, Jones, & Paggett, 2006).

Dramatic changes in hair cell basolateral membrane conductances occur during the period before hearing begins. The changes alter hair cell responses to transduction currents and transform hair cells from signal generators into exquisitely sensitive mechanoreceptors as described in the next section. The precise timing of these changes in membrane conductance is thought to be regulated in part by a number of growth factors including the neurotrophins such as brain-derived neurotrophic factor (BDNF), neurotrophin-3 (NT-3; Mou, Hunsberger, Cleary, & Davis, 1997; Pirvola & Ylikoski, 2003; Sokolowski, Csus, Hafez, & Haggerty, 1999), possibly retinoic acid (Romand, 2003), and thyroid hormone (Brandt et al., 2007; Deol, 1973; Forrest, Erway, Ng, Altschuler, & Curran, 1996; Knipper et al., 1999; Mustapha et al., 2009; Ng et al., 2001; Rüsch, Erway, Oliver, Vennström, & Forest, 1998; Sendin, Bulankina, Riedel, & Moser, 2007; Winter et al., 2006; Winter et al., 2009; Winter, Braig, Zimmerman, Engel, Rohbock, & Knipper, 2007). As in the case of vestibular hair cells, acquisition of particular channels occurs at prescribed times and each new channel modifies the hair cell response to depolarization. Next, we summarize the temporal sequence for important changes in membrane channels and underscore the functional consequences.

Inner Hair Cell Action Potentials

Beginning as early as E17.5 in the mouse (Kros, Ruppersberg, & Rüsch, 1998; Marcotti, Johnson, Holley, & Kros, 2003), by P3 to P11 in rats (Brandt et al., 2007), and at E7.5 in the bird (Fuchs & Sokolowski 1990; Levic et al., 2007), hair cells are capable of generating spontaneous repetitive action potentials. The ages for these animal models correspond to approximately weeks 10 to 13 in humans. The action potentials generated by IHCs are remarkable because mature mammalian hair cells do not produce such regenerative spiking but rather evidence graded receptor potentials that vary systematically with transduction currents (Palmer & Russell, 1986). Indeed, hair cell spiking continues in the rodent neonate but reportedly disappears prior to the onset of hearing (Kros et al., 1998; Marcotti, Johnson, Holley, et al., 2003, Marcotti, Johnson, Rüsch, & Kros, 2003). In the bird, such spiking also disappears with the onset of hearing at about E17 to E18 (Fuchs & Sokolowski, 1990; Jones, Jones, & Paggett, 2001; Jones et al., 2006; Levic et al., 2007), corresponding to week 25 or so in humans.

In mice, IHC spiking begins about one day after the acquisition of voltage gated Ca^{2+} channels ($Ca_V1.3$) at E16.5 (Marcotti, Johnson, Rusch, et al., 2003). The voltage-sensitive Ca^{2+} channels (Ca_V^{2+}) mediating spikes are mostly L-type Ca^{2+} channels (Brandt, Streissnig, & Moser, 2003) coded by the gene *Cacna1d* (Kollmar, Montgomery, Fak, Henry, & Hudspeth, 1997; Platzer et al., 2000). In the chick, $Ca_V1.3$ channels are transiently supplemented with $Ca_V3.1$ channels during periods of spontaneous hair cell spiking (~E7-E17, Levic et al., 2007). In the adult, Ca_V channels are primarily colocalized with ribbon synapses and restricted to the base of hair cells whereas only about 30 to 50% are colocalized with synaptic ribbons in the early neonate (Brandt, Khimich, & Moser, 2005; Zampini et al., 2010). Voltage gated sodium channels (Na_V) also appear transiently during embryonic and post-natal periods in rodents (Oliver, Plinkert,

Zenner, & Ruppersberg, 1997; Marcotti, Johnson, Rüsch, et al., 2003). Although Na$^+$ currents contribute, inner hair cell spike discharge depends critically on voltage sensitive Ca^{2+} channels and associated Ca^{2+} currents (Brandt et al., 2003; Kros et al., 1998; Marcotti, Johnson, Rüsch, et al., 2003b; Marcotti, Johnson, & Kros, 2004). The shape of the Ca^{2+} spike discharge changes with development, becoming larger and of shorter duration as development proceeds. Figure 7–5 represents the spiking activity present in IHCs.

In the early mouse neonate (P0 for the base, P2 for the apex), IHCs begin to express a Ca^{2+}-activated small-conductance potassium channel (SK channel; Glowatzki & Fuchs, 2000; Katz et al., 2004; Marcotti, Johnson, & Kros., 2004). The SK channel is not directly sensitive to changes in membrane voltage but rather is activated by intracellular free Ca^{2+}. Calmodulin is bound tightly to the SK channel to form a complex (Xia et al., 1998). When Ca^{2+} binds to calmodulin, the conformation of the SK channel changes forming an open diffusion path for K$^+$. Diffusion of K$^+$ then tends to hyperpolarize the cell membrane. During the onset depolarization phase of a hair cell spike, Ca^{2+} diffuses into the hair cell via Ca$_v$1.3 channels and activates the slow SK current. The slow SK current repolarizes the IHC, thereby resetting the membrane in preparation for the generation of subsequent spikes. The SK current is required for sustained trains of Ca^{2+} spikes in IHCs (Johnson, Adelman, & Marcotti, 2007; Marcotti et al., 2004) as well as for olivocochlear efferent function (see below; Glowatzki & Fuchs, 2000; Kong, Adelman, & Fuchs, 2008; Murthy, Mason, et al., 2009). The appearance of SK channels in IHCs is transient and coincides with the arrival and formation of abundant olivocochlear efferent synaptic terminals. The efferent terminals, acetylcholine receptors and SK channels colocalize in basal regions of the IHC membrane during the early prehearing period (see Figure 6–12 for innervations patterns; e.g., Katz et al., 2004; Marcotti et al., 2004; Oliver et al., 2000). SK channels disappear from IHCs during the second postnatal week in the mouse when efferent neurites reorganize to innervate OHCs. SK channels finally appear in OHCs late in the prehearing period (second postnatal week in the mouse) as efferent innervation of OHCs increases dramatically. SK channels are prominent on OHCs in mature mammals (Dulon, Luo, Zhang, & Ryan, 1998) and birds (Fuchs & Murrow, 1992).

Role of Cochlear Efferent Neurons

Apparently, the dominant functional role of the cochlear efferent system switches

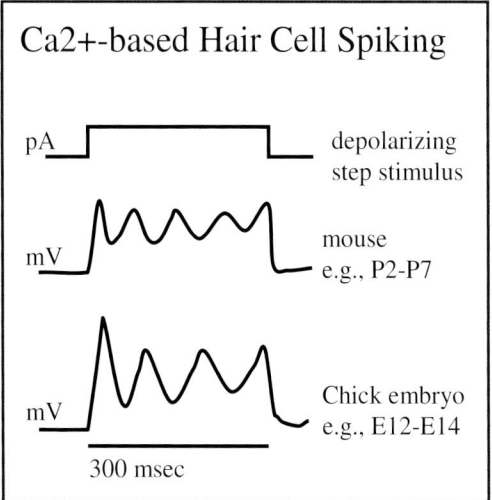

Figure 7–5. Examples of Ca^{2+}-dependent action potentials schematically represented based on data reported by Fuchs and Sokolowski (1990) and Kros et al. (1998).

from controlling primarily IHC activity (e.g., Ca^{2+} spike patterns) to one of controlling OHC activity during the prehearing period of neural reorganization. Surgical interruption and subsequent degeneration of efferent projections (Walsh, McGee, McFadden, & Liberman, 1998), genetic mutation or knockout of efferent receptors (Murthy, Taranda, Elgoyhen, & Vetter, 2009b; Turcan, Slonim, & Vetter, 2010; Vetter et al., 1999, 2007), or deletion of SK2 channels (Johnson et al., 2007; Kong et al., 2008; Murthy, Maison, et al., 2009) prevents the normal maturation of auditory afferent fiber projections and cochlear functional characteristics suggesting an important role for efferent control during development.

The principal neurotransmitter of the olivocochlear efferent system in mature animals is acetylcholine (ACh; Eybalin, 1993). There is convincing evidence that this is also true in the prehearing neonatal animals (Glowatzki & Fuchs, 2000; Marcotti, Johnson, Holley, et al., 2003; Morley & Simmons, 2002; Vetter et al., 2007). Evidence for this includes the observation that a cholinergic agent (presumably ACh) is released onto the hair cell membrane when olivocochlear efferent terminals are activated in both mature (OHC) and prehearing (IHC) cochleas (Fuchs & Murrow, 1992; Glowatzki & Fuchs, 2000; Katz et al., 2004; Oliver et al., 2000;). The action of ACh is inhibitory, that is, the membrane potential ultimately becomes more negative. The receptor mediating the inhibitory effect is α9α10 nicotinic ACh receptor (nAChR; Elgoyhen, Johnson, Boulter, Vetter, & Heineman, 1994; Elgoyhen et al., 2001; Gomez-Casati, Fuchs, Elgoyhen, & Katz, 2005). Functional nAChRs are transiently expressed in regions near efferent synapses on developing IHCs and appear in IHCs only during the early prehearing period (e.g., Glowatzki & Fuchs, 2000; Katz et al., 2004; Marcotti et al., 2004; Morley & Simmons 2002; Vetter et al., 2007). Inhibitory efferent function is eliminated when the α9α10 nAChr or the SK2 channel is deleted (Kong et al., 2008; Murthy, Maison, et al., 2009; Vetter et al., 1999, 2007).

The mechanism for ACh inhibition involves two steps and is unusual (Blanchet, Erostegui, Sugasawa, & Dulon, 1996; Evans, 1996; Fuchs & Murrow, 1992; Glowatzki & Fuchs, 2000; Nenov, Norris, & Bobbin, 1996; Oliver et al 2000). The release of ACh from efferent terminals [or the application of ACh or other agonists to fluids bathing immature IHCs] leads to the activation of α9α10 nAChRs and associated nonselective cationic channels (Weisstaub, Vetter, Elgoyhen, & Katz, 2002). Opening these channels allows Ca^{2+} to diffuse into the cell locally where it binds and activates SK channel currents that produce a slow hyperpolarization as noted above. Evidence suggests that this Ca^{2+}-dependent process in mature OHCs is augmented by Ca^{2+}-dependent Ca^{2+} release from intracellular endoplasmic reticulum located adjacent to nAChRs (Lioudyno et al., 2004). Ca^{2+} concentration within the cell can also be elevated locally by activation of $Ca_v1.3$ channels (e.g., Marcotti et al., 2004). Therefore, immature IHC SK channels can be activated by at least two mechanisms: (1) IHC depolarization (e.g., depolarizing Ca^{2+} spike), which activates $Ca_v1.3$ channels to elevate intracellular free Ca^{2+}; or (2) efferent release of ACh, which activates α9/α10 nAChRs to elevate intracellular Ca^{2+}. Whereas the indirect voltage-linked activation of SK channels serves to facilitate Ca^{2+} spike trains, as noted above, the efferent cholinergic mechanism could serve to modulate the overall pattern of discharge activity;

conceivably even extinguish spike trains to produce long periods of silence characteristic of prehearing endogenous activity.

Channels During Late Prehearing Periods: Transitions

The ability of IHCs to generate Ca^{2+} spikes disappears with the onset of hearing. The disappearance of IHC spiking coincides with the appearance of a fast activating large Ca^{2+}-activated potassium conductance at about P10 to P12 in the mouse (called $g_{K,f}$, or $g_{K(Ca)}$, or BK channels; Fuchs & Sokolowski, 1990; Kros et al., 1998; Li et al., 2009; Marcotti, Johnson, Rüsch, et al., 2003). BK channels are found in mature hair cells (Dulon, Sugusawa, Blanchet, & Erostegui, 1995; Engel et al., 2006; Hafidi, Beurg, & Dulon, 2005; Housley & Ashmore, 1992) and are thought to be essential for the precise timing of primary afferent activation and neural discharge that is tightly synchronized to transduction currents at frequencies below about 5 kHz (Oliver, Taberner, Thurm, Sausbier, Arntz, Ruth, Fakler, & Liberman, 2006). Prior to P10 in the mouse, BK channels are not present in hair cells. The upregulation of BK channels in the neonatal mouse is associated with a significant reduction in the IHC membrane time constant. In contrast to the sluggish kinetics found prior to the acquisition of BK channels, hair cell currents develop very rapidly in the presence of BK channels. This rapid response to transduction currents prevents Ca^{2+} spikes and transforms the immature spiking IHC into a reliable high-frequency encoder (Kros et al., 1998). The newly acquired short membrane time constant ensures that membrane voltages faithfully reflect transduction currents even at relatively high frequencies. Stimulus induced fluctuations in membrane voltage modulate the diffusion of Ca^{2+} into the cell (e.g., via $Ca_v1.3$ channels) and this provides the basis for synchronizing Ca^{2+}-dependent release of neurotransmitter with transduction currents. These sharper temporal characteristics mark the first evidence of the transformation of IHCs from signal generators to detectors and encoders of sound vibrations.

BK channels also appear at regions near the base of OHCs during the second neonatal week in the mouse (Rüttiger et al., 2004). Additionally, both IHCs and OHCs acquire a slowly activating delayed rectifier current ($I_{K,n}$), which activates at negative voltages near the resting membrane potentials (Kharkovets et al., 2000; Marcotti et al., 2003b; Marcotti & Kros, 1999; Oliver et al., 2003). These channels serve to prevent Ca^{2+} spiking and strengthen the direct relationship between transduction currents and membrane potential. As noted above, SK channels (essential for trains of Ca^{2+} spikes) are down-regulated in IHCs during the late prehearing period, which presumably restricts the possible patterns of spike discharge that occur immediately before hearing onset. These maturational changes are key steps that occur immediately before and with the onset of hearing.

Synaptic Transmission

Figure 7–6 provides a schematic illustration of the key steps for synaptic transmission from the IHC to the single terminal swelling of a spiral ganglion neuron (Fuchs, Glowatzki, & Moser, 2003; Glowatzki & Fuchs, 2002; Nouvian, Beutner, Parsons, & Moser, 2006). Depolarizing transduction currents in IHCs lead to the opening of voltage sensitive L-type Ca^{2+} channels (Brandt et al., 2003, 2005; Kollmar et al.,

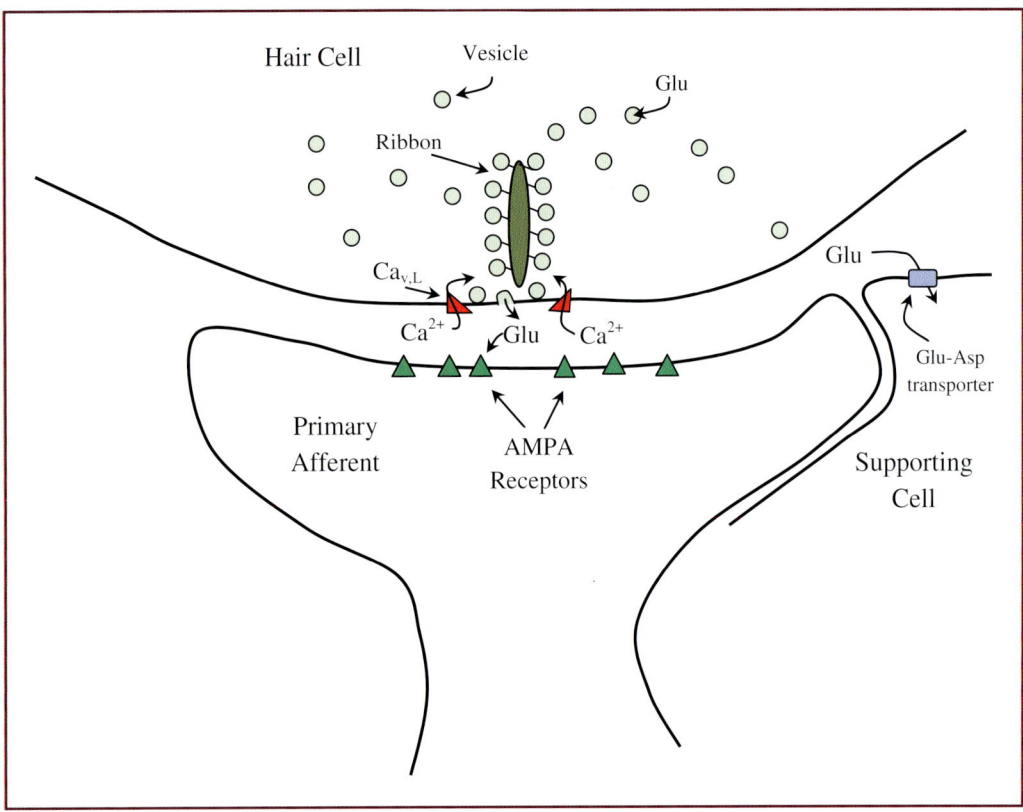

Figure 7–6. Schematic representation of mature synaptic transmission between the hair cell and spiral ganglion neuron. Glu = Glutamate; Glu-Asp = Glutamate-Aspartate transporter.

1997; Marcotti, Johnson, & Rüsch, et al., 2003; Moser & Beutner, 2000; Platzer et al., 2000; Spassova, Eisen, Saunders, & Parsons, 2001; Zampini et al., 2010; Zidanic & Fuchs, 1995) located near the presynaptic active zones of ribbon synapses. Ca^{2+} is required for the release of neurotransmitter and this is accomplished through a process called exocytosis. Exocytosis involves the fusion of synaptic vesicles with the plasma membrane, which results in the release of its contents (i.e., glutamate) into the synaptic space. The site of vesicle fusion and release is often called the "active zone." Spent vesicles (those that have fused with the membrane) are rapidly replaced with other nearby free migrating vesicles as the synaptic ribbon "traps" and holds them near the active zone. Many other vesicles in the active zone are directly apposed to the presynaptic membrane and are said to be "docked" and ready for release upon a Ca^{2+} signal. The general pool of free vesicles is continually supplied by a fast vesicle-generating process located at the apex of the hair cell (Griesinger, Richards, & Ashmore, 2005). New vesicles distribute throughout the hair cell. This vesicle recycling represents a major process that is required for continuous encoding of sound information in the adult. Shortly after birth in the mouse (equates to about week 14 in humans) and thereafter, exocytosis in IHCs (and OHCs) becomes dependent on the presence of the calcium sensor otoferlin (Beurg et al.,

2008, 2010; Dulon, Safeiddine, Jones, & Petit, 2009; Johnson & Chapman 2010; Roux et al., 2006), which when defective causes DFNB9, a human nonsyndromic hearing loss (Johnson & Chapman, 2010; Yasunaga et al., 1999). In the mature cochlea, each primary afferent terminal on an IHC is generally associated with a single ribbon and active zone (Liberman, 1980, 1982). In the early neonatal rodent (human week 12), however, individual afferent terminals are often paired with two or three ribbons (Pujol, Lavigne-Rebillard, & Lenoir, 1998; Sobkowicz, 1992; Sobkowicz et al, 1980).

Release of neurotransmitter from the hair cell synapse active zone leads to the activation of the postsynaptic spiral ganglion cell. Glutamate is thought to be the neurotransmitter released during exocytosis (Nordang, Oestreicher, Arnold, & Anniko, 2000; Puel, 1995) and postsynaptic responses are mediated by α-amino-3-hydroxy-5-methylisoxazole-4-proprionic acid (AMPA) glutamate receptors (e.g., Glowatzki & Fuchs, 2002; Irons-Brown & Jones, 2004; Keen & Hudspeth, 2006; Ruel, Bobbin, Vidal, Pujol, & Puel, 2000). Usually, only one synaptic ribbon is tethered by the scaffolding protein bassoon to the plasma membrane at the active zone (Altrock et al., 2003; Frank et al., 2010; Khimich et al., 2005). The presynaptic membrane of the active zone appears thickened and faces the postsynaptic density of the afferent bouton where AMPA receptors are located. Glutamate is synthesized in the hair cell and stored in synaptic vesicles (Ruel et al., 2008; Seal et al., 2008). Numerous vesicles containing glutamate are tethered to the ribbon, others are docked against the presynaptic membrane. In the presence of Ca^{2+}, synaptic vesicles fuse with the plasma membrane and release their contents into the synapse. Glutamate then binds with postsynaptic AMPA receptors to activate the primary afferent neurons. The process of vesicle fusion with the membrane increases the surface area of the hair cell membrane. This increases the membrane capacitance of the hair cell, which can be measured and used to monitor the rate of ongoing exocytosis (Parsons, Lenzi, Almers, & Roberts, 1994). Glutamate is removed from the synapse in part by glutamate-aspartate transporters located on the membranes of supporting cells (e.g., Furness & Lawton 2003; Furness & Lehre 1997; Glowatzki et al., 2006; Nordang et al., 2000).

As noted above, free intracellular Ca^{2+} acts to control exocytosis and the release of glutamate from hair cells. However, Ca^{2+} also acts as a trigger for numerous other intracellular functions including control of the open probability of BK channels, mediation of efferent inhibition via α9α10 nAChRs and SK channels, and initiation of intracellular signal cascades. Given the many functions linked to the Ca^{2+} signal, one wonders how the cell manages to control each function separately in a single hair cell. One way is to ensure that the free Ca^{2+} that enters the cell through the plasma membrane acts locally and is short lived. As local Ca^{2+} concentration increases, some of the Ca^{2+} acts immediately on a local target whereas the remaining is either rapidly bound by intracellular proteins and/or pumped out of the cytosol into internal compartments (e.g., endoplasmic reticulum) or out of the cell. Under these circumstances, it is the location of the Ca^{2+} channels that largely determines which function will be affected. Colocalizing cell membrane-bound Ca^{2+} channels in clusters within target synaptic active zones or with particular calcium-activated channels helps ensure discrete action. Given that different types of Ca^{2+} channels are activated by distinctly different primary signals (e.g. membrane voltage versus specific chemical ligands), it follows that many different external signals can act to produce different Ca^{2+}-mediated responses in a cell. Examples of this colocalization strategy are found in the hair cells of nonmammalian species such as turtles, zebrafish, frogs, and birds where BK channels, Ca_v channels, and ribbon synapses are located together in basolateral membranes of the hair cell (Kim et al., 2010; Li et al., 2009; Roberts, Jacobs, & Hudspeth, 1990; Tucker & Fettiplace, 1995; Zenisek, Davila, Wan, & Almers, 2003). Presumably the juxtapositioning of these three elements serves resonant electrical tuning in nonmammalian hair cells (Art & Fettiplace, 1987; Fuchs, Nagai, & Evans, 1988; Hudspeth & Lewis 1988). The clustering of Ca^{2+} channels also has been reported in mammalian IHC active zones (Brandt et al., 2005); however, mammalian IHCs lack electrical tuning and express some channels at different locations compared to nonmammals. For example, IHC BK channels are largely found clustered in the apical membrane below the cuticular plate (Hafidi et al., 2005; Pyott, Glowatzki, Trimmer, & Aldrich, 2004; Pyott et al., 2007; Rüttiger et al., 2004). Small numbers of $Ca_v1.3$ channels are found in apical regions overlapping BK channel locations (Hafidi & Dulon, 2004), but most are clustered in basolateral regions in association with synaptic ribbons and active zones (Brandt et al., 2005; Frank, Khimich, Neef, & Moser, 2009; Hafidi & Dulon, 2004; Meyer et al., 2009). The distribution of $Ca_v1.3$ channels changes in development such that only about 30% of $Ca_v1.3$ channels are colocalized within synaptic active zones in immature IHCs (e.g., P7, Zampini et al., 2010) whereas most are colocalized with synaptic active zones in mature IHCs. A wider distribution in the prehearing period would seem to better support regenerative Ca^{2+} spiking and the much larger Ca^{2+} currents reported prior to the onset of hearing (Beutner & Moser, 2001; Marcotti, Johnson, Holley, et al., 2003).

Maturational changes in exocytosis in the neonatal mouse also have been characterized. The rate of hair cell exocytosis in response to a fixed step depolarization increases dramatically during the first neonatal week (week 14 to 18 in humans) and then decreases to approach mature levels with the onset of hearing (Beutner & Moser, 2001). The changes in exocytosis over this period parallel changes in Ca^{2+} current levels (Beutner & Moser, 2001; Beurg et al., 2010) although the effectiveness of Ca^{2+} current in eliciting exocytosis increases during the same period (Johnson, Franz, Knipper, & Marcotti, 2009 ; Johnson, Marcotti, & Kros, 2005). Ca^{2+} effectiveness is higher in the base compared to the apex in mature cochleas, but not in neonatal gerbils, suggesting that there is a refinement of the link between Ca^{2+} and neurotransmitter release along the tonotopic gradient during development (Johnson, Forge, Knipper, Munkner, & Marcotti, 2008; Johnson et al., 2009). Some of these changes may be due to the consolidation of synaptic active zones where immature synapses often have multiple round ribbons associated with one active zone whereas mature hair cells generally have a single thin ribbon at each active zone, as noted previously.

Endogenous Patterned Activity in Primary Afferent Neurons

The question arises as to whether IHC Ca^{2+} spiking leads to the activation of primary afferent neurons during the prehearing period? The depolarizing current producing the IHC spike in neonates is carried in large part by Ca^{2+} and thus the

Ca^{2+} spike leads to transient increases in intracellular free Ca^{2+}. As we have noted, immature synaptic ribbons and primary afferent synapses are present in both IHCs and OHCs well before the onset of hearing (e.g., Pujol et al., 1998; Sobkowicz et al., 1982). Moreover during this period, an IHC spike or an external artificial depolarizing current applied to IHCs produces Ca^{2+}-dependent exocytosis as evidenced by increases in membrane capacitance (e.g., Beurg et al., 2010; Beutner & Moser, 2001; Johnson et al., 2005; Marcotti, Johnson, Holley, et al., 2003; Spassova et al., 2001;). Such data show that mechanisms responsible for Ca^{2+}-dependent vesicle fusion with the IHC plasma membrane are present and functional. However, alone they do not demonstrate the release of a neurotransmitter or activation of postsynaptic membrane.

A more direct approach involves recording excitatory postsynaptic currents (EPSCs) from single primary afferent bouton terminals on IHCs (Glowatzki & Fuchs, 2002; Goutman & Glowatzki, 2007). EPSCs generally represent local postsynaptic membrane currents that are elicited when a neurotransmitter, having been released from the presynaptic hair cell, binds with the postsynaptic receptor. Glowatzki and colleagues showed that depolarizing neonatal IHCs increased the rate of EPSC discharge, presumably due to the resultant increase in IHC intracellular Ca^{2+}. EPSCs were eliminated by selectively blocking postsynaptic AMPA glutamate receptors. Finally, these investigators noted that following IHC depolarization, EPSC "bursting" patterns occasionally were produced at a time corresponding to IHC spiking, suggesting that IHC spikes can lead to EPSCs. These findings demonstrate that glutamate (or glutamate-like substance) is released during exocytosis and such release results in excitation of primary afferent terminals in the isolated cochlea of the prehearing neonate. They also suggested the possibility that primary afferent neurons may be activated by IHC spikes.

Evidence for repetitive IHC spiking noted above was obtained from cells or organs isolated from their natural fluid and tissue environment. One may ask, does repetitive IHC spiking occur spontaneously in the intact animal (in vivo)? Secondly, what is the evidence that such activity actually leads to the activation of primary afferent neurons in vivo? The presence of rhythmic and intermittent bursting discharge patterns in cochlear primary afferents in prehearing birds (Jones et al., 2001) and mammals (Jones, Leake, Snyder, Stakhovskaya, & Bonham, 2007) and in central auditory relays (Gummer & Mark, 1994; Rubsamen & Schafer 1990; Sonntag, Englitz, Kopp-Scheinpflug, & Rubsamen, 2009; Tritsch, Rodrigues-Contreras, et al., 2010) provides strong evidence that these discharge processes are initiated in the neuroepithelium of the cochlea, are spontaneously generated (independent of extrinsic stimulation), and ultimately lead to the activation of central auditory projections in the intact prehearing animal.

Recent evidence suggests that the spontaneous release of ATP from supporting cells in the cochlear neuroepithelium may trigger the initiation of IHC spike activity in vivo (Tritsch, Yi, Gale, Glowatzki, & Bergles, 2007; Tritsch, Zhang, Ellis-Davies, & Bergles, 2010). ATP, acting through P2 purinergic receptors on IHCs, depolarizes the IHCs by opening nonselective cation channels. Presumably such IHC depolarization triggers the Ca^{2+} spiking as previously described. Spontaneous localized transient release of ATP is likely effective in coordinating the discharge pattern of groups of IHCs located within

100 to 200 microns of the ATP source. Such processes activate adjacent IHCs in a spatial pattern running along the tonotopic gradient thus producing spatially-patterned and correlated discharge activity in a large number of neurons innervating these IHCs (5 to 30 neurons per hair cell). As noted earlier, the overall pattern of spontaneous rhythmic bursting may be influenced by efferent discharge, which can affect ongoing activity of IHCs specifically during early prehearing periods.

An example of endogenous spontaneous discharge in the P5 kitten is illustrated in the bottom tracing of Figure 7–7. The top recording (Figure 7–7A) shows the virtually continuous spontaneous discharge of a cochlear ganglion cell from an older kitten (P36). This should be contrasted with the lower tracing (Figure 7–7B) where the neuron produces occasional bursts of action potentials separated by long silent periods (Jones et al., 2007). Similar bursting of primary afferent neurons is present in the chick at ages E12 to E17 (Jones et al., 2001). Figure 7–8 shows the discharge pattern of auditory ganglion cells over a much longer time frame (greater than 5 min). The top tracing shows the spontaneous discharge pattern of an older kitten (P36) where the mean discharge rate is virtually constant at approximately 88 spikes/sec. An example of endogenous spontaneous activity recorded from a P5 kitten is shown in the lower tracing. In the lower tracing, intermittent bursts of discharge activity are seen as sudden large rate increases and these bursts are separated by long periods where discharge activity is relatively low or absent.

Bursting rates have been measured and generally vary between 1 and 30 bursts per minute in both the cat and the chick. This activity is reminiscent of spontaneous electrophysiological activity found

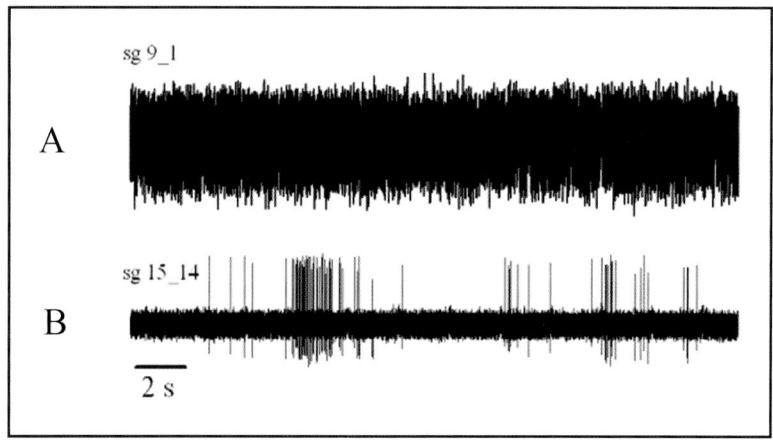

Figure 7–7. Spontaneous electrical discharge of spiral ganglion cells from kittens at P36 (top trace) and P5 (bottom trace). The activity shown reflects approximately 30 seconds of a 6-minute recording. The mature discharge rate was 88 spikes/sec. The neonatal cell discharged slowly on average (mean spike rate = 1.8 spikes/sec) with repeated periods of intense activity separated by long silent periods. From Jones et al. (2007). Used with permission of the American Physiological Society.

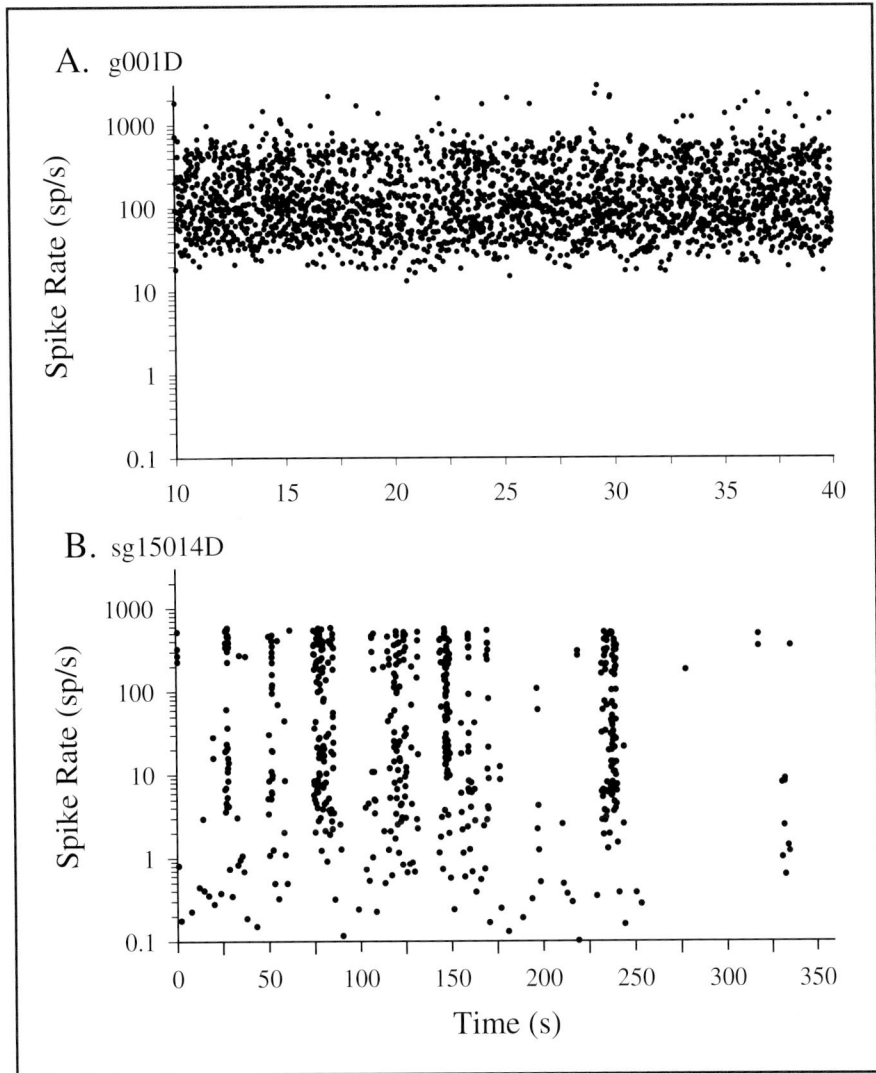

Figure 7–8. Spontaneous spike discharge rates for the same animals shown in Figure 7–7. At P36 (**A**) when cochlear function is mature, the discharge rate is high with some variation around the mean rate. For the neonatal kitten at P5 (**B**), the discharge rate shows wide variability along with intermittent bursts of activity separated by intervals with little or no activity. Note the different time scales for the two ages. The mature data reflect 30 seconds of recording while the neonatal recording lasted nearly 6 minutes. Modified from Jones et al. (2007). Used with permission of the American Physiological Society.

in widespread areas of the brain during development (see reviews by Spitzer, 2006; Blankenship & Feller, 2010). In the nervous system, endogenous discharge signaling plays a role in diverse developmental processes including neural differentiation, neurotransmitter selection, neuronal migration, and dendrite formation. In the cochlea, spontaneous hair cell discharges may serve to modulate the release of neurotrophins and influence synaptogenesis and remodeling of primary afferent and efferent terminals prior to the onset of hearing. Moreover, the patterned activity also may play a role in the refinement and segregation of cochlear neural projections.

Endogenous Signaling Occurs in Concert with Neuroepithelial Reorganization

Endogenous patterned activity may be important for the reorganization of the sensory neuroepithelium during prehearing periods. In Chapter 6, we described an initial sprouting of cochlear afferent neurites within the neuroepithelium for the early prehearing period (first week in neonatal mammal, E12 to E15 in the chick, weeks 12 to 17 human), which resulted in the expansion of radial fibers to innervate both IHCs and OHCs. Also during this early prehearing period, we noted extensive elaboration of efferent terminals that reached IHCs with minimal projections to OHCs. This immature neural configuration is found in mammals and birds, is transient in that it is reorganized over the subsequent prehearing period of days (weeks in humans) to form the mature configuration, and it occurs simultaneously with IHC spiking and bursting discharge activity in primary afferent neurons. During the late prehearing period of reorganization (approximately P7 to P10 mouse, E15 to E17 chick, 18 to 24 weeks human), type I spiral ganglion cell neurites prune contacts on OHCs and withdraw excess terminations from multiple IHCs whereas type II neurites eliminate contacts on IHCs and elaborate terminals on OHCs. Meanwhile, efferent terminal neurites of the medial olivocochlear bundle begin to establish synaptic contacts on OHCs while eliminating contacts on IHCs. This occurs just as lateral olivocochlear (LOC) neurites arrive at IHCs and form synaptic contacts on afferent terminals of type I primary afferents.

As noted above, these dramatic changes in peripheral neural circuitry are accompanied by a remarkable reconfiguration of basolateral membrane channels in hair cells. For example, the $\alpha 9/\alpha 10$ nAChR appears exclusively in IHCs during the early prehearing period when olivocochlear efferents form abundant functional synapses on IHCs (Glowatzki & Fuchs, 2000). These channels are downregulated in IHCs and upregulated in OHCs during the subsequent late prehearing period as efferent terminals retract from IHCs and form on OHCs. Ca^{2+} spiking occurs throughout the early prehearing period but begins to change during late prehearing periods as BK and $K_v7.4$ channels are upregulated in IHCs and OHCs and as SK channels are downregulated in IHCs and upregulated in OHCs. Unraveling the precise mechanisms operating to orchestrate this complex reorganization is an exciting but challenging task for future research.

Thyroid Hormone and the Timing of Developmental Events

The timing of changes during prehearing neonatal reorganization appears to be controlled in part by thyroid hormone. Thyroid hormone is required for normal cochlear development (e.g., Deol, 1973; Forrest et al., 1996; Hebert, Langlois, & Dussault, 1985; Uziel, Gabrion, Ohresser, & Legrand, 1981). The critical period for thyroid hormone action in the cochlea includes the period of neuroepithelial reorganization before hearing begins. In rodents, this timing extends from late embryonic to postnatal ages (Bock & Steel 1983; Ehret, 1976; Hebert et al., 1985; Knipper et al., 2000) and begins at about E10 in the chick (Bargman & Gardner, 1967). In the human, this occurs between approximately week 10 (onset of fetal thyroid function) to about week 26 to 27 coinciding with the onset of hearing (Birnholz & Benacerraf, 1983; Starr, Amlie, Martin, & Sanders, 1977). Excess thyroid hormone during the critical period accelerates maturation of the cochlea (Freeman, Geal-Dor, Shimoni, & Sohmer, 1993). The absence or reduction of thyroid hormone or its cognate receptors can lead to significantly altered cochlear morphology, substantially elevated ABR thresholds or profound deafness, and delayed maturation of the neuroepithelium (Brandt et al., 2007; Knipper et al., 2000; Mustapha et al 2009; Ng et al., 2001; Rüsch, Erway, et al., 1998; Winter et al., 2006, 2007, 2009). The delay in development includes a failure to eliminate excess ribbon synapses and a delay in the normal maturation of ribbon synapses (Sendin et al., 2007).

Thyroid hormone acts to upregulate or downregulate gene transcription through two nuclear receptors (TRα & TRβ, these receptors are thus transcription factors). Reduced thyroid hormone function delays the up-regulation of $K_v7.4$ in OHCs and BK channels in IHCs and OHCs, as well as neurite pruning and efferent synaptogenesis. The various different components of the cochlea affected by reduced thyroid hormone function suggest that it has a wide influence on numerous processes in different cells. It is interesting that, despite many cochlear deficits, recent studies have provided no evidence of vestibular dysfunction in hypothyroidism. Congenital thyroid disorders often are associated with human deafness such as Pendred syndrome and resistance to thyroid hormone. Understanding the mechanisms of thyroid hormone action and the identification of specific mediators may lead to new strategies that can be used for the prevention or reversal of hypothyroid-dependent auditory dysfunction.

Putative Roles for Endogenous Activity and Molecular Signaling

Auditory primary afferent neurons exert a powerful neurotrophic influence on neurons of the cochlear nucleus (e.g., Levi-Montalcini 1949; Parks 1979; for reviews: Friauf & Lohman, 1999; Rubel & Fritzsch 2002). The neurotrophic influence begins before hearing onset (by birth in the mammal and by E12 in the chick). In the chick this influence begins just after synapses between the auditory nerve and the cochlear nucleus become functional (Jackson, Hackett, & Rubel, 1982). The key evidence for a trophic function for cochlear input is the substantial loss of neurons in the cochlear nucleus following removal of the cochlea. Sensitivity to the withdrawal of cochlear input continues during development for varying periods of time depending on the species. In mammals this tends to end abruptly near the onset of hearing (e.g., Tierney, Russell, & Moore, 1997). There is evidence that it is the loss of neural discharge activity itself that plays a role in preventing apoptosis following cochlear removal (see Friauf & Lohman 1999; Harris et al., 2008). Because the susceptibility to deafferentation in the bird and mammal begins during prehearing periods and ends during the prehearing period in the mammal, it follows that spontaneous endogenous cochlear activity (hair cell spiking and bursting neural discharges) likely provides the required activity regulating cochlear nucleus apoptosis during these periods. Thus, endogenous cochlear signaling may mediate activity-dependent neurotrophic effects on neurons of the cochlear nucleus.

Endogenous cochlear activity may also be required for the normal reorganization of peripheral afferent and efferent projections during early neonatal periods. Unfortunately, there is little known about the mechanisms responsible for these processes. We do know that spontaneous Ca^{2+} action potential discharges (spiking) results in transient increases in intracellular Ca^{2+} levels in hair cells. Such activity, therefore, could in part regulate free Ca^{2+} levels, engage Ca^{2+} signal cascades, and lead to modification of gene expression in hair cells and primary afferents. Such mechanisms could help shape the reorganization of terminal afferent fields (e.g., Spitzer 2002). Evidence to support such notions has appeared recently. Deletion of the β_2 subunit of the $Ca_v1.3$ voltage-gated channel substantially reduced the number of $Ca_v1.3$ channels clustered in active zones of IHCs and dramatically decreased calcium currents (I_{Ca}) during membrane depolarization (by as much as 70%) when tested in mice older than P15 (Neef et al., 2009). Not surprisingly, the animals had a profound hearing loss. But most striking were the abnormalities in hair cell development. BK channels failed to form abundant discrete clusters at the apical neck as they normally do by the onset of hearing. Moreover, SK channels failed to downregulate normally even at ages of 4 weeks. Presuming that the weak calcium currents present in the $Ca_v\beta_2$ knockout mouse would fail to support normal IHC Ca^{2+} spiking, these investigators proposed that the absence of IHC spiking led to the developmental abnormalities. This is an interesting hypothesis. It will be important to examine this mouse model in more detail, including a complete characterization of the patterns of IHC Ca^{2+} spiking in the neonate should they exist.

On the other hand, remodeling of the cochlear neuroepithelium also may be guided by so-called "Hebbian" activity-dependent mechanisms. Hebb (1949) postulated that learning in synaptic circuits involved the strengthening of neural synapses as a result of coincident presynaptic and postsynaptic activation. A catchy well-known restatement of the rule is: "Cells that fire together, wire together." Today, the converse is also noted. Asynchronous activation of pre- and postsynaptic cells will weaken or eliminate a new synapse. These rules suggest that the status of a synapse can change depending on the activity present. Such changes are examples of synaptic plasticity. Synaptic plasticity involves the addition, strengthening, weakening, or elimination of synapses. One example of activity-dependent synaptic plasticity includes long-term potentiation, which is thought to underlie the formation of memories. In the case of the developing cochlear neuroepithelium, new synapses are formed and, as we have noted, these new synapses undergo considerable change during maturation. Such developmental plasticity may depend in part on activity-dependent Hebbian rules that govern whether newly forming synapses are stabilized and retained or destabilized and eliminated.

A combination of molecular (activity-independent) and activity-dependent mechanisms likely are responsible for neuroepithelial reorganization in the prehearing animal (e.g., Cline, 2003). Recent evidence suggests that P2X purine receptors may mediate the pruning of afferent terminals through the interference of BDNF signaling (Greenwood et al., 2007). P2X receptors are expressed in both hair cells and spiral ganglion neurons including distal neurites in synaptic regions (Haung, Ryan, Cockayne, & Housley, 2006). Thus conceivably, ATP released from supporting cells could act via P2X receptors in two ways: (1) to initiate hair cell spikes, which would strengthen synaptic contacts between an active hair cell and its cohort of afferent boutons (Hebbian processes); and (2) to inhibit BDNF signaling and in turn reduce the elaboration of neurites and branching of afferent terminals (molecular processes). Although at present we do not know all the players in this process, the interaction of endogenous activity and molecular signaling is likely to play a role. This hypothesis remains to be critically tested. Currently, we know very little about the specific mechanisms controlling the reorganization of the cochlear neuroepithelium in the prehearing animal.

Endogenous patterned activity also may be required for normal developmental refinements in central tonotopic (cochleotopic) neural projections. The activity in IHCs and in turn primary afferent neurons is likely to be spatially patterned owing to the coordinated activation of adjacent IHCs by ATP release (see above). Patterned activity such as this is well suited to drive Hebbian-based refinements in central cochleotopic projections. Similar patterned activity originating in the retina has been shown to be critical for retinotopic refinements in the visual system (e.g., Chandrasekaran, Plas, Gonzalez, & Crair, 2005; Katz & Shatz, 1996; McLaughlin, Torborg, Feller, & O'Leary, 2003; Ruthazer & Cline, 2004; Stafford, Shur, Litke, & Feldheim, 2009; Stellwagen & Shatz, 2002; Torborg & Feller, 2005).

There is evidence to support the proposition that patterned activity in the cochlea is critical for refinement of central tonotopic projections. Elimination of endogenous cochlear activity in the neonate before hearing begins: (1) disrupts the formation of aural dominance bands in the inferior colliculus (Gabriele, Brunso-Brechtold, & Henkel, 2000); (2) prevents the normal refinement and segregation of fibers of the medial nucleus of the trapezoid body (MNTB) into isofrequency bands (Sanes & Takacs, 1993); and (3) leads to decreased topographic precision in projections of cochlear spiral ganglion neurons onto the cochlear nucleus (Leake, Hradek, Chair, & Snyder, 2006). Similar results have been found in congenitally deaf mice that lack spontaneous cochlear activity (Leao et al., 2006). In this case, the normal topographical gradient of potassium channel kinetics among MNTB cells was found to be absent. In most of these studies, the activity-dependent refinement of projections occurred before hearing onset and thus could not be explained by the loss of extrinsic acoustic stimulation (Gabriele et al., 2000; Leake et al., 2006; Leao et al., 2006). Therefore, endogenous spontaneous activity may be responsible for the refinement and segregation of central cochleotopic circuits (reviewed by Kandler, Clause, & Noh, 2009). These experiments do not rule out a role for molecular mechanisms; indeed, as noted above, it is likely that a combination of activity-dependent and molecular processes operate to refine central circuits.

The Emergence and Maturation of Responses to Ambient Sound

Over the last four decades, there has been considerable progress in understanding the ontogeny of cochlear function. Examples of measures that have been used to evaluate function include the endocochlear potential (a measure of the voltage between the endolymphatic space and perilymph), cochlear microphonic (a compound electrical response to sound generated primarily by OHCs), OHC electromotility (mechanism underlying the mammalian cochlear amplifier), otoacoustic emissions (a measure of OHC function), summating potential (a compound measure of the direct current component of IHC and OHC receptor potentials), compound action potential (CAP, the compound response of a large number of spiral ganglion cells), discharge activity of single primary afferent neurons, and behavior. A number of comprehensive reviews that cover early work exploring the onset of auditory function are available (e.g., Romand, 1983; Rubsamen & Lippe, 1998; Walsh & Romand, 1992; Werner & Gray 1998). Here, we summarize the salient findings of important early studies and focus on work reported over the last decade.

Onset of Hearing

Birnholtz and Benacerraf (1983) observed human fetuses using ultrasound and evaluated the behavioral response to sound using the auropalpebral reflex (blink-startle reflex). Responses to sound could be seen at weeks 24 to 25 and were consistently observed at week 28. Auditory brainstem response (ABR) recordings have demonstrated the presence of neural responses to sound by gestation week 25 in humans (Despland & Galambos 1980; Hafner, Pratt, Joachims, Feinsod, & Blazer, 1991; Krumholz, Felix, Goldstein, & McKenzie, 1985; Ponton, Eggermont, Coupland, & Winkelaar, 1993; Starr et al., 1977). Although these studies reveal the period when hearing likely emerges in the human, it is difficult to get a sense about the functional progression and nature of sensory processing during development based solely on such studies. Studies in animals provide an effective way to explore these questions in some detail and provide a means to better understand the genetic mechanisms underlying normal and abnormal development.

Mice respond behaviorally to tones beginning on P11 to P12 with thresholds at 65 to 75 dB SPL (Ehret, 1976), whereas similar threshold responses for rats appear on P12 to P14 (Kelly, 1992) and P10 to P11 for the cat (Ehret & Romand, 1981). The initial responsiveness is restricted to a range of frequencies between 5 and 30 kHz for the mouse and 0.2 to 10 kHz in the cat. Over the next 12 days, thresholds drop 30 to 40 dB SPL and responses to higher frequencies appear (up to approximately 70 kHz for the mouse and 50 kHz for cat). Even at P30, response thresholds are not fully mature and decrease substantially over time. These findings reflect the series of changes found generally in animals where responses to sound are first restricted to middle to low frequencies and then spread to higher and lower frequencies with maturation. Table 7–4 summarizes examples of estimated ages of hearing onset for various species using an assortment of measures of auditory function. Note that, for most nonhuman mammals, hearing onset appears in the second or third postnatal week and for the human week 24 to 25 as noted above. The corresponding age for the chick is E16 to E17, where hearing onset is taken to be the age at which primary afferent threshold responses to airborne sound fall below approximately 85 dB SPL over a range of frequencies.

Table 7–4. Examples of Age at Hearing Onset for Various Species

Species	Hearing Onset	Measure	References
Human	24–28 wks	Behavior/Reflex	Birnholtz & Benacerraf (1983)
Human	25 wks	ABR	Despland & Galambos (1980); Hafner, Pratt, Joachims, Feinsod, & Blazer (1991); Krumholz, Felix, Goldstein, & McKenzie, (1985); Ponton, Eggermont, Coupland, & Winkelaar (1993); Starr et al. (1977)
Chick	E16/E17	Single Primary Afferent Recordings	Jones et al. (2006)
Rat	P12–P14	Behavior	Kelly (1992)
Rat	P14	ABR	Blatchley, Cooper, & Coleman (1987)
Rat	P11–P12	CAP	Uziel, Romand, & Marot (1981)
Hamster	P16	ABR	Schweitzer (1987)
Gerbil	P13–P14	Single Cochlear Nucleus Neurons	Woolf & Ryan (1985)
Mouse	P11–P12	Behavior	Ehret (1976)
Mouse	P12–P13	ABR	Shnerson & Pujol (1982).
Cat	P10–P11	Behavior	Ehret & Romand (1981)
Cat	P10–P12	Single Primary Afferent Recordings	Walsh & McGee (1987); Dolan, Teas, & Walton (1985); Romand (1984)

Equivalent incubation days (E), postnatal days (P) or weeks gestation (wks) are given.
ABR = Auditory brainstem response, CAP = Eighth nerve compound action potential.

Several investigators have sought to determine the age at which individual cochlear ganglion cells begin to respond to sound. The most extensive work has been done in the cat (Dolan, 1985; Romand, 1984, 1987; Walsh & McGee, 1986, 1987) and the chick (Jones et al., 2001; Jones et al., 2006). There are some remarkably similar observations for these species. The earliest electrophysiological responses to airborne sound can be obtained from single primary afferent fibers using midrange and lower frequencies (100 to 3000 Hz) delivered at extremely high stimulus levels (most thresholds between 100 and 140 dB SPL). This period occurs during the first postnatal week in the cat. During the corresponding developmental period in the chick (E12 to E16), 73% of primary afferent fibers are unresponsive to airborne tones at or below 100 dB SPL and the remaining cells exhibit thresholds greater than 90 dB SPL. Moreover, almost half of the primary afferent neurons fail to respond to direct mechanical stimulation of the stapes footplate at these ages. Despite the fact that some auditory primary afferent neurons can be activated by intense airborne sound, it is important to remember that such stimuli are well above the natural levels of ambient sound and the period is regarded as a prehearing period in both the mammal and the bird (corresponding human ages would be 13 to 20 weeks). Recall from the previous discussion that primary afferent neurons are not silent during this prehearing period, but are undergoing spontaneous bursting activity while IHCs are exhibiting Ca^{2+} spiking.

In addition to having high thresholds during the prehearing period, the few responsive primary afferents demonstrate little or no frequency tuning (Echteler, Arjmand, & Dallos, 1989; Jones et al., 2006; Romand 1984, 1987; Walsh & McGee, 1986, 1987). A frequency tuning curve (FTC) is used to evaluate the frequency selectivity of a given neuron. To produce an FTC, one determines the lowest stimulus level required to increase the neural discharge rate at each test frequency. The lowest stimulus level is the threshold at each frequency. The FTC when plotted represents a neuron's threshold as a function of frequency. The lowest threshold usually is defined as the best or characteristic frequency (CF) and this has been shown to be determined by the cochlear position innervated by the neuron. Primary afferents during the prehearing period, if responsive at all, tend to respond to wide ranging frequencies with little change in threshold (Figure 7–9, stages 41 to 42), thus precluding them from encoding a unique characteristic frequency.

The high thresholds of primary afferent neurons during prehearing periods can be attributed to either an immature middle ear and/or an immature organ of Corti. The fact that some cells cannot be activated at any stimulus level points to an immature cochlear apparatus. Although

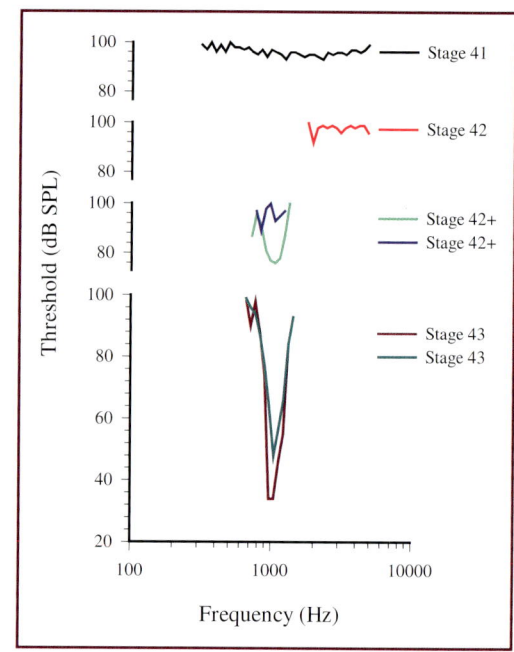

Figure 7–9. Representative frequency tuning curves from chick cochlear primary afferents at various stages of development. Stage 41 corresponds to E15 to E16. Stage 42 corresponds to E16 to E17. Stage 42+ equates to E17. Stage 43 is E17 to E18. Modified from Jones et al. (2006).

transduction channels and synapses are clearly functional (see sections above), hair cells as well as supporting and ancillary structures are not mature during the prehearing period. Hair cells have not acquired the full mature complement of basolateral ion channels and the final remodeling of the afferent and efferent terminals has yet to take place. Thus, despite being spontaneously active at this early age, the cochlea is immature and most primary afferents are unresponsive to ambient sound as a result. Nonetheless, thresholds above 100 dB SPL cannot be explained entirely by cochlear immaturity. When the middle ear is bypassed by mechanical stimulation of the stapes, there is a drop in primary afferent thresholds on the order of 30 to 40 dB SPL (Jones et al., 2006). This suggests that the middle ear plays a role in the elevated thresholds. Indeed, there is a gradual modest improvement in airborne sound transfer throughout the prehearing period (E12–E16) in the chick. A dramatic improvement in sound transfer (on the order of 20 to 25 dB) occurs just before hatching. This is largely due to the clearing of fluid from the middle ear space (Saunders, Coles, & Gates, 1973). A middle ear effect on the appearance of CM responses has also been reported in the neonatal mammal (Woolf & Ryan, 1988). Thus, during the prehearing period, evidence indicates that the middle ear presents a significant barrier to the transfer of ambient sound energy to the cochlea. Elevated cochlear response thresholds are due to both cochlear and middle ear immaturity. It is tempting to speculate that the immature middle ear may serve to protect the cochlea from external stimuli inasmuch as it is conceivable that stimulation could disrupt endogenous spontaneous patterned activity and thereby alter normal development.

The onset of hearing begins as frequency selectivity emerges and thresholds to sound stimulation drop below 80 to 90 dB SPL over a modest range of middle to low frequencies. This period corresponds reasonably well with the onset of behavioral responses to sound. In the sections below, we summarize the time line for development of key functional elements that accompany the emergence of hearing.

Endocochlear Potential

We considered the appearance of transduction channels and basolateral channels of hair cells above. In order for transduction channels to effectively depolarize hair cells in response to acoustic stimuli, the K^+ electrochemical gradient between the endolymphatic compartment and the intracellular space of hair cells must be sufficient to drive transduction currents. This requires the maintenance of elevated K^+ concentration in the cochlear endolymph (scala media) and the presence of the cochlear endolymphatic potential (endocochlear potential), which were discussed briefly in Chapter 1. The endocochlear potential (EP) normally has a magnitude on the order of +80 mV and it can be recorded by placing microelectrodes within the scala media. Figure 7–10 illustrates the specialized epithelium of the stria vascularis, the structure responsible for establishing the large cochlear endolymphatic electrochemical gradient.

Several membrane transport elements are present in the inner ear supporting cells and stria vascularis including gap junctions, ion channels, cotransporters, and pumps (Lang, Vallon, Knipper, & Wangemann, 2007; Zbedik, Wangemann, & Jentsch, 2009). Table 7–5 lists many of the transport mechanisms responsible for ionic homeostasis, the genes that encode them and the human disorders resulting from genetic mutations in these genes. K^+ pumps (Na^+-K^+ ATPase and Na^+-K^+-$2Cl^-$ cotransporter) located in the basolateral membrane (intrastrial side) of marginal cells of the stria vascularis are responsible for pumping K^+ out of the intrastrial fluid space into marginal cells at the expense of considerable metabolic energy. Blocking these pumps with pharmacological agents such as ouabain or furosemide reduces or eliminates the EP. The pumps ensure that K^+ concentration within marginal cells remains high and K^+ levels within the intrastrial fluids remain very low. The intrastrial space and intermediate cells are entirely enclosed between marginal and basal cell barrier layers where the cells of

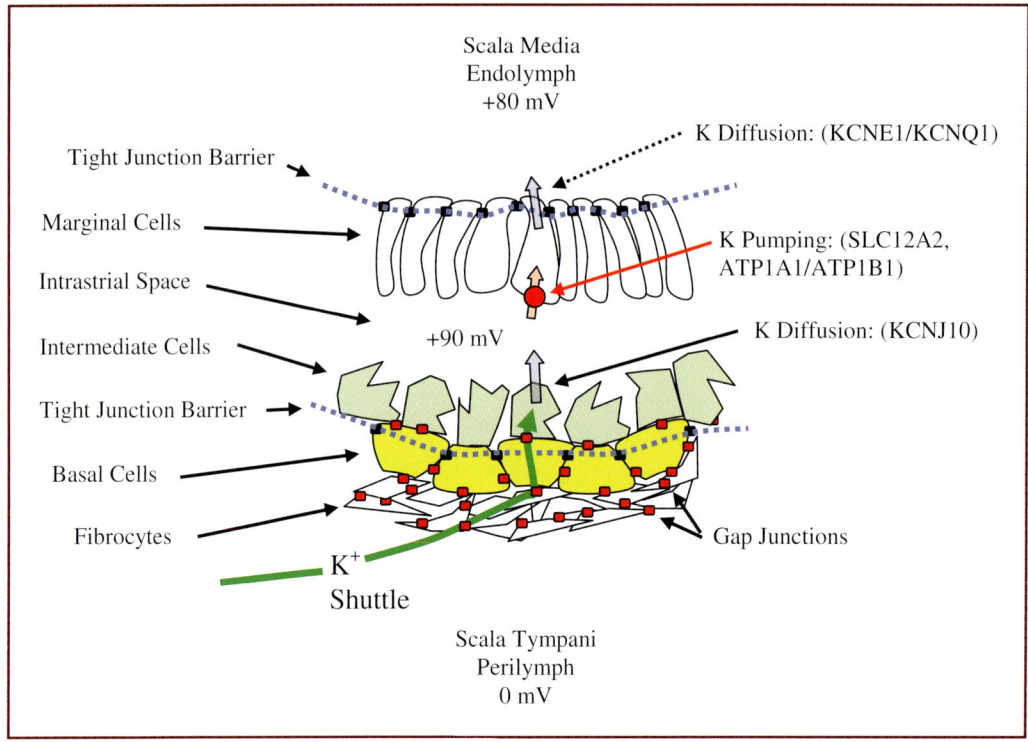

Figure 7–10. The stria vascularis has a variety of channels, pumps, and cotransporters to maintain ionic homeostasis in the inner ear.

each layer are linked together by tight junctions. The tight junctions act to prevent electrolytes from entering the intrastrial fluid from surrounding fluid spaces (i.e., from endolymph and perilymph). Intermediate cells are coupled to basal cells by gap junctions, which provide a means to supply ample K^+ to intermediate cells. Most intermediate cells are pigmented melanocytes derived from the neural crest (Hilding & Ginzberg 1977; Steel & Barkway, 1989). Intermediate cells express the inward rectifier K^+ channel $K_{ir}4.1$ (encoded by the gene *Kcnj10*). These K^+ channels permit the diffusion of K^+ into the intrastrial space. Diffusion of K^+ down its concentration gradient out of the intermediate cells into the intrastrial space generates a large diffusion potential (approximately +90 mV), which can be measured within the intrastrial space and within the adjacent marginal cell itself (Nin et al., 2008; Salt, Melichar, & Thalmann, 1987). This diffusion potential forms the basis of the EP. Elimination of intermediate cells (Steel & Barkway, 1989) or knocking out *Kcnj10* (Marcus, Wu, Wangemann, & Kofuji, 2002) eliminates the EP. Marginal cells also have K^+ channels located only on the endolymphatic border ($K_v7.1$ coded by *Kcnq1/Kcne1*). These channels permit K^+ diffusion out of the marginal cells into the endolymphatic compartment. The high concentration of K^+ combined with the large positive +90 mV intracellular potential in marginal cells serve to drive K^+ into the endolymphatic compartment and, in the adult,

Table 7–5. Transport Mechanisms That Maintain Inner Ear Fluid Homeostasis and the Endolymphatic Potential

Transport Mechanism	Gene	Locus	Human Disorder
Gap Junctions			
Gap Junction	Gja1	6q21-q23.2	Oculodentaldigital Dysplasia
Gap Junction	Gjb1	Xq13.1	Charcot-Marie Tooth-Disease
Gap Junction	Gjb2	13q11-q12	DFNB1A, DFNA3A, Vohlwinkel Syndrome, KID Syndrome
Gap Junction	Gjb3	1p35.1	DFNA2B
Gap Junction	Gjb6	13q12	DFNB1B, DFNA3B
Channels			
$K_v7.1$	Kcne1	21q22.1-q22.2	Jervell and Lange Nielsen Syndrome 2, Long QT Syndrome 5
$K_v7.1$	Kcnq1	11p15.5	Jervell and Lange Nielsen Syndrome 1, Long QT Syndrome 1
$K_{ir}4.1$	Kcnj10	1q23.2	SeSAME Syndrome[1]
	Bsnd	1p31	Bartter Syndrome Type 4A
CLCNKA	Clcnka	1p36	Bartter Syndrome Type 4B[2]
CLCNKB	Clcnkb	1p36	Bartter Syndrome Type 3[3]
SCNN1	Scnn1a, 1b, 1g	12p13, 16p13-p12	Pseudohypoaldosteronism Type 1[4], Liddle Syndrome[5]
Transporters			
Co-transporter	Slc12a2	5p23.3	
Co-transporter	Slc12a6	15q13-q14	
Co-transporter	Slc12a7	5p15.3	
Anion Transporter	Slc26a4	7q31	Pendred Syndrome, Enlarged Vestibular Aqueduct
Pumps			
Na^+-K^+ ATPase	Atp1a1	1p13-p11	
Na^+-K^+ ATPase	Atp1b1	1q22-q25	
Ca^{2+} ATPase	Atp2b1	12q21-q23	
H^+ transporting ATPase	Atp6v1b1	2q	Renal Tubular Acidosis with Progressive Deafness[6]

Note: The genes listed are expressed in nonsensory cells including various supporting cells, stria vascularis, Reissner's membrane, or the dark cells in the vestibular system. Disorders not referenced with a superscript are referenced in Table 6–9 and Chapter 6.

SeSAME = Seizure, Sensorineural deafness, Ataxia, Mental retardation, Electrolyte imbalance; KID = Keratitis-Ichthyosis-Deafness

[1]Yang et al. (2009); Scholl et al. (2009); Bockenhauer et al. (2009); [2]Schlingmann et al. (2004); [3]Simon et al. (1997); [4]Chang et al. (1996); [5]Hanson et al. (1995); [6]Karat et al. (1999).

generate K⁺ concentrations in excess of 115 millimolar (mM) and an endolymphatic electrical potential on the order of +80 mV. This endolymphatic electrochemical K⁺ gradient then drives K⁺ currents associated with hair cell transduction. Ultimately, the K⁺ that mediates transduction reaches supporting cells, perilymph, and connective tissue of the organ of Corti and is recycled to the intermediate cells by the standing transduction current, non-sensory cell K⁺ pumps, and diffusion through gap junctions. This important recycling process is sometimes referred to as the K⁺ shuttle and is briefly described in Chapter 1 (also see reviews by Lang et al., 2007; Zdebik et al., 2009).

The EP must be present before sensitive hearing can begin. Small EP voltages (<20 mV) are found in the scala media early during the prehearing periods of neonatal rodents and cats (approximately P5 to P10; Bosher & Warren, 1971; Fernandez & Hinojosa, 1974; McGuirt, Schmiedt, & Schulte, 1995; Rybak, Whitworth, & Scott, 1992; Steel & Barkway, 1989; Woolf, Ryan, & Harris, 1986). EP magnitude increases systematically over several days thereafter to reach adult values of 75 to 90 mV between P15 in the rat and mouse, P20 in the gerbil, and P30 in the cat. In cats, the response of individual primary afferent neurons to sound levels below 100 dB SPL begins to appear between P10 and P12 when the EP is only 30 to 40 mV (Romand 1984; Walsh & McGee, 1987). Thus, the endocochlear potential begins shortly before the onset of hearing and continues to mature as neural responses to sound develop and primary afferent thresholds decrease. Not surprisingly, the threshold of auditory CAPs decreases systematically with developmental increases in the EP (e.g., McGuirt et al., 1995; Rybak et al 1992; Walsh, McGee, & Javel, 1986).

Other factors that likely contribute to the improvement of CAP thresholds include improvements in middle ear transfer of sound and maturation of active basilar membrane mechanics.

Outer Hair Cell Electromotility and OAEs

In mammals, OHCs are known to respond to depolarizing current with a change in length (Ashmore, 1987; Brownell, Bader, Bertrand, & Ribaupierre, 1985; Evans & Dallos, 1993). The change in shape in response to electric current is called electromotility. Thus, during presentation of a sound, OHCs undergo sinusoidal depolarization and corresponding changes in cell length. Electromotility is thought to be the principal component of the mammalian cochlear amplifier. It is mediated by prestin and is seen by many as the primary means of injecting energy into the otherwise passive vibrations of the basilar membrane (e.g., Dallos, 2008, 2010; Santos-Sacchi, 2010). OHC electromotility begins to appear in the late prehearing period (e.g., P8 for gerbil and rat), and near the onset of hearing (P12 for mice) all OHCs demonstrate motility (Belyantseva, Adler, Curi, Frolenkov, & Kachar, 2000; He, Evans, & Dallos, 1994). However, the amplitude of electromotility continues to improve until the age of about P17 in rodents. Although this improvement occurs after the onset of hearing, the time course of the maturation of electromotility parallels the maturation of the EP magnitude and the sharpness of frequency selectivity (see below) which matures at the base by P18 in the gerbil (Echteler et al., 1989; Müller 1996; Woolf & Ryan, 1985).

The sharp tuning and high sensitivity of mature primary afferent neurons is due in part to the cochlear amplifier, which depends on normal OHC function. For a given tonotopic location, the cochlear amplifier actively boosts basilar membrane motion in response to weak (but not strong) vibrations presented at the corresponding best frequency. This process requires energy and is thought to sharpen tuning and lower thresholds. It also introduces a nonlinear mechanical response to sound that forms the basis for the generation of otoacoustic emissions (OAEs). OAEs in mammals appear to be critically dependent on the protein prestin and electromotility in OHCs (Dallos et al., 2008).

Nonmammalian vertebrates also show evidence of an active amplifier underlying sharp frequency selectivity and high sensitivities (Köppl, 2011; Manley, 2001), but the nature of the cochlear amplifier in birds may differ. The normal audibility curves for birds reveals that the highest frequencies represented are below 12 kHz and, for many bird species (e.g., the chicken), below 5 kHz. In contrast, most mammalian species (excepting the human) can hear at frequencies well above 20 kHz, extending at least to 80 kHz in some species. The basis for this frequency disparity likely is due in part to differences in middle ear transfer at high frequencies in the three ossicles of mammals versus a single process columella in the bird (Manley, 2001). However, differences also may be due in part to differences in the nature of the cochlear amplifier itself. For example, prestin is not found in the cochlear hair cells of birds and there is no evidence of shape changes in the cell body (somatic motility) for avian hair cells (e.g., He, Beisel, Chen, Ding, Jia, Fritzsch, & Salvi, 2003). Instead, nonmammalian hair cell active processes may involve hair-bundle motility in concert with electrical hair cell resonances (Duncan & Fuchs, 2003; Fettiplace, 2006; Fettiplace & Fuchs, 1999; Hudspeth, 2008; Peng & Ricci, 2010). Differences in passive mechanics of the cochlear partition also are likely.

One feature common to mammals and nonmammalian species is that cochlear amplifiers give rise to OAEs. OAEs can be measured and used in neonatal animals to characterize the development of OHC active processes. OAEs in developing animals appear at about the onset of hearing but continue to mature for a rather prolonged period up to two or three weeks after birth in rodents (e.g., Henley, Owings, Stagner, Martin, & Lonsbury-Martin, 1989; Lenoir & Puel, 1987; Mills, Norton, & Rubel, 1994; Mills & Rubel, 1996; Norton, Bargones, & Rubel, 1991). In the human, OAE amplitudes increase over the time period from gestation weeks 31 to 33 until the time of birth (Abdala, Oba, & Ramanathan, 2008). Amplitudes remain relatively constant from birth to 6 months of age and are 4 to 12 dB larger than adult OAE amplitudes. OAE suppression tuning characteristics also differ between infants and adults, but the majority of these differences have been attributed to outer and middle ear immaturities (Abdala & Keefe, 2006).

Frequency Selectivity in Primary Afferent Neurons

A number of investigators have characterized the maturation of frequency selectivity by measuring primary afferent FTCs in developing and mature animals (Dolan, Teas, & Walton, 1985; Echteler et al., 1989; Jones & Jones, 1995a, 1995b; Jones et al., 2006; Manley, Brix, & Kaiser, 1987; Müller, 1996; Romand, 1984, 1987; Walsh & McGee, 1986, 1987). The emergence of auditory responses to sound is manifest with a series of events common to all species studied including birds and mammals. As noted above, FTCs during prehearing periods have very high thresholds, are initially relatively flat across frequencies, and the range of frequencies is limited compared to an adult. Toward the end of the prehearing period (cat: approximately P9 to P10; mouse: P9 to P10, gerbil: P11 to P12; chick: E16; human week 20), FTCs begin to show narrow frequency regions within the full response bandwidth where the threshold suddenly decreases by 10 to 20 dB SPL (see Figure 7–9, stage 42+) and a distinctive CF can be identified. This is the first sign of incipient frequency selectivity but thresholds otherwise remain high generally above 90 to 100 dB SPL. Shortly thereafter, the tuning curves begin to sharpen and CF thresholds decrease dramatically over several days (presumably weeks for the human) and begin to approach mature levels (see Figure 7–9, stages 43–44).

As thresholds decrease, the range of CFs widens to include higher as well as lower frequencies. These changes in FTC shape and decreases in CF thresholds correspond to the onset of hearing and in mammals appear in parallel with the maturation of the amplitudes of OHC motility, basilar membrane stiffness (Emadi & Richter, 2008), and an increasing EP. During this period, the acquisition of high frequency CFs (above 6 kHz) at a given position in the mammalian cochlear base is accompanied by a shift in CF of up to 1.5 octaves, whereas CFs are relatively stable at more apical positions that mediate mid to low frequencies (Arjmand, Harris, & Dallos; Echteler et al., 1989; Harris & Dallos, 1984; Müller, 1996; Yancy & Dallos, 1985). Because animals tend to respond behaviorally first to middle and lower frequency sounds, it follows that responses at the onset of hearing are mediated by the middle and apical regions of the cochlea, as is the case in the adult. The shifts in CF at basal regions are associated with the maturation of OHC electromotility, the appearance of OAEs, and increases in passive stiffness of the basilar membrane (Emadi & Richter, 2008). In the chick, this developmental period corresponds to E17 to E19 (Jones & Jones, 1995a, 1995b; Jones et al., 2006).

Maturation of frequency selectivity in the chick coincides with the acquisition of functional BK channels (Duncan & Fuchs, 2009), a key component of electrical resonance in nonmammalian hair cells. Unlike mammals, in the chick there is no evidence of a shift in CF at a given position along the cochlea. The full range of CFs is present (100 to 5000 Hz) beginning at E17 to E19 and they are represented at cochlear positions matching those of the adult (Chen, Salvi, & Shero, 1994; Jones & Jones, 1995a, 1995b; Jones et al., 2006; Manley, Brix, & Kaiser, 1987). This period of hearing onset and maturation corresponds to the end of the second postnatal week in rats, mice, and cats. During this time period, the remodeling of the neuroepithelium is essentially completed and the efficiency of synaptic activity matures. Efferent neurons complete their switch from modulating IHC activity to mediating feedback control over OHC activity. In the human, this presumably occurs during weeks 20 to 28 of gestation.

Although all elements of the cochlear apparatus appear to be in place by the end of the second or third postnatal week in altricious mammals, hearing is not as acute as that demonstrated behaviorally in adult animals. Depending on the species, full sensitivities are often not achieved until weeks or months later depending on the species. The basis for continued improvement in thresholds may include maturation of middle and outer ear structures. However, it is also probable that refinement and maturation of central auditory neural circuits plays a major role in improving behavioral thresholds. The human neonate is born with mature, or nearly mature, cochlear function. However, the outer and middle ear continue to mature into childhood (reviewed in Chapter 5), and neural maturation and central refinements will also take place for several months or years after birth.

GENERAL RESOURCES

Hille, B., (1992). *Ionic channels of excitable membranes* (2nd ed.). Sunderland, MA: Sinauer Associates.

Kaufman, M. H. (1992). *The atlas of mouse development* Amsterdam, The Netherlands: Elsevier Academic Press.

Romand, R. (1992). *Development of auditory and vestibular systems 2*. Amsterdam, The Netherlands: Elsevier.

Romand, R., & Varela-Nieto, I. (2003). Development of auditory and vestibular systems 3, Molecular development of the inner ear. In G. P. Schatten (Ed.), *Current topics in developmental biology* (Vol. 57). Amsterdam, The Netherlands: Elsevier Academic Press.

Rubel, E. W., Popper, A. N., & Fay, R. R. (1998). *Development of the auditory system*. Springer Handbook of Auditory Research. New York, NY: Springer-Verlag.

LITERATURE CITED

Abdala, C., & Keefe, D. H. (2006). Effects of middle-ear immaturity on distortion product otoacoustic emission suppression tuning in infant ears. *Journal of the Acoustical Society of America, 129,* 3832–3842.

Abdala, C., Oba, S. I., & Ramanathan, R. (2008). Changes in the DP-gram during the preterm and early postnatal period. *Ear and Hearing, 29,* 512–523.

Altrock, W. D., tom Dieck, S., Sokolov, M., Meyer, A. C., Sigler, A., Brakebusch, C., . . . Gundelfinger, E. D.. (2003). Functional inactivation of a fraction of excitatory synapses in mice deficient for the active zone protein bassoon. *Neuron, 37,* 787–800.

Anniko, M. (1983). Embryonic development of vestibular sense organs and their innervation. In R. Romand (Ed.), *Development of auditory and vestibular systems* (pp. 375–423). New York, NY: Academic Press.

Arjmand, E., Harris, D., & Dallos, P. (1988). Developmental changes in frequency mapping of the gerbil cochlea: Comparison of two cochlear locations. *Hearing Research, 32,* 93–96.

Art, J. J., & Fettiplace, R. (1987). Variation in membrane properties in hair cells isolated from the turtle cochlea. *Journal of Physiology, 385,* 207–242.

Ashmore, J. F. (1987). A fast motile response in guinea-pig outer hair cells: The cellular basis of the cochlea amplifier. *Journal of Physiology, 388,* 323–347.

Baig, S. M., Koschak, A., Lieb, A., Gebhart, M., Dafinger, C., Nürnberg, G., . . . Bolz, H. J. (2009). Loss of Ca(v)1.3 (CACNA1D) function in a human channelopathy with bradycardia and congenital deafness. *Nature Neuroscience, 14,* 77–84.

Bakondi, G., Por, A., Kovács, I., Szücs, G., & Rusznák, Z. (2009). Hyperpolarization-activated, cyclic nucleotide-gated, cation nonselective channel subunit expression pattern of guinea-pig spiral ganglion cells. *Neuroscience, 158,* 1469–1477.

Bargman, G. J., & Gardner, L. I. (1967). Otic lesions and congenital hypothyroidism in the developing chick. *Journal of Clinical Investigation, 45,* 1828–1839.

Belyantseva, I. A., Adler, H. J., Curi, R., Frolenkov, G. I., & Kachar, B. (2000). Expression and localization of prestin and the sugar transporter GLUT-5 during development of electromotility in cochlear outer hair cells. *Journal of Neuroscience, 20,* RC116, 1–5.

Beurg, M., Michalski, M., Safieddine, S., Bouleau, Y., Schneggenberger, R., Chapman,

E. R., . . . Dulon, D. (2010). Control of exocytosis by synaptotagmins and otoferlin in auditory hair cells. *Journal of Neuroscience, 30*(40), 13281–13290.

Beurg, M., Safieddine, S., Roux, I., Bouleau, Y., Petit, C., & Dulon, D. (2008). Calcium- and otoferlin- dependent exocytosis by immature outer hair cells. *Journal of Neuroscience, 28*(8), 1798–1803.

Beutner, D., & Moser, T. (2001). The presynaptic function of mouse cochlear inner hair cells during development of hearing, *Journal of Neuroscience, 21*(13), 4593–4599.

Birnholz, J. C., & Benacerraf, B. R. (1983). The development of human fetal hearing. *Science, 222*, 516–518.

Blanchet, C., Erostegui, C., Sugasawa, M., & Dulon, D. (1996). Acetylcholine-induced potassium current of guinea pig outer hair cells: Its dependence on a calcium influx through nicotinic-like receptors. *Journal of Neuroscience, 16*(8), 2574–2584.

Blankenship, A. G., & Feller, M. B. (2010). Mechanisms underlying spontaneous patterned activity in developing neural circuits *Nature Reviews Neuroscience, 11,* 18–29.

Blatchley, B. J., Cooper, W. A., & Coleman, J. R. (1987). Development of auditory brainstem response to tone pip stimuli in the rat. *Brain Research, 429,* 75–84.

Bock, G. R., & Steel, K. P. (1983). Inner ear pathology in the deafness mutant mouse. *Acta Oto-Laryngologica, 96,* 39–47.

Bockenhauer, D., Feather, S., Stanescu, H. C., Bandulik, S., Zdebik, A. A., Reichold, . . . Kleta, R. (2009). Epilepsy, ataxia, sensorineural deafness, tubulopathy, and KCNJ10 mutations. *New England Journal of Medicine, 360,* 1960–1970.

Bosher, S. K., & Warren, R. L. (1971). A study of the electrochemistry and osmotic relationships of the cochlear fluids in the neonatal rat at the time of the development of the endocochlear potential. *Journal of Physiology, 212,* 739–761.

Brandt, A., Khimich, D., & Moser, T. (2005). Few Ca_v 1.3 channels regulate the exocytosis of a synaptic vesicle at the hair cell ribbon synapse. *Journal of Neuroscience, 25*(50), 11577–11585.

Brandt, A., Streissnig, J., & Moser, T. (2003). Ca_v 1.3 channels are essential for development and presynaptic activity of cochlear inner hair cells. *Journal of Neuroscience, 23*(34), 10832–10840.

Brandt, N., Kuhn, S. Münker, S., Braig, C., Winter, H., Blin, N., . . . Engel, J. (2007). Thyroid hormone deficiency affects postnatal spiking activity and expression of Ca^{2+} and K^+ channels in rodent inner hair cells. *Journal of Neuroscience, 27*(12), 3174–3186.

Brownell, W. E., Bader, C. R., Bertrand, D., & de Ribaupierre, Y. (1985). Evoked mechanical responses of isolated outer hair cells. *Science, 227,* 194–196.

Casselbrant, M. L., Mandel, E. M., Sparto, P. J., Perera, S., Redfern, M. S., Fall, P. A., . . . Furman, J. (2010). Longitudinal posturography and rotational testing in children three to nine years of age: Normative data. *Otolaryngology Head and Neck Surgery, 142,* 708–714.

Chambard, J. M., Chabbert, C., Sans, A., & Desmadryl, G. (1999). Developmental changes in low and high voltage-activated calcium currents in acutely isolated mouse vestibular neurons. *Journal of Physiology, 518,* 141–149.

Chandrasekaran, A. R., Plas, D. T., Gonzalez, E., & Crair, M. C. (2005). Evidence for an instructive role of retinal activity in retinotopic map refinement in the superior colliculus of the mouse. *Journal of Neuroscience, 25*(29), 6929–6938.

Chang, S. S., Grunder, S., Hanukoglu, A., Rosler, A., Mathew, P. M., Hanukoglu, I., . . . Lifton, R. P. (1996). Mutations in subunits of the epithelial sodium channel cause salt wasting with hyperkalaemic acidosis, pseudohypoaldosteronism type 1. *Nature Genetics, 12,* 248–253.

Charpiot, A., Tringali, S., Ionescu, E., Vital-Durand, F., & Ferber-Viart, C. (2010). Vestibulo-ocular reflex and balance maturation in healthy children aged from six to twelve years. *Audiology and Neurotology, 15,* 203–210.

Chen, L., Salvi, R., & Shero, M. (1994). Cochlear frequency-place map in adult chickens: intracellular biocytin labeling. *Hearing Research, 81,* 130–136.

Chen, Q., Kirsch, G. E., Zhang, D., Brugada, R., Brugada, J., Brugada, P., . . . Wang, Q. (1998). Genetic basis and molecular mechanism for idiopathic ventricular fibrillation. *Nature, 392,* 293–295.

Chen, W. C., & Davis, R. L. (2006). Voltage-gated and two-pore-domain potassium channels in murine spiral ganglion neurons. *Hearing Research, 222,* 89–99.

Chen, Y., Lu, J., Pan, H., Zhang, Y., Wu, H., Xu, K., . . . Wu, X. (2003). Association between genetic variation of CACNA1H and childhood absence epilepsy. *Annals of Neurology, 54,* 239–243.

Cioni, G., Favilla, M., Ghelarducci, B., & La Noce, A. (1984). Development of the dynamic characteristics of the horizontal vestibuloocular reflex in infancy. *Neuropediatrics, 15*(3), 125–130.

Cline, H. (2003). Sperry and Hebb: Oil and vinegar? *Trends in Neuroscience, 26,* 655–661.

Correia, M. F., Weng, T., Prusak, D., & Wood, T. G. (2008). Kv 1.1 associates with Kvα1.4 in Chinese hamster ovary cells and pigeon type II vestibular hair cells and enhances the amplitude, inactivation and negatively shifts the steady-state inactivation range. *Neuroscience, 152,* 809–820.

Correia, M. J., Rennie, K. J., & Koo, P. (2001). Return of potassium ion channels in regenerated hair cells. Possible pathways and the role of intracellular calcium signaling. *Annals of the New York Academy of Sciences, 942,* 228–240.

Coucke, P. J., Van Hauwe, P., Kelley, P. M., Kunst, H., Schatteman, I., Van Velzen, D., . . . Van Camp, G.. (1999). Mutations in the KCNQ4 gene are responsible for autosomal dominant deafness in four DFNA2 families. *Human Molecular Genetics, 8,* 1321–1328.

Curran, M. E., Splawski, I., Timothy, K. W., Vincent, G. M., Green, E. D., & Keating, M. T. (1995). A molecular basis for cardiac arrhythmia: HERG mutations cause long QT syndrome. *Cell, 80,* 795–803.

Curthoys, I. S. (1983). The development of function of primary vestibular neurons. In R. Romand (Ed.), *Development of auditory and vestibular systems* (pp. 425–461). New York, NY: Academic Press.

Cyr, D. G., Brookhouser, P.E., Valente, L.M., & Grossman, A. (1985). Vestibular evaluation of infants and preschool children. *Journal of Otolaryngology-Head and Neck Surgery, 93,* 463–468.

Dallos, P. (2008). Cochlear amplification, outer hair cells, and prestin. *Current Opinion in Neurobiology, 18,* 370–376.

Dallos, P. (2010). Feedback in the cochlea. *Hearing Research,* doi:10.1016/j.heares.2009.12.009.

Dallos, P., Wu, X., Cheatham, M. A., Gao, J., Zheng, J., Anderson, C. T., . . . Zuo, J. (2008). Prestin-based outer hair cell motility is necessary for mammalian cochlear amplification. *Neuron, 58,* 333–339.

Dechesne, C. J. (1992). The development of vestibular sensory organs in human. In R. Romand (Ed.), *Development of auditory and vestibular systems 2* (pp. 419–447). Amsterdam, The Netherlands: Elsevier.

Denman-Johnson, K., & Forge, A. (1999). Establishment of hair bundle polarity and orientation in the developing vestibular system of the mouse. *Journal of Neurocytology, 28,* 821–835.

Deol, M. S. (1973). An experimental approach to the understanding and treatment of hereditary syndromes with congenital deafness and hypothyroidism. *Journal of Medical Genetics, 10,* 235–242.

Desmadryl, G., Raymond, J., & Sans, A. (1986). In vitro electrophysiological study of spontaneous activity in neonatal mouse vestibular ganglion neurons during development. *Developmental Brain Research, 25,* 133–136.

Despland, P. A., & Galambos, R. (1980). The auditory brainstem response (ABR) is a useful diagnostic tool in the intensive care nursery. *Pediatric Research, 14,* 1154–1158.

Dickman, J. D., Huss, D., & Lowe, M. (2004). Morphometry of otoconia in the utricle and saccule of developing Japanese quail. *Hearing Research, 188,* 89–103.

Dolan, T., Teas, D. C., & Walton, J. P. (1985). Postnatal development of physiological responses in auditory nerve fibers. *Journal of the Acoustical Society of America, 78,* 544–554.

Dulon, D., Luo, L., Zhang, C., & Ryan, A. F. (1998). Expression of small conductance calcium-activated potassium channels (SK) in outer hair cells of rat cochlea. *European Journal of Neuroscience, 10,* 907–915.

Dulon, D., Safieddine, S., Jones, S. M., & Petit, C. (2009). Otoferlin is critical for a highly sensitive and linear calcium-dependent exocytosis at vestibular hair cell ribbon synapses. *Journal of Neuroscience, 29*(34), 10474–10487.

Dulon, D., Sugusawa, M., Blanchet, C., & Erostegui, C. (1995). Direct measurements of Ca^{2+}-activated K^+ currents in inner hair cells of the guinea-pig cochlea using photolabile Ca^{2+} chelators. *Pflügers Archiv European Journal of Physiology, 430,* 365–373.

Duncan, R. K., & Fuchs, P. A. (2003). Variation in large-conductance, calcium-activated potassium channels from hair cells along the chicken basilar papilla. *Journal of Physiology, 547,* 357–371.

Eatock, R. A., & Hurley, K. M. (2003). Functional development of hair cells. In R. Romand & I. Vareal-Nieto (Eds.), *Development of auditory and vestibular systems 3* (pp. 389–448). Amsterdam, The Netherlands: Elsevier.

Eatock, R. A., & Rüsch, A. (1997). Developmental changes in the physiology of hair cells. *Seminars in Cell & Developmental Biology, 8,* 265–275.

Eatock, R. A., Xue, J., & Kalluri, R. (2008). Ion channels in mammalian vestibular afferents may set regularity of firing. *Journal of Experimental Biology, 211,* 1764–1774.

Echteler, S. M., Arjmand, E., & Dallos, P. (1989). Developmental alterations in the frequency map of the mammalian cochlea. *Nature, 341,* 147–149.

Ehret, G., (1976). Development of absolute auditory thresholds in the house mouse (Mus musculus). *Journal of the American Audiology Society, 1*(5), 179–184.

Ehret, G., & Romand, R. (1981). Postnatal development of absolute auditory thresholds in kittens. *Journal of Comparative and Physiological Psychology, 95,* 304–311.

Elgoyhen, A. B., Johnson, D. S., Boutler, J., Vetter, D. E., & Heinemann, S. (1994). α9: An acetylcholine receptor with novel pharmacological properties expressed in rat cochlear hair cells. *Cell, 79,* 705–715.

Elgoyhen, A. B., Vetter, D. E., Katz, E., Rothlin, C. V., Heinemann, S., & Boulter, J. (2001). α10: A determinant of nicotinic cholinergic receptor function in mammalian vestibular and cochlear mechanosensory hair cells. *Proceedings of the National Academy of Sciences, 98,* 3501–3506.

Emadi, G., & Richter, C. P. (2008). Developmental changes of mechanics measured in the gerbil cochlea. *Journal of the Association for Research in Otolaryngology, 9,* 22–32.

Engel, J., Braig, C., Rüttiger, L., Kuhn, S., Zimmerman, U., . . . Knipper, M. (2006). Two classes of outer hair cells along the tonotopic axis of the cochlea. *Neuroscience, 143,* 837–849.

Evans, M. G. (1996). Acetylcholine activates two currents in guinea-pig outer hair cells. *Journal of Physiology, 491,* 563–578.

Evans, B. N., & Dallos, P. (1993). Stereocilia displacement induced somatic motility of cochlear outer hair cells. *Proceedings of the National Academy of Sciences, 90,* 8347–8351.

Eybalin, M. (1993). Neurotransmitter and neuromodulators of the mammalian cochlea. *Physiological Reviews, 73*(2), 309–373.

Fermin, C. D., & Igarashi, M. (1985). Development of otoconia in the embryonic chick. *Acta Anatomica, 123,* 148–152.

Fermin, C. D., & Igarashi, M. (1986). Review of statoconia formation in birds and original research in chick. *Scanning Electron Microscopy, 4,* 1649–1655.

Fernandez, C., & Hinojosa, R. (1974). Postnatal development of endocochlear potential and stria vascularis in the cat. *Acta Oto-laryngologica*, *78*, 173–186.

Fettiplace, R. (2006). Active hair bundle movements in auditory hair cells. *Journal of Physiology*, *576*, 29–36.

Fettiplace, R., & Fuchs, P. A. (1999). Mechanisms of hair cell tuning. *Annual Reviews in Physiology*, *61*, 809–834.

Forrest, D., Erway, L. C., Ng, L., Altschuler, R., & Curran, T. (1996). Thyroid hormone receptor β is essential for development of auditory function. *Nature Genetics*, *13*, 354–357.

Frank, T., Khimich, D., Neef, A., & Moser, T. (2009). Mechanisms contributing to synaptic Ca^{2+} signals and their heterogeneity in hair cells. *Proceedings of the National Academy of Sciences*, *106*, 4483–4488.

Frank, T., Rutherford, M.A., Strenke, N., Neef, A., Pangrsic, T., Khimich, D., . . . Moser, T. (2010). Bassoon and the synaptic ribbon organize Ca^{2+} channels and vesicles to add release sites and promote refilling. *Neuron*, *68*, 724–738.

Freeman, S., Geal-Dor, M., Shimoni, Y., & Sohmer, H. (1993). Thyroid hormone induces earlier onset of the auditory function in neonatal rats. *Hearing Research*, *69*, 229–235.

Fuchs, P. A., Glowatzki, E., & Moser, T. (2003). The afferent synapse of cochlear hair cells. *Current Opinion in Neurobiology*, *13*, 452–458.

Fuchs, P. A., & Murrow, B. W. (1992). Cholinergic inhibition of short (outer) hair cells of the chick's cochlea. *Journal of Neuroscience*, *12*, 800–809.

Fuchs, P. A., Nagai, T., & Evans, M. G. (1988). Electrical tuning in hair cells isolated from the chick cochlea. *Journal of Neuroscience*, *8*(7), 2480–2487.

Fuchs, P. A., & Sokolowski, B. H. A. (1990). The acquisition during development of Ca^{++} activated potassium currents by cochlear hair cells of the chick. *Proceedings of the Royal Society of London B*, *241*, 122–126.

Furness, D. N., & Lawton, D. M. (2003). Comparative distribution of glutamate transporters and receptors in relation to afferent innervation density in the mammalian cochlea. *Journal of Neuroscience*, *23*(36), 11296–11304.

Furness, D. N., & Lehre, K. P. (1997). Immunocytochemical localization of a high-affinity glutamate-aspartate transporter, GLAST, in the rat and guinea-pig cochlea. *European Journal of Neuroscience*, *9*, 1961–1969.

Furness, D. N., Richardson, G. P., & Russell, I. J. (1989). Stereociliary bundle morphology in organotypic cultures of the mouse cochlea. *Hearing Research*, *38*, 95–110.

Gabriele, M. L., Brunso-Brechtold, J. K., & Henkel, C. K. (2000). Plasticity in the development of afferent patterns in the inferior colliculus of the rat after unilateral cochlear ablation. *Journal of Neuroscience*, *20*(18), 6939–6949.

Galicia, S., Cotes, C., & Galindo, F. (2010) Development of spontaneous activity and response properties of primary lagenar neurons in the chick. *Cell and Molecular Neurobiology*, *30*, 327–331.

Géléoc, G. S. G., & Holt, J. R. (2004). Developmental acquisition of sensory transduction in hair cells of the mouse inner ear. *Nature Neuroscience*, *6*(10), 1019–1020.

Géléoc, G. S. G., Lennan, G. W. T., Richardson, G. P., & Kros, C. J. (1997). A quantitative comparison of mechanoelectrical transduction in vestibular and auditory hair cells of neonatal mice. *Proceedings of the Royal Society of London B*, *264*, 611–621.

Géléoc, G. S. G., Risner, J. R., & Holt, J. R. (2004). Developmental acquisition of voltage-dependent conductances and sensory signaling in hair cells of the embryonic mouse inner ear. *Journal of Neuroscience*, *24*(49), 11148–11159.

Glowatzki, E., Cheng, N., Hiel, H., Yi, E., Tanaka, K., Ellis-Davies, G. C. R., . . . Bergies, D. E. (2006). The glutamate–aspartate transporter GLAST mediates glutamate uptake at inner hair cell afferent synapses in the mammalian cochlea. *Journal of Neuroscience*, *26*(29), 7659–7664.

Glowatzki, E., & Fuchs, P. A. (2000). Cholinergic synaptic inhibition of inner hair cells in the neonatal mammalian cochlea. *Science, 288,* 2366–2368.

Glowatzki, E., & Fuchs, P. A. (2002). Transmitter release at the hair cell ribbon synapse. *Nature Neuroscience, 5,* 147–154.

Gomez-Casati, M. E., Fuchs, P. A., Elgoyhen, A. B., & Katz, E. (2005). Biophysical and pharmacological characterization of nicotinic cholinergic receptors in rat cochlear inner hair cells. *Journal of Physiology, 566,* 103–118.

Goutman, J. D., & Glowatzki, E. (2007). Time course and calcium dependence of transmitter release at a single ribbon synapse. *Proceedings of the National Academy of Sciences, 104,* 16341–16346.

Greenwood, D., Jagger, D. J., Huang, L. C., Hoya, N., Thorne, P. R., Wildman, S. S., . . . Housely, G. D. (2007). P2X receptor signaling inhibits BDNF-mediated spiral ganglion neuron development in the neonatal rat cochlea. *Development, 134,* 1407–1417.

Griesinger, C. B., Richards, C. D., & Ashmore, J. F. (2005). Fast vesicle replenishment allows indefatigable signaling and the first auditory synapse. *Nature, 435,* 212–215.

Gummer, A. W., & Mark, R. F. (1994). Patterned neural activity in brainstem auditory areas of a prehearing mammal: The tammar wallaby *Macropus eugenii. NeuroReport, 5,* 685–688.

Hafidi, A., Beurg, M., & Dulon, D. (2005). Localization and developmental expression of BK channels in mammalian cochlear hair cells. *Neuroscience, 130,* 475–484.

Hafidi, A., & Dulon, D. (2004). Developmental expression of $Ca_v1.3$ (α1D) calcium channels in the mouse inner ear. *Developmental Brain Research, 150,* 167–175.

Hafner, H., Pratt, H., Joachims, Z., Feinsod, M., & Blazer, S. (1991). Development of auditory brainstem evoked potentials in newborn infants: A three-channel Lissajous' trajectory study. *Hearing Research, 51,* 33–47.

Hamburger, V., & Hamilton, H. L. (1951). A series of normal stages in the development of the chick embryo. *Journal of Morphology, 88,* 49092.

Hansson, J. H., Nelson-Williams, C., Suzuki, H., Schild, L., Shimkets, R., Lu, Y., . . . Lifton, R. P..(1995). Hypertension caused by a truncated epithelial sodium channel gamma subunit: Genetic heterogeneity of Liddle syndrome. *Nature Genetics, 11,* 76–82.

Hard, E., & Larsson, K. (1975). Development of righting in rats. *Brain and Behavioral Evolution, 11,* 53–59.

Harris, D. M., & Dallos, P. (1984). Ontogenetic changes in frequency mapping of a mammalian ear. *Science, 225,* 741–743.

Harris, J. A., Iguchi, F., Seidl, A. H., Lurie, D. I., & Rubel, E. W. Afferent deprivation elicits a transcriptional response associated with neuronal survival after a critical period in the mouse cochlear nucleus. *Journal of Neuroscience, 28*(43), 10990–11002.

Haung, L., Ryan, A. F., Cockayne, D. A., & Housley, G. (2006). Developmentally regulated expression of the P2X3 receptor in the mouse cochlea. *Histochemistry and Cell Biology, 125,* 681–692.

He, D. Z. Z., Beisel, K. W., Chen, L., Ding, D., Jia, S., Fritzsch, B., & Salvi, R. (2003). Chick hair cells do not exhibit voltage-dependent somatic motility. *Journal of Physiology, 546,* 511–520.

He, D. Z. Z., Evans, B. N., & Dallos, P. (1994). First appearance and development of electromotility in neonatal gerbil outer hair cells. *Hearing Research, 78,* 77–90.

Hebb, D. O. (1949). *The organization of behavior.* New York, NY: Wiley & Sons.

Hebert, R., Langlois, J. M., & Dussault, J. H. (1985). Permanent defects in rat peripheral auditory function following perinatal hypothyroidism: Determination of a critical period. *Developmental Brain Research, 23,* 161–170.

Henley, C. M., Owings, M. H., Stagner, B. B., Martin, G. K., & Lonsbury-Martin, B. L. (1989). Postnatal development of 2f1-f2 otoacoustic emission in pigmented rat. *Hearing Research, 43,* 141–148.

Herman, R., Maulucci, R., & Stuyck, J. (1982). Development and plasticity of visual and vestibular generated eye movements. *Experimental Brain Research, 47,* 69–78.

Hilding, D. A., & Ginzberg, R. D. (1977). Pigmentation of the stria vascularis. The contribution of neural crest melanocytes. *Acta Oto-laryngologica*, *84*, 24–37.

Housely, G. D., & Ashmore, J. F. (1992). Ionic currents of outer hair cells isolated from the guinea pig-cochlea. *Journal of Physiology*, *448*, 73–98.

Hudspeth, A. J. (1982). Extracellular current flow and the site of transduction by vertebrate hair cells. *Journal of Neuroscience, 2*(1), 1–10.

Hudspeth, A. J. (1985). The cellular basis of hearing: The biophysics of hair cells. *Science*, *230*, 745–752.

Hudspeth, A. J. (2008). Making an effort to listen: Mechanical amplification in the ear. *Neuron, 59*, 530–545.

Hudspeth, A. J., & Lewis, R. S. (1988). A model for electrical resonance and frequency tuning in saccular hair cells of the bullfrog, *Rana castesbieana*. *Journal of Physiology, 400*, 275–297.

Hurley, K., M., Gaboyard, S., Zhong, M., Price, S. D., Wooltorton, J. R. A., Lysakowski, A., . . . Eatock, R. K. (2006). M-like K+ currents in type I hair cells and calyx afferent endings of the developing rat utricle. *Journal of Neuroscience, 26*(40), 10253–10269.

Irons-Brown, S., & Jones, T. A. (2004). Effects of selected pharmacological agents on avian auditory and vestibular compound action potentials. *Hearing Research, 195*, 54–66.

Iwasaki, S., Chihara, Y., Komuta, Y., Ito, K., & Sahara, Y. (2008). Low-voltage-activated potassium channels underlie the regulation of intrinsic firing properties of rat vestibular ganglion cells. *Journal of Neurophysiology*, *100*, 2192–2204.

Jackson, H., Hackett, J. T., & Rubel, E. W. (1982). Organization and development of brainstem auditory nuclei in the chick: Ontogeny of postsynaptic responses. *Journal of Comparative Neurology, 210*, 80–86.

Johnson, C. P., & Chapman, E. R. (2010). Otoferlin is a calcium sensor that directly regulates SNARE-mediated membrane fusion. *Journal of Cell Biology, 191*, 187–197.

Johnson, S. L., Adelman, A. P., & Marcotti, W. (2007). Genetic deletion of SK2 channels in mouse inner hair cells prevents the developmental linearization in the Ca^{2+} dependence of exocytosis. *Journal of Physiology*, *583*, 631–646.

Johnson, S. L., Forge, A., Knipper, M., Münkner, S., & Marcotti, W. (2008). Tonotopic variation in the calcium dependence of neurotransmitter release and vesicle pool replenishment at mammalian auditory ribbon synapses. *Journal of Neuroscience*, *28*(30), 7670–7678.

Johnson, S. L., Franz, C., Knipper, M., & Marcotti, W. (2009). Functional maturation of the exocytotic machinery at gerbil hair cell ribbon synapses. *Journal of Physiology, 587*, 1715–1726.

Johnson, S. L., Marcotti, W., & Kros, C. J. (2005). Increase in efficiency and reduction in Ca^{2+} dependence of exocytosis during development of mouse inner hair cells. *Journal of Physiology, 563*, 177–191.

Jones, S. M., Erway, L. C., Bergstrom, R. A., Schimenti, J. C., & Jones, T. A. (1999). Vestibular responses to linear acceleration are absent in otoconia-deficient C57BL/6JEi-*het* mice. *Hearing Research*, *135*, 56–60.

Jones, S. M., & Jones, T. A. (1995a). Neural tuning characteristics of auditory primary afferents in the chicken embryo. *Hearing Research, 82*, 139–148.

Jones, S. M., & Jones, T. A. (1995b). The tonotopic map in the embryonic chicken cochlea. *Hearing Research, 82*, 149–157.

Jones, S. M., & Jones, T. A. (2000a). Ontogeny of vestibular compound action potentials in the domestic chicken. *Journal of the Association for Research in Otolaryngology, 1*, 232–242.

Jones, T. A., & Jones, S. M. (2000b). Spontaneous activity in the statoacoustic ganglion of the chicken embryo. *Journal of Neurophysiology, 83*, 1452–1468.

Jones, T. A., Jones, S. M., & Hoffman, L. F. (2008). Resting discharge patterns of macular primary afferents in otoconia-deficient mice.

Journal of the Association for Research in Otolaryngology, 9, 490–505.

Jones, T. A., Jones, S. M., & Paggett, K. (2001). Primordial rhythmic bursting in embryonic cochlear ganglion cells. *Journal of Neuroscience, 21,* 8129–8135.

Jones, T. A., Jones, S. M., & Paggett, K. (2006). Emergence of hearing in the chicken embryo. *Journal of Neurophysiology, 96,* 128–141.

Jones, T. A., Leake, P. A., Snyder, R. L., Stakhovskaya, O., & Bonham, B. (2007). Spontaneous discharge patterns in cochlear spiral ganglion cells before the onset of hearing in cats. *Journal of Neurophysiology, 98,* 1898–1908.

Kalluri, R., Xue, J., & Eatock, R. A. (2010). Ion channels set spike timing regularity of mammalian vestibular afferent neurons. *Journal of Neurophysiology, 104,* 2034–2051.

Kandler, K., Clause, A., & Noh, J. (2009). Tonotopic reorganization of developing auditory brainstem circuits. *Nature Neuroscience, 12,* 711–717.

Karet, F. E., Finberg, K. E., Nelson, R. D., Nayir, A., Mocan, H., Sanjad, S. A, . . . Lifton, R. P. (1999). Mutations in the gene encoding B1 subunit of H(+)-ATPase cause renal tubular acidosis with sensorineural deafness. *Nature Genetics, 21,* 84–90.

Katz, E., Elgoyhan, E. B., Gomez-Casati, M. E., Knipper, M., Vetter, D. E., Fuchs, P. A., & Glowatski, E. (2004). Developmental regulation of nicotinic synapses on cochlear inner hair cells. *Journal of Neuroscience, 24(36),* 7814–7820.

Katz, L. C., & Schatz, C. J. (1996). Synaptic activity and the construction of cortical circuits. *Science, 274,* 1133–1138.

Keen, E. C., & Hudspeth, A. J. (2006). Transfer characteristics of the hair cell's afferent synapse. *Proceedings of the National Academy of Sciences, 103,* 5537–5542.

Kelly, J. B. (1992). Behavioral development of the auditory orientation response. In R. Romand (Ed.). *Development of auditory and vestibular systems 2* (pp. 391–418), Amsterdam, The Netherlands: Elsevier.

Kharkovets, T., Hardelin, J. P., Safieddine, S., Schweizer, M., El-Amraoui, I., Petit, C., & Jentsch, T. J. (2000). KCNQ4, a K1 channel mutated in a form of dominant deafness, is expressed in the inner ear and the central auditory pathway. *Proceedings of the National Academy of Sciences, 97,* 4333–4338.

Khimich, D., Nouvian, R., Pujol, R., tom Dieck, S., Egner, A., Gundelfinger, E. D., . . . Moser, T.. (2005). Hair cell synaptic ribbons are essential for synchronous auditory signaling. *Nature, 434* 888–894.

Kido, T. (1997). Otoconial formation in the chick: Changing patterns of tetracycline incorporation during embryonic development and after hatching. *Hearing Research, 105,* 191–201.

Kido, T., Sekitani, T., Yamashita, H., Endo, S., Okami, K., Ogata, Y., & Hara, H. (1993). The otolithic organ in the developing chick embryo. *Acta Oto-Laryngologica, 113,* 128–136.

Kim, J. M., Beyer, R., Morales, M., Chen, S., Liu, L. Q., & Duncan, R. K. (2010). Expression of BK-type calcium-activated potassium channel splice variants during chick cochlear development. *Journal of Comparative Neurology, 518,* 2556–2569.

Knipper, M., Gestwa, L., Cate, W. J. T., Lauterman, J., Brugger, H., Maier, H., . . . Zenner, H. P. (1999). Distinct thyroid hormone dependent expression of TrkB and p75[NGFR] in nonneuronal cells during the critical TH-dependent period of the cochlea. *Journal of Neurobiology, 38,* 338–356.

Kollmar, R., Montgomery, L. G., Fak, J., Henry, L. J., & Hudspeth, A. J. (1997). Predominance of the α_{1D} subunit in L-type voltage gated Ca^{2+} channels of hair cells in the chicken cochlea. *Proceedings of the National Academy of Sciences, 94,* 14883–14888.

Kong, I. H., Adelman, J. P., & Fuchs, P. A. (2008). Expression of the SK2 calcium-activated potassium channel is required for cholinergic function in mouse cochlear hair cells. *Journal of Physiology, 586,* 5471–5485.

Köppl, C. (2011). Birds—same thing, but different? Convergent evolution in the avian and mammalian auditory systems provides informative comparative models. *Hearing Research, 273,* 65–71.

Kros, C. J., Ruppersberg, J. P., & Rüsch, A. (1998). Expression of a potassium current in inner hair cells during development of hearing in mice. *Nature, 394,* 281–284.

Kros, C. J., Rüsch, A., & Richardson, G. P. (1992). Mechano-electrical transducer currents in hair cells of the cultured neonatal mouse cochlea. *Proceedings of the Royal Society of London B, 249,* 185–193.

Krumholz, A., Felix, J. K., Goldstein, P. H., & McKenzie, E. (1985). Maturation of the brainstem auditory evoked potentials in pre-mature infants. *Electroencephalography and Clinical Neurophysiology, 62,* 124–134.

Kubisch, C., Schroeder, B. C., Friedrich, T., Lutjohann, B., El-Amraoui, A., Marlin, S., . . . Jentsch, T. J. (1999). KCNQ4, a novel potassium channel expressed in sensory outer hair cells, is mutated in dominant deafness. *Cell, 96,* 437–446.

Lang, F., Vallon, V., Knipper, M., & Wangemann, P. (2007). Functional significance of channels and transporters expressed in the inner ear and kidney. *American Journal of Physiology Cell Physiology, 293,* C1187–C1208.

Laouris, Y., Kalli-Laouri, J., & Schwartze, P. (1990). The postnatal development of the air-righting reaction in albino rats. Quantitative analysis of normal development and the effect of preventing nec-torso and torso-pelvis rotations. *Behavioural Brain Research, 37,* 37–44.

Lavigne-Rebillard, M., & Pujol, R. (1986). Development of the auditory hair cell surface in human fetuses. A scanning electron microscopy study. *Anatomy & Embryology, 174,* 360–377.

Leake, P. A., Hradek, G. T., Chair, L., & Snyder, R. L. (2006) Neonatal deafness results in degraded topographic specificity of auditory nerve projections to the cochlear nucleus in cats. *Journal of Comparative Neurology, 497,* 13–31.

Leao, R. N., Sun, H., Svahn, K., Berntson, A., Youssoufian, M., Paolini, A. G., . . . Walmsley, B. (2006). Topographic organization in the auditory brainstem of juvenile mice is disrupted in congenital deafness. *Journal of Physiology, 571,* 563–578.

Lenoir, M., & Puel, J. L. (1987). Development of the 2f1–f2 otoacoustic emission in the rat. *Hearing Research, 29,* 265–271.

Levic, S., Nie, L., Tuteja, D., Harvey, M., Sokolowski, B. H. A., & Yamoah, E. N. (2007). Development and regeneration of hair cells share common functional features. *Proceedings of the National Academy of Sciences, 104,* 19108–19113.

Li, G. Q., Meredith, F. L., & Rennie, K. J. (2010). Development of K^+ and Na^+ conductances in rodent postnatal semicircular canal type I hair cells. *American Journal of Physiology – Regulatory Integrative and Comparative Physiology, 298,* R351–R358.

Li, Y., Atkin, G., M., Morales, M. M., Liu, L. Q., Tong, M., & Duncan, R. K. (2009). Developmental expression of BK channels in chick cochlear hair cells. *BMC Developmental Biology, 9,* 67. doi:10.1186/1471–213X-9–67.

Liberman, M. C. (1980). Morphological differences among radial afferent fibers in the cat cochlea: An electron-microscopic study of serial sections. *Hearing Research, 3,* 45–63.

Liberman, M. C. (1982). Single neuron labeling in the cat auditory nerve. *Science, 216,* 1239–1241.

Lim, D. J. (1984). The development and structure of otoconia. In I. Friedman & J. Ballantyne (Eds.), *Ultrastructural atlas of the inner ear* (pp. 245–269). London, UK: Butterworth.

Lim, D. J., & Anniko, M. (1985). Developmental morphology of the mouse inner ear. *Acta Oto-Laryngologica Supplement, 422,* 1–69.

Lioudyno, M., Heil, H., Kong, J. H., Katz, E., Waldman, E., Parameshwaran-Iyer, S., . . . Fuchs, P. A. (2004). A "synaptoplasmic cistern" mediates rapid inhibition of cochlear hair cells. *Journal of Neuroscience, 24*(49), 11160–11164.

Lippe, W. R. (1994). Rhythmic spontaneous activity in the developing avian auditory system. *Journal of Neuroscience, 14*(3), 1486–1495.

Manley, G. A. (2001). Evidence for an active process and a cochlea amplifier in nonmammals. *Journal of Neurophysiology, 86,* 541–549.

Manley, G. A., Brix, J., & Kaiser, A. (1987). Developmental stability of the tonotopic organization of the chick's basilar papilla. *Science, 237,* 665–666.

Marcotti, W., Johnson, S. J., Holley, M. C., & Kros, C. J. (2003). Developmental changes in the expression of potassium currents of embryonic, neonatal, and mature mouse inner hair cells. *Journal of Physiology, 548,* 383–400.

Marcotti, W., Johnson, S. J., Rüsch, A. R., & Kros, C. J. (2003). Sodium and calcium currents shape action potentials in immature mouse inner hair cells. *Journal of Physiology, 552,* 743–761.

Marcotti, W., Johnson, S. L., Kros, C. J. (2004). A transiently expressed SK current sustains and modulates action potential activity in immature mouse inner hair cells. *Journal of Physiology, 560,* 691–708.

Marcotti, W., & Kros, C. J. (1999). Developmental expression of the potassium current $I_{K,n}$ contributes to maturation of mouse outer hair cells. *Journal of Physiology, 520,* 653–660.

Marcus, D. C., Wu, T., Wangemann, P., & Kofuji, P. (2002). KCNJ10 (Kir4.1) potassium channel knockout abolishes endocochlear potential. *American Journal of Physiology Cell Physiology, 282,* C403–C407.

Masetto, S., Bosica, M., Correia, M. J., Ottersen, O. P., Zucca, G., Perin, P., . . . Valli, P. (2003). Na$^+$ currents in vestibular type I and type II hair cells of the embryo and adult chicken. *Journal of Neurophysiology, 90,* 1266–1278.

Masetto, S., Perin, P., Malusà, A., Lucca, G., & Valli, P. (2000). Membrane properties of chick semicircular canal hair cells in situ during embryonic development. *Journal of Neurophysiology, 83,* 2740–2756.

McGuirt, J. P., Schmiedt, R. A., & Schulte, B. A. (1995). Development of cochlear potentials in neonatal gerbil. *Hearing Research, 84,* 52–60.

McLaughlin, T., Torborg, C. L., Feller, M. B., & O'Leary, D. D. M. (2003). Retinotopic map refinement requires spontaneous retinal waves during a brief critical period of development. *Neuron, 40,* 1147–1160.

Meyer, A. C., Frank, T., Khimich, D., Hoch, G., Riedel, D., Chapochnikov, N. M., . . . Moser, T. (2009). Tuning of synapse number, structure, and function in the cochlea. *Nature Neuroscience, 12,* 444–453.

Milanesi, R., Baruscotti, M., Gnecchi-Ruscone, T., & DiFrancesco, D. (2006). Familial sinus bradycardia associated with a mutation in the cardiac pacemaker channel. *New England Journal of Medicine, 354,* 151–157.

Mills, D. M., Norton, S. J., & Rubel, E. W. (1994). Development of active and passive mechanics in the mammalian cochlea. *Auditory Neuroscience, 1,* 77–99.

Mills, D. M., & Rubel, E. W. (1996). Development of the cochlear amplifier. *Journal of the Acoustical Society of America, 11,* 1–15.

Morley, B. J., & Simmons, D. E. (2002). Developmental mRNA expression of the α10 nicotinic acetylcholine receptor subunit in the rat cochlea. *Developmental Brain Research, 139,* 87–96.

Moser, T., & Beutner, D. (2000). Kinetics of exocytosis and endocytosis at the cochlear inner hair cell afferent synapse of the mouse. *Proceedings of the National Academy of Sciences, 97,* 883–888.

Mou, K., Hunsberger, C. L., Cleary, J. M., & Davis, R. L. (1997). Synergistic effects BDNF and NT-3 on postnatal spiral ganglion neurons. *Journal of Comparative Neurology, 386,* 529–539.

Müller, M. (1996). The cochlear place frequency map of the developing and mature Mongolian gerbil. *Hearing Research, 94,* 148–154.

Murthy, V., Maison, S. F., Taranda, J., Haque, N., Bond, C. T., Elgoyhen, A. B., . . . Vetter, D. E. (2009). SK2 channels are required for function and long term survival of efferent synapses on mammalian outer hair cells. *Molecular and Cellular Neurosciences, 40,* 39–49.

Murthy, V., Taranda, J., Elgoyhen, A. B., & Vetter, D. E. (2009). Activity of nAChRs containing α9 subunits modulates synapse stabilization via bidirectional signaling programs. *Developmental Neurobiology, 69,* 931–949.

Mustapha, M., Fang, Q., Gong, T. W., Dolan, D. F., Raphael, Y., Camper, S., & Duncan, R. K. (2009). Deafness and permanently reduced potassium channel gene expression and function in hypothyroid *Pit^{dw}* mutants. *Journal of Neuroscience, 29*(4), 1212–1223.

Nakahara, H., & Bevelander, G. (1979). An electron microscope study of crystal calcium carbonate formation in the mouse otolith. *Anatomical Record, 193,* 233–242.

Nayak, G. D., Ratnayaka, H. S. K., Goodyear, R. J., & Richardson, G. P. (2007). Development of the hair bundle and mechanotransduction. *International Journal of Developmental Biology, 51,* 597–608.

Neef, J., Gehrt, A., Bulankina, A. V., Meyer, A. C., Riedel, D., Gregg, R. G., . . . Moser, T. (2009). The Ca^{2+} channel subunit β2 regulates Ca^{2+} channel abundance and function in inner hair cells and is required for hearing. *Journal of Neuroscience, 29*(34), 10730–10740.

Nenov, A. P., Norris, C., & Bobbin, R. P. (1996). Acetylcholine response in guinea pig outer hair cells. II. Activation of a small conductance Ca^{2+}-activated K$^+$ channel. *Hearing Research, 101,* 149–172.

Ng, L., Rüsch, A., Amma, L. L., Nordström, K., Erway, L. C., & Vennström, B. (2001). Suppression of the deafness and thyroid dysfunction in *Thrb*-null mice by an independent mutation in the *Thra* thyroid hormone receptor α gene. *Human Molecular Genetics, 10*(23), 2701–2708.

Nin, F., Hibino, H., Doi, K., Suzuki, T., Hisa, Y., & Kurachi, Y. (2008). The endocochlear potential depends on two K$^+$ diffusion potentials and an electrical barrier in the stria vascularis of the inner ear. *Proceedings of the National Academy of Sciences, 105,* 1751–1756.

Nof, E., Luria, D., Brass, D., Marek, D., Lahat, H., Reznik-Wolf, H., . . . Glikson, M. (2007). Point mutation in the HCN4 cardiac ion channel pore affecting synthesis, trafficking, and functional expression is associated with familial asymptomatic sinus bradycardia. *Circulation, 116,* 463–470.

Nordang, L., Oestreicher, E., Arnold, W., & Anniko, M. (2000). Glutamate is the afferent neurotransmitter in the human cochlea. *Acta Oto-Laryngologica, 120,* 359–362.

Norton, S. J., Bargones, J. Y., & Rubel, E. W. (1991). Development of otoacoustic emissions in gerbil: Evidence for micromechanical changes underlying development of the place code. *Hearing Research, 51,* 73–92.

Nouvian, R., Beutner, D., Parsons, T. D., & Moser, T. (2006). Structure and function of the hair cell ribbon synapse. *Journal of Membrane Biology, 209,* 153–165.

Oliver, D., Klocker, N., Schuck, J., Baukrowitz, T. Ruppersberg, J. P., & Fakler, B. (2000). Gating of Ca^{2+}-activated K+ channels controls fast inhibitory synaptic transmission at auditory outer hair cells. *Neuron, 26,* 595–601.

Oliver, D., Knipper, M., Derst, C., & Fakler, B. (2003). Resting potential and submembrane calcium concentration of inner hair cells in the isolated mouse cochlea are set by KCNQ-type potassium channels. *Journal of Neuroscience, 23*(6), 2141–2149.

Oliver, D., Plinkert, P., Zenner, H. P., & Ruppersberg, J. P. (1997). Sodium current expression during postnatal development of rat outer hair cells. *Pflügers Archivs European Journal of Physiology, 434,* 772–778.

Oliver, D., Taberner, A. M., Thurm, H., Sausbier, M., Arntz, C., Ruth, P., . . . Liberman, C. (2006).The role of BK$_{Ca}$ signaling in electrical signal encoding in the mammalian auditory periphery. *Journal of Neuroscience, 26*(23), 6181–6189.

Ophoff, R. A., Terwindt, G. M., Vergouwe, M. N., van Eijk, R., Oefner, P. J., Hoffman, S. M. G., . . . Frants, R. R. (1996). Familial hemiplegic migraine and episodic ataxia type-2 are caused by mutations in the Ca(2+) channel gene CACNL1A4. *Cell, 87,* 543–552.

Palmer, A. R., & Russell, I. J. (1986). Phase locking in the cochlear nerve of the guinea-pig and its relation to receptor potential of inner hair-cells. *Hearing Research, 24,* 1–15.

Parrad, P. J., & Cottereau, P. (1977). Apparition des réactions rotatoires chez le rat nouveau-né. *Physiology and Behavior, 18,* 1017–1020.

Parsons, T. D., Lenzi, D., Almers, W., & Roberts, W. M. (1994). Calcium-triggered exocytosis and endocytosis in an isolated presynaptic cell: Capacitance measurements in saccular hair cells. *Neuron, 13,* 875–883.

Peng, A. W., & Ricci, A. J. (2010). Somatic motility and hair bundle mechanics, are both necessary for cochlear amplification? *Hearing Research,* doi:10.1016/j.heares.2010.03.094

Pickles, J. O., Comis, S. D., & Osborne, M. E. (1984). Cross links between stereocilia in the guinea pig organ of Corti, and their possible relation to sensory transduction. *Hearing Research, 15*(2), 103–112.

Pickles, J. O., von Perger, M., Rouse, G. W., & Brix, J. (1991). The development of links between stereocilia in hair cells of the chick basilar papilla. *Hearing Research, 54,* 153–163.

Pirvola, U., & Ylikoski, J. (2003). Neurotrophic factors during inner ear development. In R. Romand (Ed.), *Development of auditory and vestibular systems 2* (pp. 207–223). Amsterdam, The Netherlands: Elsevier.

Plaster, N. M., Tawil, R., Tristani-Firouzi, M., Canun, S., Bendahhou, S., Tsunoda, A., . . . Ptácek, L. J. (2001). Mutations in Kir2.1 cause the developmental and episodic electrical phenotypes of Andersen's syndrome. *Cell, 105,* 511–519.

Platzer, J., Engel, J., Schrott-Fisher, A., Stephan, K., Bova, S., Chen, H., . . . Striessnig, J. (2000). Congenital deafness and sinoatrial node dysfunction in mice lacking class D L-type Ca^{++} channels. *Cell, 102,* 89–97.

Ponton, C. W., Eggermont, J. J., Coupland, S. G., & Winkelaar, R. (1993). The relation between head size and auditory brain stem response (ABR) interpeak latency maturation. *Journal of the Acoustical Society of America, 94,* 2149–2158.

Puel, J. L. (1995). Chemical synaptic transmission in the cochlea. *Progress in Neurobiology, 47,* 449–476.

Pujol, R., Lavigne-Rebillard, M., & Lenoir, M. (1998). Development of sensory and neural structures in the mammalian cochlea. In E. W. Rubel, A. N. Popper, & R. R. Fay (Eds.), *Development of the auditory system* (pp. 146–192). New York, NY: Springer-Verlag.

Pyott, S. J., Glowatzki, E., Timmer, J. S., & Aldrich, R. W. (2004). Extrasynaptic localization of inactivating calcium-activated potassium channels in mouse inner hair cells. *Journal of Neuroscience, 24*(43), 9469–9474.

Pyott, S. J., Meredith, A. L., Fodor, A. A., Vazquez, A. E., Yamoah, E. N., & Aldrich, R. W. (2007). Cochlear function in mice lacking the BK channel α, 1, or 4 subunits. *Journal of Biological Chemistry, 282,* 3312–3324.

Rennie, K. J., & Correia, M. J. (1994). Potassium currents in mammalian and avian isolate type I semicircular canal hair cells. *Journal of Neurophysiology, 71,* 317–329.

Roberts, W. M., Jacobs, R. A., & Hudspeth, A. J. (1990). Colocalization of ion channels involved in frequency selectivity and synaptic transmission at presynaptic active zones of hair cells. *Journal of Neuroscience, 10*(11), 3664–3684.

Romand, R. (1983). Development of the cochlea. In R. Romand (Ed.), *Development of auditory and vestibular systems* (pp. 47–88). New York, NY: Academic Press.

Romand, R. (1984). Functional properties of auditory nerve fibers during postnatal development in the kitten. *Experimental Brain Research, 56,* 395–402.

Romand, R. (1987). Tonotopic evolution during development. *Hearing Research, 28,* 117–123.

Romand, R. (2003). The roles of retinoic acid during inner ear development. In R. Romand & I. Varela-Nieto (Eds.), *Development of auditory and vestibular systems 3* (pp. 261–291). Amsterdam, The Netherlands: Elsevier.

Roux, I., Safieddine, S., Nouvian, R., Grati, M., Simmler, M.C., Bahloul, H., . . . Petit, C. (2006). Otoferlin, defective in a human deafness form, is essential for exocytosis at the auditory ribbon synapse. *Cell, 127,* 277–289.

Ruan, Q, Chen, D., Wang, Z., Chi, F., Yin, S., & Wang, J. (2008). Topological and developmental expression gradients of Kir2.1, an inward rectifier K⁺ channel, in spiral ganglion and cochlear hair cells of mouse inner ear. *Developmental Neuroscience*, *30*, 374–388.

Rubsamen, R., & Lippe, W. R. (1998). The development of cochlear function. In E. W. Rubel, A. N. Popper, & R. R. Fay (Eds.), *Development of the auditory system* (pp. 193–270). Berlin, Germany: Springer-Verlag.

Rubsamen, R., & Schafer, M. (1990). Ontogenesis of auditory fovea representation in inferior colliculus of the Sri Lankan rufous horseshoe bat *Rhinolofus rouxi*. *Journal of Comparative Physiology A*, *167*, 757–769.

Ruel, J., Bobbin, R. P., Vidal. D., Pujol, R., & Puel, J. L. (2000). The selective AMPA receptor antagonist GYKI 53784 blocks action potential generation and excitotoxicity in the guinea pig cochlea. *Neuropharmacology*, *39*, 1959–1973.

Ruel, J., Emery, S., Nouvian, R., Bersot, T., Amilhon, B., Van Rybroek, J. M., . . . Puel, J. L. (2008). Impairment of SLC17A8 encoding vesicular glutamate transporter-3, VGLUT3, underlies nonsyndromic deafness DFNA25 and inner hair cell dysfunction in null mice. *American Journal of Human Genetics*, *83*, 278–292.

Rüsch, A., & Eatock, R. A. (1996). A delayed rectifier conductance in type I hair cells of the mouse utricle. *Journal of Neurophysiology*, *76*, 995–1004.

Rüsch, A., Erway, L. C., Oliver, D., Vennström, B., & Forest, D. (1998). Thyroid hormone receptor β expression of a potassium conductance in inner hair cells at the onset of hearing. *Proceedings of the National Academy of Sciences*, *95*, 15758–15762.

Rüsch, A., Lysakowski, A., & Eatock, R. A. (1998). Postnatal development of type I and type II hair cells in the mouse utricle: Acquisition of voltage-gated conductances and differentiated morphology. *Journal of Neuroscience*, *18*(18), 7487–7501.

Russell, I. J., Cody, A. R., & Richardson, G. P. (1986). The responses of inner and outer hair cells in the basal turn of the guinea-pig cochlea and in the mouse cochlea grown in vitro. *Hearing Research*, *22*, 199–216.

Russell, I. J., Richardson, G. P., & Cody, A. R. (1986). Mechanosensitivity of mammalian auditory hair cells in vitro. *Nature*, *321*, 517–519.

Rusznák, Z., & Szücs, G. (2009). Spiral ganglion neurons: An overview of morphology, firing behavior, ionic channels and function. *Pflügers Archiv – European Journal of Physiology*, *457*, 1303–1325.

Ruthazer, E. S., & Hollis, C. T. (2004). Insights into activity-dependent map formation from the retinotectal system: A middle-of-the-brain perspective. *Journal of Neurobiology*, *59*, 134–146.

Rüttiger, L., Sausbier, M., Zimmermann, U., Winter, H., Braig, C., Engel, J., . . . Knipper, M. (2004). Deletion of the Ca^{2+}-activated potassium (BK)α-subunit but not the BKβ1-subunit leads to progressive hearing loss. *Proceedings of the National Academy of Sciences*, *101*, 12922–12927.

Rybak, L. P., Whitworth, C., & Scott, V. (1992). Development of endocochlear potential and compound action potential in the rat. *Hearing Research*, 59, 189–194.

Salamat, M. S., Ross, M. D., & Peacor, D. R. (1980). Otoconia formation in the fetal rat. *Annals of Otology, Rhinology, and Laryngology*, *89*, 229–238.

Salt, A., N., Melichar, I., & Thalmann, R. (1987). Mechanisms of endocochlear potential generation by stria vascularis. *Laryngoscope*, *97*, 984–991.

Sanes, D. H., & Takacs, C. (1993). Activity-dependent refinement of inhibitory connections. *European Journal of Neuroscience*, *5*, 570–574.

Santos-Sacchi, J. (2010). The mammalian cochlear amplifier done. *Hearing Research*, doi:10.1016/j.heares.2009.12.014.

Saunders, J. C., Coles, R. B., & Gates, G. R. (1973). The development of auditory evoked responses in the cochlea and cochlear nuclei of the chick. *Brain Research*, *63*, 59–74.

Schlingmann, K. P., Konrad, M., Jeck, N., Waldegger, P., Reinalter, S. C., Holder, M., . . . Waldegger, S. (2004). Salt wasting and deafness resulting from mutations in two chloride channels. *New England Journal of Medicine, 350,* 1314–1319.

Scholl, U. I., Choi, M., Liu, T., Ramaekers, V. T., Hausler, M. G., Grimmer, J., . . . Lifton, R. P. (2009). Seizures, sensorineural deafness, ataxia, mental retardation, and electrolyte imbalance (SeSAME syndrome) caused by mutations in KCNJ10. *Proceedings of the National Academy of Science, 106,* 5842–5847.

Schweitzer, L. (1987). Development of brainstem auditory evoked responses in the hamster. *Hearing Research, 25,* 249–255.

Schweizer, F. E., Savin, D., Suu, C., Sultemeier, E. R., & Hoffman, L. F. (2009). Distribution of high-conductance calcium-activated potassium channels in rat vestibular epithelia. *Journal of Comparative Neurology, 517,* 134–145.

Seal, R. P., Akil, O., Yi, E., Weber, C. M., Grant, L., Yoo, J., . . . Edwards, R. H. (2008). Sensorineural deafness and seizures in mice lacking vesicular glutamate transporter 3. *Neuron, 57,* 263–275.

Sendin, G., Bulankina, A. V., Riedel, D., & Moser, T. (2007). Maturation of ribbon synapses in hair cells is driven by thyroid hormone. *Journal of Neuroscience, 27*(12), 3163–3173.

Sher, A. E. (1971). The embryonic and postnatal development of the inner ear of the mouse. *Acta Oto-Laryngologica Supplement, 285,* 1–77.

Shnerson, A., & Pujol, R. (1982). Age related changes in the C57BL/6J mouse cochlea. I. Physiological findings. *Developmental Brain Research, 2,* 65–75.

Si, F., Brodie, H., Gillespie, P., Vazquez, A., & Yamoah E. (2003). Developmental assembly of transduction apparatus in chick basilar papilla. *Journal of Neuroscience, 23*(34), 10815–10826.

Simon, D. B., Bindra, R. S., Mansfield, T. A., Nelson-Williams, C., Mendonca, E., Stone, R., . . . Lifton, R. P. (1987). Mutations in the chloride channel gene, CLCNKB, cause Bartter's syndrome type III. *Nature Genetics, 17,* 171–178.

Sobkowicz, H. N. (1992). The development of innervations in the organ of Corti. In R. Romand. (Ed.), *Development of auditory and vestibular systems 2* (pp. 59–100). Amsterdam, The Netherlands: Elsevier.

Sobkowicz, H. N., Rose, J. E., Scott, G. E., Kuwada, S., Hind, J. E., Oertel, D., & Slapnick, S. M. (1980). Neuronal growth in the organ of Corti in culture. In E. Giacobini, A. Vernadkis, & A. Shahr (Eds.), *Tissue culture in neurobiology* (pp. 253–275). New York, NY: Raven Press.

Sobkowicz, H. N., Rose, J. E., Scott, G. E., & Slapnick, S. M . (1982). Ribbon synapses in the developing intact and cultured organ of Corti in the mouse. *Journal of Neuroscience, 2*(7), 942–957.

Sokolowski, B. H. A., Csus, J., Hafez, O. I., & Haggerty, H. S. (1999). Neurotrophic factors modulate hair cells and their potassium currents in chick otocyst explants. *European Journal of Neuroscience, 11,* 682–690.

Sokolowski, B. H. M., Stahl, L. M., & Fuchs, P. A. (1993). Morphological and physiological development of vestibular hair cells in the organ-cultured otocyst of the chick. *Developmental Biology, 155,* 134–146.

Sonntag, M., Englitz, B., Kopp-Scheinpflug, C., & Rubsamen, R. (2009). Early postnatal development of spontaneous and acoustically evoked discharge activity of principal cells of the medial nucleus of the trapezoid body: An *in vivo* study in mice. *Journal of Neuroscience, 29*(30), 9510–9520.

Souter, M., Nevill, G., & Forge, A. (1995). Postnatal development of membrane specializations of gerbil outer hair cells. *Hearing Research, 91,* 43–62.

Spassova, M., Eisen, M.D., Saunders, J. C., & Parsons, T. D. (2001). Chick cochlear hair cell exocytosis mediated by dihydropyridine-sensitive calcium channels. *Journal of Physiology, 535,* 689–696.

Spitzer, N. C. (2002). Activity-dependent neuronal differentiation prior to synapse formation: The functions of calcium transients. *Journal of Physiology – Paris, 96,* 73–80.

Spitzer, N. C. (2006). Electrical activity in early neuronal development. *Nature, 444,* 707–712.

Stafford, B. K., Shur, A., Litke, A. M., & Feldheim, D. A. (2009). Spatial-temporal patterns of retinal waves underlying activity-dependent refinement of retinofugal projections. *Neuron*, 64, 200–212.

Starr, A., Amlie, R. N., Martin, W. H., & Sanders, S. (1977). Development of auditory function in newborn infants revealed by auditory brainstem potentials. *Pediatrics*, 60, 831–839.

Steel, K. P., & Barkway, C., (1989). Another role for melanocytes: Their importance for normal stria vascularis development in the mammalian inner ear. *Development*, 107, 453–463.

Stellwagen, D., & Schatz, C. J. (2002). An instructive role for retinal waves in the development of retinogeniculate connectivity. *Neuron*, 33, 357–367.

Tierney, T. S., Russell, F. A., & Moore, D. R. (1997). Susceptibility of developing cochlear nucleus neurons to deafferentation-induced death abruptly ends just before the onset of hearing. *Journal of Comparative Neurology*, 378, 295–306.

Tilney, L. G., Tilney, M. S., Saunders, J. S., & DeRosier, D. J. (1986). Actin filaments, stereocilia, and hair cells of the bird cochlea. III. The development and differentiation of hair cells and stereocilia. *Developmental Biology*, 116, 100–118.

Torberg, C. L., & Feller, M. B. (2005). Spontaneous patterned retinal activity and the refinement of retinal projections. *Progress in Neurobiology*, 76, 213–235.

Tritsch, N. X., Rodríguez-Contreras, A., Crins, T. T. H., Wang, H. C., Borst, J. G. G., & Bergles, D. E. (2010). Calcium action potentials in hair cells pattern auditory neuron activity before hearing onset. *Nature Neuroscience*, 13, 1050–1052.

Tritsch, N. X., Yi, E., Gale, J. E., Glowatzki, E., & Bergles, D. E. (2007). The origin of spontaneous activity in the developing auditory system. *Nature*, 450, 50–55.

Tritsch, N. X., Zhang, Y. X., Ellis-Davies, G., & Bergles, D. E. (2010). ATP-induced morphological changes in supporting cells of the developing cochlea. *Purinergic Signaling*, 6, 155–166.

Trudeau, M. M., Dalton, J. C., Day, J. W., Ranum, L. P. W., & Meisler, M. H. (2006). Heterozygosity for a protein truncation mutation of sodium channel SCN8A in a patient with cerebellar atrophy, ataxia, and mental retardation. *Journal of Medical Genetics*, 43, 527–530.

Tucker, T., & Fettiplace, R., (1995). Confocal imaging of calcium microdomains and calcium extrusion in turtle hair cells. *Neuron*, 15, 1323–1335.

Turcan, S., Sonim, D. K., & Vetter, D. E. (2010). Lack of nAChR activity depresses cochlear maturation and up-regulates GABA system components: Temporal profiling of gene expression in α9 null mice. *PLoS ONE*, 5(2), e9058. doi: 1371/journal.pone.0009058.

Uziel, A., Gabrion, J., Ohresser, M., & Legrand, C. (1981). Effects of hypothyroidism on the structural development of the organ of Corti in the rat. *Acta Oto-Laryngologica*, 92, 469–480.

Uziel, A., Romand, R., & Marot, M. (1981). Development of cochlear potentials in rats. *Audiology*, 20, 89–100.

Valente, M. (2007). Maturational effects of the vestibular system: A study of rotary chair, computerized dynamic posturography, and vestibular evoked myogenic potentials with children. *Journal of the American Academy of Audiology*, 18(6), 461–481.

Veenhof, V. B. (1969). *The development of statoconia in mice*. Amsterdam, The Netherlands: N.V. North-Holland Publishing Company.

Vetter, D. E., Katz, E., Maison, S. F., Taranda, J., Turcan, S., Ballestero, J., . . . Boulter, J. (2007). The α10 nicotinic Ach receptor subunit is required for normal synaptic function and integrity of the olivocochlear system. *Proceedings of the National Academy of Sciences*, 104, 20594–20599.

Vetter, D. E., Liberman, M. C., Mann, J., Barhanin, J., Boulter, J., Brown, M. C., . . . Elgoyhen, A. B. (1999). Role of α9 nicotinic Ach receptor subunits in the development and function of cochlear efferent innervations. *Neuron*, 23, 93–103.

Viener-Wacher, S. R., Toupet, F., & Narcy, P. (1996). Canal and otolith vestibulo-ocular reflexes to vertical and off vertical axis rotations in children learning to walk. *Acta Oto-Laryngologica*, 116, 657–665.

Walsh, E. J., & McGee, J. (1986). The development of function in the auditory periphery.

In R. A. Altschuler, D. W. Hoffman, & R. P. Bobbin (Eds.), *Neurobiology of hearing: The Cochlea* (pp. 247–269). New York, NY: Raven Press.

Walsh, E. J., & McGee, J. (1987). Postnatal development of auditory nerve and cochlear nucleus neuronal responses in kittens. *Hearing Research, 28,* 97–116.

Walsh, E. J., McGee, J., & Javel, E. (1986). Development of auditory-evoked potentials in the cat. I. Onset of response and development of sensitivity. *Journal of the Acoustical Society of America, 79,* 712–724.

Walsh, E. J., McGee, J., McFadden, S. L., & Liberman, M. C. (1998). Long term effects of sectioning the olivocochlear bundle in neonatal cats. *Journal of Neuroscience, 18,* 3859–3869.

Walsh, E. J., & Romand, R. (1992). Functional development of the cochlea and the cochlear nerve. In R. Romand (Ed.), *Development of auditory and vestibular systems 2* (pp. 161–220). Amsterdam, The Netherlands: Elsevier.

Weisstaub, N., Vetter, D. E., Elgoyhen, A. B., & Katz, E. (2002). The α9α10 nicotinic acetylcholine receptor is permeable to and is modulated by divalent cations. *Hearing Research, 167,* 122–135.

Werner, L. A., & Gray, L. (1998). Behavioral studies of hearing development. In E. W. Rubel, A. N. Popper, & R. R. Fay (Eds.). *Development of the auditory system* (pp. 12–79). New York, NY: Springer-Verlag.

Winter, H., Braig, C., Zimmerman, U., Engel, J., Rohbock, J., & Knipper, M. (2007). Thyroid hormone receptor α1 is a critical regulator for the expression of ion channels during final differentiation of outer hair cells. *Histochemistry and Cell Biology, 128,* 65–75.

Winter, H., Braig, C., Zimmerman, U., Geisler, H.S., Fränzer, J. T., Weber, T., . . . Knipper, M. (2006). Thyroid hormone receptors TRα1 and TR differentially regulate gene expression of *Kcnq4* and prestin during final differentiation of outer hair cells. *Journal of Cell Science, 119,* 2975–2984.

Winter, H., Rüdiger, L., Müller, M., Kuhn, S., Brandt, N., Zimmerman, U., . . . Knipper, M.

(2009). Deafness in TR mutants is caused by malformation of the tectorial membrane. *Journal of Neuroscience, 29*(8), 2581–2587.

Woolf, N. K., & Ryan, A. F. (1985). Ontogeny of neural discharge patterns in the ventral cochlear nucleus of the Mongolian gerbil. *Brain Research, 349,* 131–147.

Woolf, N. K., & Ryan, A. F. (1988). Contributions of the middle ear to the development of function in the cochlea. *Hearing Research, 35,* 131–142.

Woolf, N. K., Ryan, A. F., & Harris, J. P. (1986). Development of mammalian endocochlear potential: Normal ontogeny and effects of anoxia. *American Journal of Physiology, 250,* R493–R498.

Xia, X. M., Fakler, B., Rivard, A., Wayman, G., Johnson-Pais, T., Keen, J. E., . . . Adelman, J. P. (1998). Mechanism of calcium gating in small conductance calcium-activated potassium channels. *Nature, 395,* 503–507.

Yancey, C., & Dallos, P. (1985). Ontogenic changes in cochlear characteristic frequency at a basal turn location as reflected in the summating potential. *Hearing Research, 18,* 189–195.

Yang, T., Gurrola, J. G., II, Wu, H., Chiu, S. M., Wangemann, P., Snyder, P. M., . . . Smith, R. J. (2009). Mutations of KCNJ10 together with mutations of SLC26A4 cause digenic nonsyndromic hearing loss associated with enlarged vestibular aqueduct syndrome. *American Journal of Human Genetics, 84,* 651–657.

Yang, Y., Wang, Y., Li, S., Xu, Z., Li, H., Ma, L., . . . Shen, Y. (2004). Mutations in SCN9A, encoding a sodium channel alpha subunit, in patients with primary erythermalgia. *Journal of Medical Genetics, 41,* 171–174.

Yasunaga, S., Grati, M., Cohen-Salmon, S., El-Amraoui, A., Moustapha, M., Salem, N., . . . Petit, C. (1999). A mutation in OTOF, encoding otoferlin, a FER-like protein, causes DFNB9, a nonsyndromic form of deafness. *Nature Genetics, 21,* 363–369.

Zampini, V., Johnson, S. L., Franz, C., Lawrence, N. D., Münkner, S., Engel, J., . . . Marcotti, W. (2010). Elementary properties of

Ca$_v$ 1.3 Ca^{2+} channels expressed in mouse cochlear inner hair cells. *Journal of Physiology, 588,* 187–199.

Zampini, V., Masetto, S., & Correia, M. J. (2008). Elementary properties of Kir2.1, a strong inwardly rectifying K(+) channel expressed by pigeon vestibular type II hair cells. *Neuroscience, 155,* 1250–1261.

Zdebik, A. A., Wangemann, P., & Jentsch, T. J. (2009). Potassium ion movement in the inner ear: Insights from genetic disease and mouse models. *Physiology, 24,* 307–316.

Zenisek, D., Davila, V., Wan, L., & Almers, W., (2003). Imaging calcium entry sites and ribbon structures in two presynaptic cells. *Journal of Neuroscience, 23*(7), 2538–2548.

Zidanic, M., & Fuchs, P. A. (1995). Kinetic analysis of barium currents in chick cochlear hair cells. *Biophysical Journal, 68,* 1323–1336.

Zine, A., & Romand, R. (1996). Development of the auditory receptors of the rat: A SEM study. *Brain Research,* 721, 49–58.

Index

A

Achondroplasia, 132
Action potentials and sensory processing philosophy, 5
Alagille syndrome, 188, 189, 190
Alcohol as teratogen, 126
Altricial defined, 2
Aminoglycosides, 85
Aminoglycosides as teratogens, 126
Amniotic bands as teratogenic, 126
Anatomy. *See* Ear at maturity
Anderson syndrome, 215
Animal models for ontogeny
    developmental age comparisons: chick, mouse, human, 4
    mouse as model, 3
    overview, 2–3
Anophthalmia, 189, 190
Antibiotics as teratogens, 126
Anticoagulants as teratogens, 126
Anticonvulsants as teratogens, 126
Aortic valve disease, 191
Apert syndrome, 132, 188, 189
Ataxia, episodic, 215

B

Bacteria as teratogens, 126
Bosley-Salih-Alorainy/Athabaskan brainstem dysgenesis syndromes, 132
Branchial-oto-renal (BOR) syndrome, 124, 133, 188, 189
Branchiooculofacial syndrome, 133
Brugada syndrome, 215

C

Campomelic dysplasia, 135
Camurati-Engelmann disease, 135
Cell cycles. *See also main entry* Genetics (basic)
    cell division stage, 60, 61
        mitosis, 60
    cytokinesis
        mitosis, 60
    DNA to RNA to protein, 65–68
        transcription factors, 70–71
        transcription/translation control, 68–71
    and dominance (Mendelian), 62, 64. *See also* Mendelian inheritance patterns *main entry*
    and gamete cells (ovum/sperm), 60, 62, 63
    gametogenesis, 62, 63
    and genetic alleles, 62
    and germ cells (ovum/sperm), 60–61, 62, 63
    haploid cells from spermatogenesis/oogenesis, 62
    interphase stage, 60
    meiosis, 60–62, 64
        random assortment principle of Mendel, 64
    Mendel, Johann Gregor (1822–1884), 62–64

Cell cycles *(continued)*
  oogenesis, 62
  random assortment principle of Mendel, 64
  and recessiveness (Mendelian), 62, 64. *See also* Mendelian inheritance patterns
  segregation principle of Mendel, 62–64
  spermatogenesis, 62
Cell signaling
  and convergent extension (embryogenesis), 46–49
  hedgehog (Hh) signaling pathway, 39–40
  notch signaling pathway, 40–41
  overview, 38–39
  PCP (planar cell polarity)
    and convergent extension (embryogenesis), 46–49
    epithelial junctional complex, 43–45
    individual cell polarization, 42–45
    planar organization, polarized cells, 45–46
  Wnt signaling pathway, 41–42
Cells (nature of)
  chromosomes, 28
  cytoskeletons, 29
  ER (endoplasmic reticulum), 29
  and fluids, 27–28
  homeostasis, 28
  intermediate filaments, 30–31
  key molecular components, 28–31
  microtubule radiations, 29–30
  mitochondria, 28
  nuclei, 28
  organelles, 28–31
  ribosomes, 28–29
Centromeres defined, 56–57
Cerebellar atrophy, ataxia, and mental retardation, 215
CGD (chronic granulomatosis disease), 193
Charcot-Marie-Tooth disease, 194
Chemicals as teratogens, 126
Chemotherapeutics as teratogens, 126
Chromosomes
  abnormalities of, 72–73
    aneuploidy, 72
    Cri du Chat syndrome, 73
    deletions, 73, 74
    diGeorge syndrome, 73
    Down syndrome, 72. *See also main entry* Down syndrome
    duplications, 73, 74
    and embryogenesis, 73
    inversions, 73, 74
    monosomy, 72
    numerical anomalies, 72–73
    overview, 72
    polyploidy, 72–73
    Shprintzen syndrome, 73
    structural anomalies, 73
    translocations, 73, 74
    triploidy, 72
    trisomy 21, 72. *See also main entry* Trisomy 21
    22q11 deletion (22q11-), 73
  autosomes, 55
  and cell nature, 28
  centromeres defined, 56–57
  chromatids, 55
  and DNA (deoxyribonucleic acid), 54, 57, 58
  numbered karotype regions, 57
  overview, 54–55
  pictured by karotype, 54–57
  telomeres defined, 56
Cisplatin as teratogen, 126
Cleft lip, 125
Cleft lip and palate, 189, 190
Cleft palate, 124, 125
  and microtia, hearing impairment, 132, 135
Cleidocranial dysplasia, 135
Cochlea
  at maturity
    action potentials/hair cells, 16
    cochlear ganglion cells (primary afferents), 16
    cochlear partition tuning, 16, 18
    efferent system, 16, 18
    hair cells, 15–16, 19
    neurotransmitters, 16
    sound arrival path, 15
Congenital, 132
Craniosynostosis, 135
Cri du Chat syndrome, 73
Crouzon syndrome, 132, 188, 189

Cytogenetics. *See* chromosomes under Genetics (basic) *main entry*
Cytomegalovirus as teratogen, 126

D

Deafness
  congenital, 132, 188, 190, 193
    with inner ear agenesis, microtia, microdontia, 132
    with palmoplantar kerotoderma, 75
Diabetes
  maternal, 126, 189
  type II, 189
diGeorge syndrome, 73, 133, 188
Dilantin as teratogen, 126
Dispermy, 72–73
Down syndrome
  karyotype, 72
  phenotype, 72
Drugs as teratogens, 126

E

Ear at maturity
  acoustic reflex pathway of middle ear, 8–9
  auditory pathways, 17
  basilar membrane, 33
  bony labyrinth, 9
  cells
    ATP chemical energy functions, 28
    intermediate filaments, 30–31
  cochlea, 15–18. *See also main entry* Cochlea
  connective tissues, 32–33. *See also main entry* Tissues (nature of)
  endolymph, 9, 19
  eustachian tube, 7
  external auditory meatus, 6–7
  facial nerve and stapedius muscle, 9
  hair cell polarization, 10, 12
  hair cell sensor receptors, 10. *See also* inner hair cells *in this section;* outer hair cells *in this section*
  internal auditory artery, 12
  and motion or sound encoded for brain processing, 10, 12
  and sensory neuroepithelium, 10, 12, 15
  inner ear, 9–23
    overview, 9–15
  inner hair cells, 15–16
  membranous labyrinth, 9
  middle ear, 7–9
    sound transmission structures, 7–8
  mucosa of middle ear innervation, 9
  muscle types, 38
  neural connections to cochlear hair cells, 15
  ossicles (malleus, incus, stapes), 7, 8–9
  otoconial organs, 20–23. *See also main entry* Otoconial organs
  outer ear, 6–7
  outer hair cells, 15, 16
  overview, 5, 23
  perilymph, 9
  potassium (K^+) as inner ear key ion, 12–15
  receptor organs, inner ear, 9–10
  saccule, 10, 20, 21. *See also* Otoconial organs
  Scarpa's ganglion (VIIIth nerve vestibular branch), 21–22
  semicircular canals, 18–20. *See also main entry* Semicircular canals
  sensory epithelia of inner ear, 34
  sensory systems of inner ear, 37
  stapedius muscle, 8
  stereociliary bundles, 10, 11, 20
  tensor tympani muscle, 8
    trigeminal nerve innervation, 9
  tympanic membrane, 6, 7
    innervation, 9
    layers, 8
  utricle, 10, 20, 21. *See also* Otoconial organs
  vagus nerve
    and tympanic membrane, 9
Ears during development
  inner ear
    POU domain transcription factors, 71
  inner hair cells as neurons, 36
Embryology (basic), 92
  bird model neurulation illustrated, 102
  branchio-oto-renal (BOR) syndrome, 124
  cell lineage, 90

Embryology (basic) *(continued)*
  cleavage
    blastocyst hatching, 92
    blastocoele: embryoblast/future embryonic cells, 92
    blastocoele: from morula in uterus, 92
    blastocoele: trophoblast/nutrient exchange tissues, 92
    compaction, 92
    on embryogenesis timetable, 107
    illustrated, 93
    journey to uterus, 92–93
    morula from embryo in uterus, 92, 107
    specialized cell division, 92
  cleft palate, 124
  convergent extension, cellular and PCP (planar cell polarity), 46–49
  cranial placode development, 121–122
  egg gametogenesis illustrated, 91
  embryogenesis timetable, 107
  embryonic disc anatomical landmarks, 97, 98
  environmental influences, 126
    of chemicals, 126
    ear development, 127
    and embryo's genetic makeup, 127
    of infectious agents, 126
    of maternal health, 126
    of mechanical factors, 126
    multiple factors, 127
    of physical agents, 126
    teratogen categories, 125
  fertilization, 89–90
  fetal period
    developmental milestones, 125
    9th week beginning averages, 124–125
  gastrulation
    AVE (anterior visceral endoderm) formation, 100
    mesodermal flanks to midline, 101
    as morphogenesis beginning, 97
    notochordal process (chordamesoderm) elongation, 101
    primary embryonic germ layer formation, 101
    primitive groove formation/growth, 100, 101
    primitive streak appearance, 97, 99, 107
    primitive streak elongation, 101
    and TGF-b (transforming growth factor beta), 97
  hearing impairment, 124
  holoprosencephaly, 124
  Hox genes
    hindbrain/spinal cord formation, 115
    illustrated, 108–110
  implantation
    attachment of blastocyst into uterine wall, 93–95, 107
    bilaminar embryonic disc, 96, 107
    blastogenesis completion, 97
    early structure changes, 93, 96
    embryonic disc anatomical landmarks, 97
    exchange with maternal tissue/blood supply, 93
    preplacodal domain appears, 107
  Kallman syndrome 6, 124
  microtia, 124
  nervous system regionalization, 108–114
  neurulation
    and anatomic structures from germ layers, 103, 104–105
    begins, 103, 107
    bird model illustrated, 102
    neural plate formation, 104
    neural tube closure, 104
    neural tube forming, 107
    pharyngeal arches develop, 105, 107
  overview, 127
  pharyngeal apparatus development
    overview, 122
    pharyngeal arch development, 122–123
    pharyngeal groove development, 123
    pharyngeal pouch development, 123
    syndromic hearing loss, 123
  regionalization of nervous system, 108–114
    anterior-posterior patterning: diencephalon, metencephalon, mesencephalon, 116–117
    anterior-posterior patterning: forebrain, 117–118
    anterior-posterior patterning: hindbrain, 114–115

anterior-posterior patterning: metencephalon, mesencephalon, diencephalon, 116–117
dorsal-ventral patterning: forebrain, 119–121
dorsal-ventral patterning: overview, 118–119
segmentation. *See also* nervous system regionalization *in this section*
    anterior, 105–106
    axial, 105
    homeosis, 108
    neuromere model, 110
    orchestration of, 106, 108
somites: embryonic development staging, 105, 106, 107
timetable for embryogenesis, 107
totipotency, early embryogenesis, 90
Treacher Collins syndrome, 124
zygote formation, 89–90
    cell developmental change mechanisms, 90
    and cell lineage, 90
    cell molecular mechanisms, 90
    gastrulation germ layer development, 90
    signaling pathways, 90
    totipotency, early embryogenesis, 90
    zygote formation: cell molecular mechanisms, 90
Endocrine deficits, maternal, as teratogen, 126
Enlarged vestibular aqueduct, 194
Epilepsy, childhood susceptibility, 215
Erythermalgia, 215

F

Familial hemiplegic migraine, 215
Fechtner syndrome, 192
Folic acid deficit, maternal, as teratogen, 126

G

Gene disorders, single
    framesense mutations, 75
    GJB2 gene mutation numbers, 75
    and hearing loss, 75
    and Mendelian inheritance patterns, 73. *See also* Mendelian inheritance patterns *main entry*
    missense mutations, 75
    mutation varieties, 73
    nonsense mutations, 75
    SNPs (single nucleotide polymorphisms), 75
Genetics (basic)
    amino acids, 57–59
    cell cycles, 60–64. *See also main entry* Cell cycles
    centromeres defined, 56–57
    cerumen inheritance studies, 62, 64
    chromatin defined, 54
    chromosomes
        abnormalities of, 72–73. *See also main entry* Chromosomes
        autosomes, 55
        centromeres defined, 56–57
        chromatids, 55
        and DNA (deoxyribonucleic acid), 54, 57, 58
        overview, 54–55
        pictured by karotype, 54–57
        telomeres defined, 56
    cytokinesis defined, 60. *See also main entry* Cell cycles
    DNA defined, 57, 59–60
    dominance (Mendelian), 62, 64. *See also* Mendelian inheritance patterns *main entry*
    gene-environment interactions, 85
    gene expression for RNA production, 66–67
    genes defined, 57
    genetic alleles defined, 62
    genotype defined, 62
    haploid cells defined, 54
    HOX (homeobox) genes, 71
    human somatic cells defined, 54
    karyotype defined, 54–55, 60
    Mendel, Johann Gregor (1822–1884), 62–64
    mitosis defined, 60. *See also main entry* Cell cycles
    modifier genes, 84–85
    multifactorial inheritance, 85
    nomenclature, 64–65

Genetics (basic) *(continued)*
  overview, 85
  phenotype defined, 62
  proteins, 57–59
  random assortment principle of Mendel, 64
  recessive (Mendelian), 62, 64. *See also* Mendelian inheritance patterns *main entry*
  RNAs (ribonucleic acids) defined, 57, 59–60
  segregation principle of Mendel, 62–64
  twins, 62
Genome, human, 57
Gray's Anatomy (Standring), 137

H

HDR (hypoparathyroidism, deafness (sensorineural), renal disease), 189
Hearing defined, 4
Hearing impairment, 124
  and cleft palate, microtia, 132, 135
Hearing loss
  and aminoglycosides, 85
  compound heterozygosity, 83–84
  deafness
    congenital, 132, 188, 190, 193
    with inner ear agenesis, microtia, microdontia, 132
    with palmoplantar kerotoderma, 75
  DFNB1: autosomal recessive syndromic, 76–77, 82–83
  difficulties of genetic/phenocopy identification, 77
  dominant inheritance, Mendelian, 76. *See also* Mendelian inheritance patterns *main entry*
  environmental factors, 77
  genetic: as heterogeneous, 77
  Hereditary Hearing Loss Homepage, 77
  inheritance, 76
  Jervell and Lange-Nielsen syndrome, 83–84
  keratitis-ichthyosis-deafness (KID) syndrome, 75
  and MELAS (*m*itochondrial myopathy, *e*ncephalopathy, *l*actic *a*cidosis, *s*troke-like episodes), 82
  and MERRF (*m*yoclonus *e*pilepsy associated with *r*agged *r*ed *f*ibers), 82
  and modifier genes, 85
  and multifactorial inheritance, 85
  nomenclature, 75–77
  ototoxicity, 85
  phenocopic, 77
  recessive inheritance, Mendelian, 76. *See also* Mendelian inheritance patterns
  syndromic *versus* nonsyndromic, 75–76
  Vohlwinkel syndrome, 75
  X-linked inheritance, Mendelian, 76. *See also* Mendelian inheritance patterns
  Y-linked inheritance, Mendelian, 76, 81. *See also* Mendelian inheritance patterns
Hereditary Hearing Loss Homepage, 77
Herpes simplex as teratogen, 126
Hirschsprung disease, 133
Holoprosencephaly, 124
Human genome, 57
Hyperthermia as teratogen, 126
Hypoparathyroidism, 189
Hypothyroidism
  congenital, 188
  maternal, as teratogen, 126

I

Infectious agents, maternal, as teratogen, 126
Inheritance patterns. *See* Mendelian inheritance patterns; Non-Mendelian inheritance
Inner ear agenesis, 188, 190, 193
Inner ear embryogenesis
  cochlear formation, 163–165. *See also* ZNP (zone of nonproliferation) *in this section*
  axes/landmarks of developing mouse cochlea, 164
  hair cell differentiation, 167–169, 173
  hair cell patterning/supporting epithelial cells, 169–173
  hair cell patterning/supporting epithelial cells: genes involved in, 191
  stereociliary bundles, 173–180
  stereociliary bundles: cochlear, 175–178

stereociliary bundles: genes expressed in, 192
stereociliary bundles: vestibular, 178–180
germ layer derivatives of inner ear structures, 195
innervation
  cochlear afferent, 183–186
  efferent, 186–187
  statoacoustic ganglion development, 180–182
  vestibular afferent, 182–183
nonsensory cell gene involvement, 194
otic placode/otic vesicle formation, 154, 158–160
  genes involved/expressed, 188
  illustrated, 158
overlying membrane gene involvement, 193
overview, 154
pillar cell gene involvement, 193
prosensory patches formation
  dorsal sensory patches, 162–163
  genes for prosensory domain declamination/specification, 189
  genes for prosensory patches and terminal mitosis, 190
  overview, 162
  ventral sensory patches, 163
statoacoustic ganglion declamination/ formation, 160–162
time line, 155–157
ZNP (zone of nonproliferation) formation, 165–167
  elongation/reorganization, 165–167
  illustrated, 166
Inner ear function emergence during ontogeny
  auditory function development
    ambient sound: endocochlear potential, 239–242
    ambient sound: frequency selectivity in primary afferent neurons, 243–245
    ambient sound: hearing onset, 236–239
    ambient sound: outer hair cell electromotility/OAEs, 242–243
    ambient sound emergence/ maturation responses, 235–245

overview, 221
prehearing endogenous activity/ molecular signaling roles, 233–235
prehearing endogenous patterned activity in primary afferent neurons, 228–232
prehearing endogenous signaling period, 221–235
prehearing endogenous signaling with neuroepithelial reorganization, 232
prehearing inner hair cell action potentials, 222–223
prehearing late period channel transitions, 225
prehearing synaptic transmission, 225–228
cochlea developmental time line, 211
genes of ion channels in sensory cells/ primary afferents, 215
ion channels in sensory cells/primary afferents, 214
overview
  functional stimulation adequacy, 208–209
  hearing onset definition, 211
  of hearing process, 208
  studies/conventions, 209
vestibular function development
  hair cell transduction current response, 213–214, 216–217
  head motion behavioral response, 217–218
  overview, 212
  primary afferent function, 218–221
  transduction channels/associated membrane currents, 212–213

J

Jervell and Lange-Nielsen syndrome, 83–84
Jervell and Lange Nielsen syndrome I, 194

K

Kallmann syndrome, 6, 124, 133
Karotype regions, numbered, 57
Keratitis-ichthyosis-deafness (KID) syndrome, 75, 194

L

LADD (lacrimo-auriculo-dento-digital) syndrome, 132, 188, 190
Lead as teratogen, 126
Long QT syndrome, 3, 5, 194, 215
Long QT syndrome type 2, 215
Lymphedema-distichiasis syndrome, 135

M

May-Hegglin anomaly, 192
MELAS (*m*itochondrial myopathy, *e*ncephalopathy, *l*actic acidosis, *s*troke-like episodes), 82
Mendel, Johann Gregor (1822–1884), 62–64
Mendelian inheritance patterns. *See also* Non-Mendelian inheritance
   autosomal dominant inheritance, 77–79
   autosomal recessive inheritance, 79–80
   expressivity defined, 82
   overview, 77
   pedigree defined, 77, 78
   penetrance defined, 81–82
   Punnett square risk analysis, 79
   sex-linked genetic disorders, 80–81
   Waardenburg syndrome type I (WSI), 81–82
   X-linked genetic disorders, 81
   Y-linked genetic disorders, 81
Mercury as teratogen, 126
MERRF (*m*yoclonus *e*pilepsy associated with *r*agged *r*ed *f*ibers), 82
Microdontia, 188, 190, 193
Microphthalmia, 189, 190
Microphthamia syndrome 6, 135
Microtia, 124, 133, 188, 190, 193
   and cleft palate, hearing impairment, 135
   and hearing impairment, cleft palate, 132
Middle ear development, 123
   ligament development, 146
   muscle development, 145–146
   neural development, 146
   ossicles' development
      origin theories illustrated, 144
Middle ear embryogenesis
   eustachian tube development, 146
   overview, 132, 134, 146–147
   temporal bone development
      weeks to 7 weeks, 137
   overview, 134
   and skull base development, illustrated, 137
   time line, 136
   time line, 134
   tympanic cavity development, 146
Multiple endocrine neoplasia, type IV, 190
Multiple synostosis syndrome, 135

N

Neural tube defects, 125
Nicotine as teratogen, 126
Non-Mendelian inheritance. *See also* Mendelian inheritance patterns
   digenic, 82
   expressivity, 82
   heteroplasmic mitochondrial, 82
   homoplasmic mitochondrial, 82
   MELAS (*m*itochondrial myopathy, *e*ncephalopathy, *l*actic acidosis, *s*troke-like episodes), 82
   MERRF (*m*yoclonus *e*pilepsy associated with *r*agged *r*ed *f*ibers), 82
   mitochondrial, 82, 83
   penetrance, 82
Nutritional deficits, maternal, as teratogen, 126

O

Oculo-auricular syndrome, 132
Oculodentaldigital dysplasia, 194
Oligohydramnios as teratogenic, 126
Ontogeny defined, 1–2. *See also* Animal models for ontogeny
Orofacial cleft, 133, 135
OTIS (Organizaton of Teratology Information Specialists), 126
Otoconial organs
   at maturity
      as gravity receptors, 20
      head position/motion perception, 22
      macular hair cells, 20–21
      Scarpa's ganglion (VIIIth nerve vestibular branch), 21–22
      stereociliary bundles, 20
      vestibular efferent fibers, 23

Otoconial organs at maturity and gravitoinertial stimuli, 20
Outer ear embryogenesis
  eustachian tube development, 146
  external auditory meatus (EAM)/tympanic membrane development, 140–143
    illustrated, 141
  overview, 132, 134, 146–147
  pinna development, 137, 139–140
    malformations, 140
    morphogenesis, 139
  tympanic cavity development, 146

P

Pallister Hall syndrome, 189
Parasites as teratogens, 126
Parietal foramina, 135
PCBs (polychlorinated biphenyls) as teratogens, 126
Pendred syndrome, 194
Perception defined, 4–5
Pfeiffer syndrome, 188, 189
Phenylketonuria, maternal, as teratogen, 126
Physical agents as teratogens, 126
Physiology. *See* Ear at maturity
Polyploidy, 72–73
Precocial defined, 2

R

Radiation as teratogen, 126
Renal coloboma syndrome, 135, 188, 190
Renin-angiotensin inhibitors as teratogens, 126
Rubella as teratogen, 126

S

SANDD (sinoatrial node dysfunction and deafness) syndrome, 215
Sebastian syndrome, 192
Semicircular canals
  at maturity
    ampullae, 18
    anatomical position/angular rotation, 18, 19
    angular acceleration, 20
    brain distribution of vectored motion information, 20
    and head rotation, 19–20
SeSAME (seizures, sensorineural deafness, ataxia, mental retardation, electrolyte imbalance), 194
Shprintzen syndrome, 73
Sick sinus syndrome, 215
SMMD (spondylo-megaepiphyseal-metaphyseal-dysplasia), 133
Spina bifida, 85, 125
Spondylocostal dysostosis, 188, 189, 190
Stapes ankylosis syndrome without symphalangism, 135
Stickler syndrome type I, 193
Symphalangism, proximal, 135
Syndromic hearing loss, 123
Synpolydactyly, 135
Syphilis as teratogen, 126

T

Telomeres defined, 56
Time lines
  cochlear development, 211
  inner ear embryogenesis, 155–157
  middle ear embryogenesis, 134
  temporal bone development, 136
Tissues (nature of)
  connective, 31–33
  epithelia, 33–34
  muscles
    cardiac, 38
    overview, 37–38
    skeletal cells, 38
    smooth, 38
  neural tissue
    and action potentials, 35–36
    excitability of neurons, 35–36. *See also* action potentials *in this section*
    and functional components of nervous system, 36–37
    neurons, 33–37
    primary afferent neurons passing sensory information from periphery, 37
    sensory neurons, 34
    and thalmic relays, 37
  types in vertebrates, 31

TOXNET (Toxicology Data Network), 126
Toxoplasmosis as teratogen, 126
Treacher Collins syndrome, 124
Trisomy 21
   karyotype, 72
   phenotype, 72
Turner syndrome phenotype, 72
22q11 deletion (22q11-), 73

U

Umbilical cord construction as teratogenic, 126
Usher 2A syndrome, 192
Usher 1B, syndrome, 191, 192
Usher 1C syndrome, 192
Usher 2C syndrome, 192
Usher 1D syndrome, 192
Usher 2D syndrome, 192
Usher 1F syndrome, 192
Usher 1G syndrome, 192

V

Varicella zoster as teratogen, 126
Velocardiofacial syndrome, 73, 133, 188

Vestibular impairment
   compound heterozygosity, 83–84
   Jervell and Lange-Nielsen syndrome, 83–84
   and modifier genes, 85
Viruses as teratogens, 126
Vohlwinkel syndrome, 75, 194

W

Waardenburg syndrome type I (WSI)
   and Mendelian penetrance/expressivity, 81–82
Warfarin as teratogen, 126
Web sites
   Atlas and Database on Human Developmental Anatomy, 137
   Hereditary Hearing Loss Homepage, 77
   HUGO (Human Genome Organisation), 65
   Online Mendelian Inheritance in Man (OMIM), 188
   OTIS (Organizaton of Teratology Information Specialists), 126
   TOXNET (Toxicology Data Network), 126
Witkop syndrome, 133, 135